THE EVOLUTION OF CAPITALISM

THE EVOLUTION OF CAPITALISM

REVOLUTION IN GLASSMAKING

WARREN C. SCOVILLE

ARNO PRESS
A NEW YORK TIMES COMPANY
New York • 1972

Reprint Edition 1972 by Arno Press Inc.

Reprinted from a copy in
The State Historical Society of Wisconsin Library

The Evolution of Capitalism
ISBN for complete set: 0-405-04110-1
See last pages of this volume for titles.

Manufactured in the United States of America

- - - - - - - - - - - - - -

Library of Congress Cataloging in Publication Data

Scoville, Warren Candler, 1913-
Revolution in glassmaking.

(The Evolution of capitalism)
Original ed. published in Studies in economic history series.
Bibliography: p.
1. Glass manufacture--U. S. I. Title.
II. Series. III. Series: Studies in economic history.
[TP853.A29 1972] 666'.1'0973 71-39485
ISBN 0-405-04144-6

STUDIES IN ECONOMIC HISTORY

REVOLUTION IN GLASSMAKING

Prepared under the direction of the Committee on Research in Economic History, Social Science Research Council, in collaboration with the Committee on Technological Change, Massachusetts Institute of Technology

(*Courtesy of Owens-Illinois*)

EDWARD DRUMMOND LIBBEY, 1854–1925

REVOLUTION IN GLASSMAKING

Entrepreneurship and Technological Change In the American Industry

1880–1920

WARREN C. SCOVILLE

ASSOCIATE PROFESSOR OF ECONOMICS
UNIVERSITY OF CALIFORNIA, LOS ANGELES

HARVARD UNIVERSITY PRESS
Cambridge, Massachusetts
1948

PRINTED BY THE HARVARD UNIVERSITY PRINTING OFFICE
CAMBRIDGE, MASSACHUSETTS
U.S.A.

To

HELEN

AND

OUR TWO SONS

PREFACE

The decades falling between 1880 and 1920 witnessed an extremely rapid growth in American industry. The value added to all products by manufacturing processes increased roughly twelvefold, and by 1919 the number of persons engaged in industrial pursuits had surpassed the number engaged in agriculture. There was a marked tendency for the size of individual enterprises to increase. The manufacture of glass participated in the industrial expansion, even though it never became one of the largest in the American economy. According to the 1880 census returns, among 331 separate industries listed, it ranked twenty-eighth in the number of persons engaged, and fifty-third in value of output. In 1921 it ranked twenty-ninth in the number of wage earners employed and forty-eighth in value of output. It fell short of such industries as iron and steel, cotton goods, woolens and worsteds, boots and shoes, paper and wood pulp, and chemicals. The development of glassmaking nevertheless conformed in many ways to the general pattern of industrial expansion. The individual firm enlarged its scale of operations, combinations of separate producing units appeared and met with varying degrees of success, new techniques superseded older processes, and the total output of glass increased in both quantity and quality.

The development of glass manufacture in other respects was unusual. In 1880 it was essentially a handicraft industry employing large numbers of highly skilled workmen; by 1920 a technological revolution destined to displace nearly all skilled laborers was well under way. In 1880 the physical equipment of most glass plants permitted them to move from place to place while searching for better and cheaper fuel; by 1920 modern buildings and larger and more costly furnaces had anchored them to one location. And the corporation, with its intricate ownership and managerial controls, rapidly supplanted the older forms of business which had been predominantly individual or family affairs. These developments were obviously not peculiar to the glass industry alone, but the sudden break with the past and the rapidity of change made the expansion of glassmaking unusual. The bottleneck created by handicraft techniques and strong labor unions was not effectively cracked until the turn of the century; and in this process a few men at Toledo, Ohio, seem to have played a more important role than any other single group. The industry did not feel the full impact of their activities until after 1920, but their pioneering work was essentially completed before then.

Because of both its usual and unusual features the expansion of glassmaking offers interesting possibilities to scholars interested in ascertaining the factors responsible for industrial growth. To what extent did the industry ride the wave of increased economic activity which swept across the country in 1880–1920? To what extent was its progress due to the revolutionary techniques developed, and were these technological advances the result of increased inventive activity in industry generally? Or did they emerge in the favorable environment primarily because inventive genius was at work on them? Can the growth of glassmaking be explained solely or even adequately in terms of extraneous or impersonal factors such as market developments, increased availability of raw materials, discovery of natural gas sources, high protective tariffs, the presence or absence of competitive conditions, business cycles and war, and the scarcity or abundance of skilled labor? Or was it perhaps that the influence exercised by personalities — the inventor and the entrepreneur — was the compelling factor in creating the favorable environment and then in making the most of the opportunities offered? The basic question is whether expansion in glassmaking was more or less inevitable at the time or whether individuals or groups of men played a truly active and purposive role in its accomplishment.

The opportunity to embark upon a study of the glass industry came in the fall of 1942. A short time earlier the Industrial Relations Section at the Massachusetts Institute of Technology had initiated several exploratory excursions in the field of technological change, with the purpose of ascertaining in so far as possible both the direct and indirect causes for technological progress. Different men began to work on the incandescent and fluorescent lighting industry, the radio and television industry, the kraft-paper industry, and the glass-container industry. Although these studies were primarily concerned with recent technological change, it soon became apparent that in each case some attention would have to be accorded historical antecedents if the proper perspective was to be attained. At this point, Professor W. Rupert Maclaurin, Head of the Industrial Relations Section, approached the Committee on Research in Economic History of the Social Science Research Council with the proposal of jointly sponsoring a historical investigation of the technological and industrial expansion in at least one manufacturing field. The Committee on Research in Economic History was interested, among other things, in promoting investigation of the entrepreneur's role in American industry. It was finally decided that the joint project should involve a study of the relation of entrepreneurial activity to technological and industrial growth.

At the outset, I had planned to concentrate upon the growth in the American glass industry during the nineteenth century, but a canvass of the older firms revealed that they had preserved few original records prior to 1880. Several tons of old documents dating back to the early 1820's had been turned in during the 1943 "paper-salvage campaign" by one of the oldest glass factories in New Jersey about four weeks before I contacted the firm's officials. Enough material, however, was available to demonstrate that 1880–1920 was the really crucial period as far as the revolution in the manner of making glass was concerned. Additional inquiry revealed that the Owens-Illinois Glass Company and the Libbey-Owens-Ford Glass Company, both of whose home offices are at Toledo, had kept many important records pertaining to the entrepreneurial and promotional activities of the men who pioneered the technological advances. In the light of these facts and also because the Toledo companies showed an unselfish willingness to co-operate by making available all of their extant pre-1921 records, I decided to concentrate upon the role of a group of men at Toledo (headed by E. D. Libbey and Michael J. Owens) in the mechanization and expansion of glassmaking.

The number of people whose assistance has made the present study possible is legion. Professors Howard R. Bartlett and W.

Rupert Maclaurin, with whom I worked during two years at the Massachusetts Institute of Technology, co-operated in every possible way and persuaded the administration there to give me only half the normal teaching load. The group engaged in technological research under the guidance of Professor Maclaurin met from time to time and discussed problems and objectives of mutual interest. Out of these discussions certain broad questions evolved which gave a desirable and helpful focus to all the studies. No small amount of credit for this is due to the personality and intellectual caliber of the individuals who participated: Mrs. Barbara Cole Adams, Mr. Daniel C. Vandermeulen, Mr. Arthur A. Bright, Jr., and Dr. Lawrence R. Klein. I wish especially to acknowledge the assistance rendered by one member of the group. Mr. Robert L. Bishop, who is preparing a monograph on recent developments in the glass-container industry, has willingly supplied me with much of the information he has collected, and his criticism of my views has always been helpful.

The Committee on Research in Economic History liberally subsidized the research necessary to complete this study, and all of its members have constantly offered helpful suggestions for enhancing the study's effectiveness. Professor Earl J. Hamilton, one of its members and also my former teacher, has given an unusual amount of his time and boundless intellectual energy to correcting many errors in content, style, and presentation of the script in its various stages of completion. My debt to him is even greater than this would suggest. He was the first to interest me in economic-historical problems and for more than a decade has never failed to counsel and guide my intellectual development. Dr. Arthur H. Cole, chairman of the Committee, has probably been more helpful than any other single individual in my present undertaking. Through numerous conferences and endless correspondence he has aided me in grasping the entrepreneurial concept and in evaluating the functions performed by the individuals who enter into my story.

I welcome this opportunity to pay tribute to those men in the field of business who, despite the pressure of business and private affairs, have made their companies' records available and have spent long hours answering my questions about their former friends and associates who participated directly in the glassmaking revolution. Among these, Messrs. John H. McNerney and Henry W. Carter of Owens-Illinois, John H. Wright and Carl U. Fauster of the Libbey Glass Division of Owens-Illinois, and John D. Biggers and Herbert H. Baker of Libbey-Owens-Ford deserve special mention. Messrs. Edward J. Barry, James C. Blair, William H. Boshart, Joseph Crowley, Meyer Epstein, Richard La France, William J. Jacobs, A. J. Martin, Charles C. Oswald, Thomas Owens, Thomas W. Rowe, Charles W. Schwenzfeier, Leonard D. Soubier, and Lloyd T. Williams — all of whom knew personally the principal characters of the "Toledo Group" and associated with them as glassworkers, inventor and engineer, labor leader, plant manager, legal advisor, or fellow executive — have supplied me with much useful information which would never have been available in written form. Mr. James Gillinder, president of Gillinder Brothers, Inc., and Mr. Henry Calford Gauss of Vineland, New Jersey, have furnished much information and considerable encouragement through their interest in my work.

I am indebted to the University of Chicago Press and the editors of the *Journal of Political Economy* for permission to reproduce in slightly revised form as Chapters I and II my article entitled, "Growth of the American Glass Industry to 1880," which appeared in volume LII, pp. 193–216, 340–355. I am also grateful to Dr. Pearce Davis for permission to utilize the material presented in his doctoral dissertation, "The American Glass Industry and the Tariff" (Harvard University, April 1, 1938), before it appears in published form as a Harvard Economic Study.

Finally, mention should be made of the tremendous debt I owe my wife, Helen Cook Scoville. Although she has ever assumed the prejudiced attitude that her husband "could do no wrong," she has critically read the entire manuscript and has offered innumerable suggestions whereby its value has been increased. Most of all, she has preserved a happy household environment which has encouraged and bolstered my spirit during long hours of research and writing.

WARREN C. SCOVILLE

Los Angeles, 1947

gent, long hours answering my questions about their former friends and associates who once participated directly in the glassmaking industry. Among these Messrs. John H. McNerney and Henry S. Carter of Streator, Illinois; John F. Wright and Carl G. Fenninger of the Libbey Glass Division of Owens-Illinois; and John T. Rogers and Herbert E. Barker of Libbey-Owens-Ford deserve special mention. Messrs. Howard D. Bartow, James G. Blake, W[illegible] [illegible], Joseph Crowley, Meyer Epstein, Richard Ira Francis, William J. Jacobs, [illegible] Martin, Charles [illegible] O'Donald[?], Thomas Owens, Thomas W. Rowe, Charles W. Schwenzfeier, [illegible] D. [illegible], and Lloyd [illegible] Hahnes[?]—[illegible] of whom knew personally the principal characters of the Toledo group and associated with them as glassworkers, inventors and engineers, [illegible] managers, executives, or fellow executives—have supplied me with much useful information which would never have been available in written form. [illegible] [illegible] [illegible] 20. Henry [illegible] of [illegible] letters [illegible] [illegible] [illegible] through their interest in my work. [illegible] [illegible] [illegible] of the [illegible] for permission [illegible] [illegible] [illegible] I and II of my article on [illegible] Owens [illegible] American Glass [illegible] [illegible] which appeared in volume [illegible] [illegible] Harvard [illegible] for permission to utilize the material presented in [illegible] [illegible] dissertation, "[illegible] American Glass Industry [illegible]" [illegible] [illegible] [illegible] published [illegible]

[illegible] [illegible] [illegible] [illegible] [illegible] [illegible] the [illegible] that [illegible] [illegible] [illegible] of the entire [illegible] [illegible] [illegible] [illegible] [illegible] [illegible] [illegible] [illegible] [illegible] [illegible] [illegible] [illegible] [illegible] [illegible] and [illegible] [illegible] [illegible]

[illegible]

[illegible]

CONTENTS

TABLES

CHARTS

ILLUSTRATIONS

REVOLUTION IN GLASSMAKING

CHAPTER I

THE HISTORY OF AMERICAN GLASSMAKING PRIOR TO 1880

Although glass was among the first commodities made in the colonies, the progress achieved in its manufacture during the seventeenth, eighteenth, and nineteenth centuries was not spectacular. With the exception of pressing, the industry developed few peculiarly American features until after 1880, and its basic techniques differed little from those which had been practiced in other parts of the world four hundred years earlier.[1] Most individual firms were operated on a comparatively small scale by individual proprietors, small groups of partners, or families. The cleavage between the owner-entrepreneur and his workmen was rarely very distinct because the owner most frequently was or had been at one time a glassworker himself. A marked degree of comradeship persisted between employer and employee and, especially, among the workers. The present chapter and the one immediately following are concerned with the origin and evolution of the industry during its protracted period of adolescence and with an analysis of the factors conditioning its progress before 1880. After a brief description of its infancy during the seventeenth and eighteenth centuries, the growth of glassmaking during the nineteenth century will be portrayed in terms of the increase in the number of glasshouses and in the size of the "typical" firm as suggested by data on capitalization, numbers employed, or output. Since technology played a vital role in controll-

[1] As a matter of fact, even the development of the glassmaking art in western Europe following A.D. 1000 had been limited to a great extent to "rediscovering" the methods formerly developed by the Phoenicians, Syrians, Egyptians, and Romans. There is some evidence that before the end of the Roman era glass had been cast and pressed as well as blown, that relatively transparent "crystal" and beautifully colored — especially ruby — glass had been developed, and that hooded crucibles had been used. See, e.g., Jarves, *Reminiscences*, pp. 19, 39–40; Phillips, *Glass*, pp. 6–8; Northend, *American Glass*, p. 49; and Weeks, *Report on Manufacture of Glass*, pp. 56, 58.

ing this growth, changes in the methods followed by American glassmen will be traced in some detail. Such other factors as the availability of skilled workmen, the sources and quality of domestic supplies of basic materials, the evolution of consumer demand, and government policy, crises, and wars will then be discussed and an attempt made to evaluate their effects. Finally, the influence which the "typical" entrepreneur exercised upon the organization of the industry, the extent of competition, and the adoption of new techniques will be considered.

1. Glassmaking During the Seventeenth and Eighteenth Centuries

Apparently glass had never been made in what was to become the continental United States until the first English settlers arrived at Jamestown. In 1608, the London Company sent over eight Dutchmen or other foreigners to teach its colonists to make glass, and for a very few years they actually operated a glasshouse near the settlement. In 1621, a second group — this time predominantly Italian — arrived at Jamestown to make glass beads for the Indian trade, and its leader was cautioned to limit the quantity to the needs of trade lest the value of the beads be "vilified" through too great abundance. Furthermore, he was not to allow the natives under any circumstances to witness the process of manufacture. Undoubtedly, the new furnace made glass. Some of its glass beads have been unearthed and are now preserved in the Toledo Museum of Art and a few other places. But production continued for only a few years.[2] Less definite evidence suggests that individuals may have worked other furnaces for a short time at Salem, Massachusetts, in the 1640's;[3] near Philadelphia in the 1680's; and at two places in New Am-

[2] See Irwin, *Sandwich Glass*, pp. 9–11; McKearin, *American Glass*, pp. 75–76; Northend, *American Glass*, pp. 13–16; Gaffield Collection, Notes on Glass, III, 211; V, 92–95; Davis, "American Glass Industry," pp. 1–2.

[3] J. Gillinder, in *One Hundred Years of American Commerce*, p. 276; Lanier, "A Cloth of Glass," *Technology Rev.*, XLIV (1942), 268; McKearin, *American Glass*, p. 77; Davis, "American Glass Industry," p. 2. There is some uncertainty about the exact dates when glass was made here. In 1639 several acres of land were allotted to three glassmakers, and as late as 1669 there is some mention of the "glass house people." — Gaffield Collection, Notes on Glass, III, 212–213.

sterdam during the third quarter of the seventeenth century. Americans apparently made no glass in 1700 nor in the following four decades.

A successful brass-button manufacturer at Philadelphia named Wistar built a glasshouse in Salem County, New Jersey, in 1739, which he and his son operated until the outbreak of the Revolution. No domestic competitor of any importance appeared on the scene until 1763, when the colorful "Baron" Henry William Stiegel — both the hope and the despair of present-day collectors — diverted his attention from ironworking to glass blowing.[4] He was bankrupt in 1774; and only a small furnace lighted in that year at Glassboro, New Jersey, by a small group of runaway workers from Wistar's plant, and another built near Philadelphia in 1771–1772 survived the Revolutionary era. During the last two decades of the century, however, the industry expanded sufficiently to establish itself as an American manufacture. Furnaces were built at Temple, New Hampshire, in 1780; at Manchester, Connecticut, in 1783; at Frederick, Maryland, in 1784; near Albany, New York, in 1785; at Boston sometime between 1787 and 1792; and at Philadelphia between 1780 and 1786. In addition, two and possibly three houses in western Pennsylvania actually blew glass during the 1790's.[5] Most of these establishments were still active in 1800.[6]

Very little is definitely known about the scale upon which these early shops operated. Most of them doubtless had only one small furnace and employed a dozen or more men. On the other hand, two or three built during the eighteenth century were much larger, with two melting furnaces, many auxiliary ovens, and a working force of a hundred or more individuals.[7] But they were

[4] See Heiges, *Henry William Stiegel* and Hunter, *Stiegel Glass.*

[5] Albert Gallatin, later Secretary of the Treasury, started a factory in 1797 at New Geneva, which continued after he sold his interest in it; and James O'Hara and Isaac Craig built their plant at Pittsburgh in 1796. Some mention is made of a small window-glass shop called "Scott's" at Pittsburgh in 1795. — Thurston, *Pittsburgh*, pp. 140 ff.; J. Gillinder, in *One Hundred Years of American Commerce*, p. 275; Weeks, *Report on Manufacture of Glass*, p. 82.

[6] Two or three glasshouses, in addition to these mentioned, may have operated in other places for a short time. See McKearin, *American Glass*, pp. 584 ff.; Moore, *Old Glass*, p. 369.

[7] For example, Amelung's New Bremen works (McKearin, *American Glass*, pp.

less successful financially than many of the smaller plants. Since the country's sparse population and inadequate transportation facilities narrowly limited the size of the market for any one furnace, it is not surprising that small-scale production characterized the industry. "Most new works," a Pennsylvanian wrote in 1789, "have been begun too large in this country. . . . If we built a glass house, it was at the expense of thousands [of dollars], and calculated to cover all that part of the country with glass which was not covered by the houses." [8]

The bankruptcy of "Baron" Stiegel's grandiose undertaking at Manheim, Pennsylvania (1764–1774),[9] probably evoked this reference to glass. After having first limited himself to window and bottle glass, Stiegel soon ventured into the production of enameled and colored tableware, skillfully patterned after popular Bohemian importations. Some of this contained lead (that is, was flint glass), but most of it had a soda-lime base. Amelung's New Bremen factory in Maryland was the only other enterprise which made very large quantities of fine glassware for domestic use during the seventeenth and eighteenth centuries. Nearly all the others made bottles or window glass, both of which not only required less expense and skill but also were in much greater demand. Window glass apparently was not blown in America until Wistar fired his South Jersey furnace shortly after 1739 or before Stiegel first experimented with glass at his Elizabeth Furnace ironworks in 1763.[10] Before the century ended, production of window glass was well under way at several points along the seaboard and also west of the Alleghenies.[11]

---

100 ff.) and Stiegel's (*ibid.*, pp. 83–84; Heiges, *Henry William Stiegel*, p. 33; Hunter, *Stiegel Glass*, pp. 62–63 and Appendix; Knittle, *Early American Glass*, p. 122).

[8] Quoted in V. S. Clark, *Manufactures in U. S.*, I, 105–106.

[9] Others attempted to continue operations there for a very short time. — Heiges, *Henry William Stiegel*, p. 42; Hunter, *Stiegel Glass*, pp. 103 ff.

[10] No one seems to know exactly what sort of glass was produced by Wistar during the early years. Some writers think that his men made flint as well as window glass. But Stiegel probably was the first to make the former, and he might even have been able to claim priority in the production of glass for windowpanes. See Davis, "American Glass Industry," p. 4; McKearin, *American Glass*, p. 37; Northend, *American Glass*, pp. 24 ff.; Fowle, *Flat Glass*, p. 25.

[11] Evidently all of the furnaces in the West made window glass as well as bottles. — J. Gillinder, in *One Hundred Years of American Commerce*, p. 275; Thurston,

Although glassmaking had not flourished in America before 1800, workers had blown bottles, window glass, and both flint and ordinary glassware with varying degrees of success. Various entrepreneurs had brought over many Europeans, skilled in the art,[12] and had located sources of some of the basic materials and had tested the quality of their output. Furthermore, they had built several houses both east and west of the mountains and had operated them with enough success to warrant continued production over a period of years. These accomplishments helped to prepare the way for the industry's expansion during the next hundred years.

2. Industrial Growth During the Nineteenth Century

A particularly interesting feature of this growth was the rapid increase in the number of glasshouses. There probably were around 9 or 10 in 1800, more than twice this number a decade later,[13] and at least 33 in 1820.[14] According to the inadequate returns of the *Seventh Census* and the more reliable figures of the *Tenth*, 94 establishments were making glass at the middle of the century; and in 1880, 169 were actually at work while 42 others were being built or were standing idle.

Pittsburgh, pp. 140–141; Weeks, *Report on Manufacture of Glass*, p. 82; Macbeth-Evans Glass Company, *Glass Making*, p. 20.

[12] Stiegel himself went to Europe in search of Venetian, German, and English glassmen (Knittle, *Early American Glass*, pp. 121–122); and John Frederick Amelung brought sixty-eight workmen over from Germany when he first arrived in Baltimore in August, 1784, and an assistant accompanied by fourteen more came over later. — J. F. Amelung, *Remarks on Manufactures* (1787), mss. in the Library of the Boston Athenaeum but reprinted in part by McKearin, *American Glass*, pp. 100–101. Earlier, on December 7, 1738, Caspar Wistar had made a contract with one Captain John Marshall to bring over four expert glass blowers from Rotterdam. — *Ibid.*, p. 79; Northend, *American Glass*, p. 23.

[13] Davis, "American Glass Industry," pp. 7–8, 12–13; Coxe, *Arts and Manufactures for 1810*, Table 28.

[14] American State Papers, *Fourth Census*. This is probably an understatement since the census returns were quite incomplete. For example, the census includes only 5 works for New Jersey, whereas the careful checking done by the McKearins led them to conclude that there were about 11 operating in South Jersey alone (*American Glass*, p. 165). In their opinion, "more than 40" probably were at work throughout the United States in 1820. — *Ibid.*, p. 135.

Increase in the Number of Shops in the Window-Glass, Bottle, Tableware, and Plate-Glass Branches

These figures, however, do not mean that all branches of the industry expanded at the same rate. Nearly all the 9 or 10 houses operating in 1800 made window glass as their specialty or side line, and 18 in 1820 turned out this product exclusively. Four others produced some bottles as well.[15] The number of window-glass shops continued to increase, but the importance of their output relative to total production declined from 70 per cent in 1820 to about 40 per cent in 1831 and to approximately 24 per cent in 1880, when 49 factories were making window glass.[16]

Unfortunately, the available data will not permit a similar analysis of the bottle or container branch. In 1800 some of the shops made bottles and flasks. Two decades later 4 were exclusively engaged in this line, and 6 others turned out bottles along with either tableware or window glass. According to Weeks's report, 42 green-glass factories were producing a little less than 27 per cent of the industry's output in 1880.[17] Most of the green glass was made into bottles, and the small portion which was blown as tableware may have been more than offset by the number of containers made of flint and lime glass. Hence, bottle factories probably increased in number at a somewhat faster rate after the decades 1800–1820 than did those making window glass.

Apparently not a single shop in 1800 specialized in tableware, although the workers in most of them made a few pieces on the side. By 1820, however, this branch of the industry was securely established [18] and of the 68-odd factories built between 1820 and 1840 at least 25 were designed primarily to fashion glassware for table or domestic use.[19] The 73 houses reported in the *Tenth*

[15] American State Papers, *Fourth Census.* Weeks concludes that there were "at least" 15 window-glass plants in the United States by 1810. — *Report on Manufacture of Glass*, p. 83, n. *c*.

[16] Davis, "American Glass Industry," pp. 20, 66; Weeks, *Report on Manufacture of Glass*, Tables II and VI.

[17] Cf. Table IV with Table VI in Weeks, *Report on Manufacture of Glass*, pp. 12, 15.

[18] The census for this year reported 5 which made either fine or common glassware exclusively and 2 others which also made some bottles.

[19] McKearin, *American Glass*, p. 135.

Census as actually making flint and lime glassware in 1880 accounted for slightly more than 45 per cent of the total output.[20] Until the Civil War this growth was a reflection of the increasing popularity of flint glass. Certain Pittsburgh entrepreneurs in 1807–1808 made the first attempt of the nineteenth century[21] to produce white glass with a lead base, and several others soon followed their example.[22] At the time of the New York Tariff Convention in 1831, for example, there were at least 17 firms operating 21 flint-glass furnaces, and by 1852 the number of such shops had increased to around 25. But an improved form of lime glass, developed in 1864, was used so successfully for tableware that the relative importance of flint thereafter declined.

Rough and polished plate was not produced until the latter half of the century. In 1852 a window-glass plant at Cheshire, Massachusetts, made some rough plate for about six months and then sold its equipment to a group of men who continued production at Brooklyn, New York, for two or three years.[23] Meanwhile, two of the workmen at Cheshire went to another window-glass plant at Lenox, Massachusetts, in order to introduce there the technique of casting plate.[24] Although the company failed several times and reorganized under various firm names,[25] the Lenox

[20] Cf. Table III with Table VI in Weeks, *Report on Manufacture of Glass*, pp. 12, 15. It is true that not all their output was tableware since some druggists' and chemists' ware was made of flint glass; but, on the other hand, tumblers, dishes, pitchers, and the like were also made by a few of the 42 factories listed under "green glass." Those making lamp chimneys are included in the "tableware" branch.

[21] Evidently, no American furnace had produced flint since "Baron" Stiegel's bankruptcy in 1774 (see above, p. 6).

[22] In 1820, four or five houses were making some flint. — American State Papers, *Fourth Census*; Davis, "American Glass Industry," pp. 24–25.

[23] Gaffield Collection, Scraps, I, 162; Weeks, *Report on Manufacture of Glass*, pp. 98–99; J. Gillinder, in *One Hundred Years of American Commerce*, p. 279. There is some unsatisfactory evidence that rough plate was also made in the early 1850's at Green Point, New York (*ibid.*, p. 279; Davis, "American Glass Industry," p. 102), and at Philadelphia by the Anthracite Glass Works (Gaffield Collection, Prices Current of Window Glass, p. 103).

[24] Gaffield Collection, Scraps, I, 28–30. Later, in 1855 or 1856, the itinerant equipment at Brooklyn was brought to the Lenox plant.

[25] It was first an adjunct of the Lenox Iron Works, then was reorganized by other entrepreneurs as the National Plate Glass Company in 1855; reverted to the original owners in 1857; was separately incorporated as the Lenox Rough Plate Glass Company in May, 1864; had a disastrous experience with cryolite glass in

works were the first in the United States to make rough plate successfully. During all but a few of the years from 1853 until after 1880 they cast plate glass, and once or twice they attempted to polish it. The first factory which both cast and polished commercial quantities was built at New Albany, Indiana, between 1867 and 1870. During the following decade various enterprises erected three new plants on extensive scales,[26] and the Berkshire Glass Works at Lanesboro, Massachusetts, also began to make unfinished plate for church windows. Hence in 1880 there were three factories which made and finished plate glass, two which only cast it, and a sixth — the Berkshire Company — which, as it happened, at that time lay idle.

Increase in the Size of Shops in the Four Branches

This increase in the number of firms does not reveal adequately either the extent of expansion or the nature of the industry. Just as significant were changes in the scale of operations of individual factories.[27] A majority of those making window glass between 1800 and 1830 employed between 25 and 40 workmen, produced an annual output valued at from $12,000 to $30,000, and had fixed investments which probably ranged from $10,000 to

---

1870–1871 as the Lenox Glass Company; then passed into the hands of other individuals who retained its name; and was finally reorganized as the Lenox Glass Works in 1879. — Gaffield Collection, Glass Journal, I, 272; II, 20–21, 26–27, 87, 221; III, 281; Glass-Making, Vol. IV; Notes on Glass, IV, 165–166; Scraps, I, 28–30, 91–92, 124; Davis, "American Glass Industry," pp. 102–103, 244–245; Weeks, *Report on Manufacture of Glass*, pp. 98, 99.

[26] At Louisville, Ky., in 1870; at Crystal City, Mo., in 1872; and at Jeffersonville, Ind., in 1876. — Pound, *Salt of the Earth*, p. 27; Gaffield Collection, Correspondence on Glass, I, 21; Glass Journal, II, 224, 226; III, 131; Scraps, I, 4, 123, 125, 151, 158, 162; II, 125, 134, 162; J. Gillinder, in *One Hundred Years of American Commerce*, p. 280; Weeks, *Report on Manufacture of Glass*, p. 99.

[27] The sources which lend support to the following generalizations are far too numerous to give here. Hundreds of statements concerning the size of individual plants have been culled from the publications of various collectors and antique dealers and from the notes, journals, and newspaper clippings in Thomas Gaffield's collection. Whenever possible the figures in the census reports were first broken down into counties and branches of the industry; and then separate averages were made. This permitted some sort of approach to a "modal average" and minimized the distortion which would have been caused by unusually large or small shops in a comprehensive arithmetic average.

$25,000.[28] An insufficiency of data prevents the striking of such modal averages for bottle and tableware shops, but it appears that the former were smaller and the latter larger than window-glass plants.[29]

By the decade 1840–1850 the "typical" plant had definitely increased in size. The census returns suggest that the average laboring force numbered from 30 to 60 individuals in 1840 and from 35 to 70 in 1850. The vast majority made glass valued at $20,000 to $50,000 each year,[30] and most of the owners reported a capitalization ranging from $20,000 to $40,000.[31] Since the published returns for this period do not group the firms in accordance with the kind of glass made but only by counties or states, it is difficult to judge whether establishments in any one branch of the industry operated on a larger scale than those in another. Supplementary data, however, lend the impression that most bottle and window-glass factories were of about equal size. Those making tableware apparently were the largest.

Weeks's excellent report and the increased efficiency of the census enable one to ascertain the state of the industry in 1879 with much greater accuracy than in any previous year. The arithmetic averages given in Table 1 suggest the scale upon which plants in the four branches operated at that time. The plate-glass factories were the largest, as measured by the number of workmen and the value of their output. Furthermore, they employed over twice the amount of capital per worker relative to other fac-

[28] Several were capitalized at $100,000, but it is extremely doubtful whether anything like this amount was ever paid in by the owners. In order to make the figures on output and capitalization comparable with those for 1880, they have to be adjusted in accordance with changes in the price of glassware and of general commodities, respectively. The value of glass made each year in 1800–1830, therefore, should be multiplied by something like 0.3; capitalization, by 0.8 (see Chart II, below).

[29] All of the bottle shops listed in the 1820 census, for example, were said to provide work for only 10 to 20 individuals, and other information suggests that those specializing in tableware from 1817 to 1825 hired from 22 to 96 persons and produced an output valued at $20,000–$65,000.

[30] The modal average for 1840 was probably $25,000–$40,000; for 1850, $35,000–$50,000.

[31] The value of output for 1840–1850 should be multiplied by 0.75 and capitalization by 1.2 in order to make the figures comparable with the respective figures for 1880.

tories because the equipment necessary for casting, grinding, and polishing was very expensive.[32] Most green-glass plants (whose output was predominantly bottles and other containers) and those making flint and lime glassware primarily for household use [33] hired at least twice as many workmen as the majority of

TABLE I

AVERAGE SIZE OF ACTIVE GLASSHOUSES IN 1879*

Kind of Glass Produced	No. of Firms Included in Average	Capital Employed per Plant	Number of Workers per Plant	Value of Annual Output per Plant
Plate glass**	4	$635,500	225	$205,641
Window glass	49	95,983	79	103,006
Glassware (lime and flint)	73	94,620	173	131,076
Green glass	42	109,694	159	135,010

* Based upon data given in Weeks, *Report on the Manufacture of Glass*, Tables I–IV and table on p. 3. Forty-two establishments, either lying idle or in process of construction, have been omitted from the averages.

** A fifth factory, the Lenox Glass Works in Massachusetts, was not included because its much smaller size would vitiate the average. It had only $45,000 capital, 58 employees, and $45,843 output.

window-glass houses. This was due not only to their larger output but also to the fact that the processes involved in shaping and finishing bottles and tableware required relatively more labor and less capital per worker than the manufacture of window glass.

For the industry as a whole, the size of the "typical" plant had grown considerably since the first half of the century. In all but the window-glass branch the average working force had expanded from threefold to fivefold, with the greatest absolute increase apparently coming after the Civil War. It should be recognized, however, that such arithmetic or modal averages as have been used to trace this growth have definite limitations. It would be a mistake, for example, to conclude that the smaller shop so com-

[32] By 1880–1882 the one at Crystal City, Mo., had a capitalization of over $1,000,000, a personnel numbering between 350 and 500, and a yearly product of something like $322,000. — Weeks, *Report on Manufacture of Glass*, Table I; Gaffield Collection, Glass Journal, III, 267, 280; Scraps, II, 154, 193. Another at New Albany, Ind., boasted in 1882 that it employed 1,000 workpeople to produce its million dollar output and that it had over $1,000,000 of capital invested in plant, equipment, and stock on hand. — Gaffield Collection, Scraps, II, 132; III, 123.

[33] Some, however, specialized in druggists' and chemists' ware.

(*Courtesy of Hartford-Empire*)

FIG. 1. THE NEW ENGLAND GLASS COMPANY AT CAMBRIDGE, MASS., 1851

mon before 1830 no longer existed in 1880. Several were still operating on such a small scale that only from 15 to 40 workers were needed. It would be equally misleading to assume from the foregoing analysis that no glasshouses before 1850 were comparable in size to the larger ones at work in 1880. Two in the tableware division and one making containers were outstanding exceptions. Soon after its incorporation in 1818, the New England Glass Company was capitalized at $40,000 and used about 40 persons to make $40,000 of glassware a year. The second — the Boston and Sandwich Glass Company, founded in 1825–1826 — hired about 60 individuals and boasted of an annual output valued at $75,000. By 1852 each of these companies had a capitalization of $400,000, from 400 to 500 workpeople, and an annual product worth around $500,000.[34] The owner of the third — a bottle and vial factory called the Dyottville Glass Works in the Kensington district of Philadelphia — claimed in 1832 that he employed almost 400 individuals and made between 1,100 and 1,200 tons of glass each year.[35] Production on a scale approaching that of these three companies had been achieved before 1880 by several others in each branch of the industry. The vast majority, however, were still small compared with the typical plant following the mechanical revolution of the twentieth century.

3. Composition of Glass and Methods Adopted in Working It

The main reason why most shops remained relatively small in size until after 1880 was that the techniques they employed still necessitated handicraft production and highly skilled labor. The first glassmen in the United States naturally patterned their methods after those current in European countries, and they continued to do so throughout most of the nineteenth century. Few advances were of American origin. Technological development

[34] Jarves, "Manufacture of Glass," *Hunt's Merchants' Mag.*, XXVIII (1853), 121; Jarves, *Reminiscences*, pp. 61–63; Chipman, *Old Sandwich Glass*, p. 38; Watkins, *Cambridge Glass*, pp. 18, 28.

[35] Dyott, *Labor at the Glass Factory of Dyottville*, pp. 13, 54; J. Gillinder, in *One Hundred Years of American Commerce*, p. 274; "Report on Glass Manufacturing in 1832" in Gaffield Collection, Glass-Making, III, 124.

lagged appreciably behind that in Belgium, England, France, and Germany. In order to explain the complexion of the industry from 1800 to 1880, therefore, it is essential to understand the evolution of those techniques which were basic in melting and fashioning glass.

Lime, silica, soda or potash, and cullet were the chief basic materials, but small amounts of other chemicals such as arsenic, manganese, or cobalt were often used to lessen the greenish tint resulting from iron particles in the sand or to add color to the finished product. After the materials were thoroughly mixed by hand,[36] the resulting batch was placed in several pots ranged against the inside wall of the main furnace and melted. Usually, as the ingredients fused and the batch decreased in volume, more was added. The metal was then subjected to a sustained heat and refined,[37] after which the temperature was lowered to bring the glass to a consistency approaching that of molasses on a cold December morning. By this time the metal was ready to be worked.

Production of Window Glass

There were two distinct processes employed in 1800 for making glass suitable for windowpanes. According to one, a worker, called the "gatherer," collected on the end of his blowpipe a suitable quantity of metal from one of the pots, rolled ("marvered") it on a marble or wooden block [38] until it formed a cohesive mass of the desired shape, and then turned it over to the blower. He blew the glass into a large globe, to which an assistant attached a solid iron rod (the "pontil" or "punty") at the opposite center.

[36] Mixing was generally done by hand as late as 1880. Weeks remarked: "Many attempts have been made to introduce mechanical mixers, but, though some are used, they have been discarded in many works, and the older method is employed." —*Report on Manufacture of Glass*, p. 43.

[37] In the early days when the basic materials varied a great deal in quality, relatively large amounts of sandiver or glass gall formed on the surface as scum, and this had to be carefully ladled out of the pots. Furthermore, in order to agitate the metal and remove all "seeds" or bubbles, an Irish potato was not infrequently dropped into each pot. Its almost instant consumption produced steam which, in escaping, accomplished the desired result.

[38] Shortly before or after 1880 iron blocks, dusted with sawdust or faced with wood, were used instead of wooden marvers.— Davis, "American Glass Industry," p. 164.

Next, the blowpipe was broken away, and a workman twirled the pontil until the bowl opened out into a flat plate.[39] When the pontil was finally cracked off, a rough, thick bull's-eye remained in the center.[40] Glass made in this manner was called "crown." [41]

The other process was used in forming cylinder or sheet glass. The gathering and marvering operations were the same as with crown, but instead of a globe the workman blew a large cylinder.[42] One end of this was popped open by reheating it at the furnace, and the second, to which the pipe was attached, was cracked off by encircling it with a thread of hot glass which was then suddenly cooled with a few drops of water. Next, the cylinder was slit with a red-hot or diamond-pointed knife, reheated in the flattening oven, and gradually unfolded and smoothed by a worker with the aid of a small block of charred wood attached to the end of an iron rod. This last operation caused cylinder glass to have a much less brilliant finish than crown because the latter did not come in contact with any tool until it had hardened. Cylinder, nonetheless, had other features which more than offset this disadvantage. There was less waste in cutting windowpanes because there was no bull's-eye, and for the same reason larger pieces could be obtained.[43] Furthermore, since its thickness was more uniform than that of crown, light was less distorted.[44] These features, more

[39] Meanwhile, the glass had to be reheated several times at the furnace in order to preserve its viscous state.

[40] Exactly the same process had been monopolized in France for centuries by the members of four noble Norman families. See Scoville, "Labor in French Glass Industry," *Jour. of Modern Hist.*, XV (1943), 293.

[41] Today, "crown glass" designates an optical glass of different refractive powers from flint.

[42] "The labor is severe, and commands wages equal, if not superior, to any manual labor performed. Some of the cylinders blown are very large, being from four to five feet in length and from twelve to fourteen inches in diameter." — *Scientific American*, February 20, 1869, p. 144. By 1880, they were usually 60 inches long and 40 inches in circumference. — Weeks, *Report on Manufacture of Glass*, p. 51.

[43] Writing in 1865, Richard Pilkington, Jr., stated that, whereas squares measuring 40″ x 30″ could be cut from sheet glass, panes measuring 34″ x 22″ at the very most could be obtained from crown. — "Manufacture of Polished Sheet Glass," *Jour. of Franklin Institute*, LXXX (1865), 126.

[44] In comparison with later-day methods of drawing window glass in a continuous sheet directly from the furnace, the cylinder process produced vastly inferior panes. Since the inside circumference of the cylinders was less than the outside circumference, the pieces naturally rippled when flattened.

than a difference in cost of production, were responsible for the increasing popularity of cylinder glass during the nineteenth century.[45]

Although "Baron" Stiegel may have made both kinds of window glass before the Revolution, the first American factory primarily engaged in the production of crown was the very successful Boston house established on Essex Street between 1787 and 1792. But even this one and the six others which had been started in the meantime either had failed by 1830 or had switched entirely to sheet. In 1831–1832 only the New England Crown Glass Company in East Cambridge was active, although other works were being constructed at that time near Redford, New York. Two other houses later attempted to make a little crown glass, but shortly after 1850 none was being produced anywhere in the country.[46] The blowing and flattening of cylinders, therefore, remained the only process employed in window-glass houses during the last half of the century.

Production of Bottles

Bottles required less skill on the part of workers than did either crown or sheet. After a small gob of glass had been taken from the furnace pots on the end of the pipe and then marvered, the workman proceeded to shape the bottle, usually, but not always, with the aid of some sort of mold. Most of the molds used around 1800 were made either of clay or of wood and formed only the base and body of the container; the neck had to be drawn out by the blower. Shortly thereafter the two-piece, hinged mold which formed all but the bottle's mouth became common;[47] and as early

[45] Cf. Henderson, "Glass-Making," *Popular Science Monthly*, XXXIV (1888–1889), 611. At the end of the century crown glass came back into vogue for decorative windows.

[46] A large importing firm at Boston, of which Thomas Gaffield was a member, even ceased importing crown glass around 1857 because there was no demand for it.— Gaffield Collection, Notes on Glass, IV, 156. And the manager of the Lenox plate-glass works wrote him on December 21, 1868, that "as Crown glass is not made in this country, of course, it is not necessary to say anything about that."— Gaffield Collection, Correspondence on Sunlight Experiments, I, 95.

[47] All except "paste-molds" left seams on the bottle. Paste-molds were simply metal ones coated with carbon and then sprinkled with water or oil before each blowing. Because of the resulting lubrication, the blower could slowly turn the bottle in the mold as he blew and hence avoid all seams on his finished product.

(*Courtesy of Libbey-Owens-Ford*)

FIG. 2. WORKMEN MAKING CYLINDER WINDOW GLASS BY HAND

as 1820, apparently, three- and four-piece molds were being used in several places. Some were made of such hard woods as apple and beech, but these quickly burned away under quantity production, in spite of the fact that they were always dipped in water before each blowing. Hence, brass, copper, and particularly cast iron were increasingly employed in mold construction during the century.[48] Chilled-iron molds, introduced immediately following the Civil War, had a very smooth surface and consequently could be used without having the bottles subjected to a fire-polishing process in order to efface defects which were always imprinted by a coarser metal.

Regardless of the type of mold used, most bottles had to have their mouth formed by a workman using hand tools. After the bottle was taken from the mold and before it was cracked off the blowpipe, a pontil was stuck to the center of its base. This allowed the worker to manipulate the red-hot bottle during the finishing process. Sometime between 1850 and 1860 the pontil was replaced by a snap case, which fitted around the body of the bottle and thereby left no disfiguration at its base. The threads and mouth of screw-top jars,[49] however, had to be formed in the original mold, and such containers did not pass through a finishing process like other bottles. The jagged edge left when the pipe was broken off was simply ground away after the jar had been annealed.[50]

Production of Tableware; the Introduction of Pressing

The method of blowing tableware in 1880 demanded much more skill than any of the other techniques, and it was in this

---

Obviously, paste-molds were practicable only when the bottle or other object blown was perfectly cylindrical and of a smooth surface.

[48] Brass and copper molds were much too expensive for most bottle shops; they were used more in patterned pressed and blown tableware. See, e.g., Lee, *Sandwich Glass*, p. 87.

[49] These appeared on the market during the latter half of the century. Mason's famous patent was issued on November 30, 1858, for the present-style screw-top fruit jar. I am indebted to my former colleague, Robert L. Bishop, for lending me material on this subject which he collected while working on twentieth-century developments in the glass-container field.

[50] This procedure had been adopted even earlier than 1850 for jars whose wide mouths could not be shaped very easily with hand tools.

branch that the individuality and artistry of each workman found their greatest expression. The various processes and tools did not differ from those which the famed Venetian craftsmen had employed centuries earlier, but the finished American goblet, bowl, or pitcher was considerably less artistic and much sturdier than the pieces fashioned during the Renaissance. When glass is being worked, it has to be in a viscous state so that the artisan can blow, shear, and shape it practically at will. At the beginning of the nineteenth century all glassware for table and other domestic use was blown,[51] but within a few decades this state of affairs was altered in an almost revolutionary manner. Pressing, the most important improvement in forming glass during the period preceding 1880, was introduced in the 1820's.

The technique itself had probably been used by the ancients, and European glassmen apparently "rediscovered" it during the last years of the eighteenth and first two decades of the nineteenth centuries.[52] Various shops in the United States have been credited by different writers with introducing, if not inventing, the process, but it now appears that the first American pressed ware was made by the New England Glass Company shortly after 1820. This was several years before the legendary carpenter was supposed to have confronted Deming Jarves at the Sandwich plant with the startling proposal to press instead of blow a queerly shaped piece of glass which he needed. Whatever the origin of the principle, the side-lever press — operated by hand until much later than 1880 — was an American product; and glasshouses in the United States were unquestionably responsible for the perfection of the technique [53] and its widespread use in quantity production.

[51] If no pattern or form mold was used, the object was designated as "off-hand"; otherwise, it was said to be "mold-blown."

[52] Much controversy has arisen over this point. Cf., e.g., Davis, "American Glass Industry," p. 101, and McKearin, *American Glass*, pp. 333–334, with J. Gillinder, in *One Hundred Years of American Commerce*, pp. 278–279; Weeks, *Report on Manufacture of Glass*, p. 58; and Colné, *Paris Exposition, 1878*, p. 387. Several Britishers after 1850 gave America credit for originating the process. See Dodd, *Curiosities of Industry*, p. 15, and *The Mechanics Magazine*, November 18, 1871, p. 395. But Deming Jarves, whom several people have erroneously credited with the invention, himself admitted that square feet for stemware and certain kinds of bottles, as well as ordinary salts, were pressed in European houses at an earlier date than any sort of glass in the United States. — *Reminiscences*, pp. 93 ff.

[53] The pioneer firms appear to have been the Bakewell and the James B. Lyon

The basic process was quite simple. A gob of metal was sheared off the gathering rod and allowed to drop into a mold. A plunger was then brought down by means of a hand lever and forced the glass around the interior of the mold. The main disadvantage of this method arose from the fact that glass, in a viscous state, loses much of its brilliancy when it comes in contact with metal. Hence, the molded object had usually to be reheated enough at the furnace to regain some of its sparkle but not enough to efface its form or pattern.[54] Sometimes it was possible to affirm that pressed ware with delicate lacelike designs was scarcely distinguishable from blown ware except to the experienced eye.

On the other hand, pressing had at least two distinct advantages. Since thousands of identical copies could be made from the same mold within a few weeks, more expensive and intricately engraved molds were economically feasible than when blow molds were used. Near the middle of the century Jarves estimated that American factories had up to two million dollars invested in pressing machinery and molds alone.[55] Secondly, relatively unskilled labor could be substituted for highly paid blowers.[56] As a result of these two advantages the common run of tableware was more beautiful, though more standardized, and less expensive than it had formerly been. Definite information which would permit a comparison of the costs of production is wanting, but Jarves claimed that "the tendency . . . has been so to reduce the cost . . . that it has multiplied the consumption at least tenfold." [57] Certainly, no branch of the industry prospered as much as the tableware division in 1830–1880, and Pearce Davis has concluded that it led all others in technical improvement in 1861–1890.[58]

Production of Rough and Polished Plate Glass

The process of casting plate glass — the only major *forming* technique which has yet to be considered in this chapter — was

---

companies at Pittsburgh and the New England and the Sandwich companies in Massachusetts.

[54] The technical term for this operation was "fire-polishing."

[55] *Reminiscences*, p. 94.

[56] See below, p. 33.

[57] *Reminiscences*, p. 86.

[58] "American Glass Industry," p. 220.

borrowed in its entirety from Europe. During the last quarter of the seventeenth century the French had learned to roll or cast plate on a large table,[59] and entrepreneurs in other countries had followed the method closely. No important technological change had been made by 1850. This was about the time that a few Americans first attempted to make rough plate. With the aid of grapplers, carriers, and chains the glass was taken from the melting furnace either directly in the melting pots or in smaller crucibles into which it had first been ladled; then it was poured on the smooth top of a casting table and spread out by means of a large chilled-iron or copper roller;[60] and finally it was pushed into an annealing oven and allowed to cool gradually.[61] Relatively few workmen were required for these operations, and nearly all were less skilled than blowers. Hence, at first glance, it would appear somewhat unexpected that this process of making plate was not adopted at an earlier date in the United States because industries requiring little skilled labor and much machinery were usually the ones in which American entrepreneurs were able to compete effectively with foreigners. The real explanation involved other considerations. Casting equipment occasioned a much greater outlay than was needed to make other kinds of glass;[62] glassmakers themselves could not supply sufficient capital, and outsiders were reluctant to invest in the industry. Furthermore, success seems to have hinged to a marked degree upon the organization and direction of the working force, and Americans had had no previous experience with this type of production. Probably more important than these reasons was the fact that casting was economically feasible only if operations were on a large scale, and the American

[59] See Scoville, "Large-Scale Production in French Plate-Glass Industry," *Jour. of Pol. Econ.*, L (1942), 693–695.

[60] As late as 1884 the roller in American plants was equipped with handles at both ends and was pulled over the molten metal by two strings of workmen. European factories had long discarded this backbreaking practice in favor of a windlass-and-chain mechanism. — Davis, "American Glass Industry," pp. 250–251.

[61] Three of the four American factories in 1876 used stationary casting tables and had to shove the plates onto a rickety rolling table in order to get them to the mouth of the various annealing ovens. The best European plants meantime were employing mobile casting tables which could be run on rails up to the ovens. — Gaffield Collection, Scraps, I, 123.

[62] See above, pp. 11–12.

market for unpolished plate during the first half of the nineteenth century was not very extensive.[63] As late as 1867, it was said, even the comparatively small Lenox plant could supply the entire domestic demand for rough plate.[64]

The machinery for grinding and polishing such glass required additional large investments and also demanded the services of many unskilled workmen.[65] Hence no American plant made finished plate in commercial quantities until after 1870. And then a good part of the grinding and polishing machinery was imported from Europe, even though some of the equipment used there was apparently of American origin.[66]

Production of Flint and Lime Glass

Even if glassmen in the United States made no significant contribution to the technology of casting and finishing plate, the introduction of the processes must still be ranked as an outstanding innovation. So must the development of flint manufacture. As far as its composition was concerned, flint [67] differed from ordinary glass in two ways: red lead (or litharge) was used in its batch (from 30 to 50 portions for every 100 of silica); and the other ingredients — fine white sand, potash, and lime — were of the best possible quality. The preparation of these materials required much labor, and the batch had to be melted in hooded crucibles in order to protect the metal from injurious fumes and smoke. From 1808 to 1864 American producers made flint glass in ever increasing quantities, most of it appearing on the market after 1830 as pressed tableware.

[63] Rough plate was used for skylights, sidewalk lights, church windows, and other things requiring thick, translucent panes.

[64] Gaffield Collection, Notes on Glass, IV, 166.

[65] Gaffield Collection, Glass Journal, I, 190. Very little mirror glass was made in this country because it had to be polished by hand. — V. S. Clark, *Manufactures in U. S.*, II, 502. The low wages paid her unskilled labor enabled France to control the mirror trade for a long time. — *Boston Commercial Bulletin*, June 17, 1874, in Gaffield Collection, Scraps, I, 51.

[66] Colné, *Paris Exposition, 1878*, p. 336; Davis, "American Glass Industry," p. 251.

[67] The name "flint" was really a misnomer. English glass manufacturers had first made crystal by pulverizing flints; then after the early part of the seventeenth century, when they began to use litharge as a flux along with ordinary silica, they retained the term.

In 1864, however, an improved kind of lime glass which almost immediately captured this field was developed by William Leighton, Sr., of J. H. Hobbs, Brockunier, and Company at Wheeling, West Virginia. After long experimentation Leighton succeeded in making a beautiful crystalline glass by substituting bicarbonate of soda for poor grades of soda ash and by perfecting the proportions of all materials. The resulting product had neither the weight nor resonant ring of glass containing lead, but it was almost as brilliant and considerably cheaper.[68] Since it hardened sooner than flint, employees had to work faster; this put a premium on mechanical and organizational improvement around the shop. Lime-glass production expanded at the expense of flint during the latter part of the century and far outclassed in volume any other nongreen glass made in the United States.

Finally, it should be noted, there was a tendency toward increased specialization of functions within each plant. Around 1800 each blower, regardless of what he made, had usually gathered, marvered, and shaped his own glass; and some in the window branch had even flattened and cut their cylinders. These operations became more and more sharply differentiated after 1820. The output per worker rose as a result of the accompanying division of labor, but available data are not sufficient to measure the increase. Much later, additional specialization in bottle shops raised individual efficiency a great deal more. Instead of having each pot manned by one blower and his assistants, as had been the practice until shortly after 1870, three skilled blowers, aided by four helpers, were put to work at each pot. Two blowers gathered and inflated bottles while the third finished the mouths and lips for the other two. Partially as a result of this change to the so-called "shop" system, the daily output per worker more than doubled between 1870 and 1905.[69]

[68] Lime-glass batch cost only about one-third as much as flint.—V. S. Clark, *Manufactures in U. S.*, II, 500. In 1923, M. J. Owens testified in court that lime batch cost 1 cent per pound, whereas lead batch cost 3.75 cents per pound.

[69] *Glass Industry*, pp. 14, 299. Cf. also *Productivity of Labor in the Glass Industry*, p. 28.

4. Types of Furnaces and Other Equipment Used

The technological developments considered thus far have concerned either the working of glass or its composition. Attention must now be directed to those innovations which related to fuel, furnace construction, and type of equipment.

Abundant and easily accessible forests provided fuel for all the early glassworks in this country because pine and oak, kiln-dried and split into small sticks, produced an intense fire at extremely low cost. A majority of the melting furnaces continued to burn wood until about the middle of the nineteenth century,[70] and annealing ovens did so much later. Sometimes resin or pitch, either alone or with wood, was burned during the melting or the working process. As had happened centuries earlier in England, local supplies of wood eventually failed, and increasing numbers of entrepreneurs had to cease work or turn to mineral fuel.[71]

The first American glasshouse to use soft coal was built by Isaac Craig and James O'Hara at Pittsburgh in 1796–1797, and until after 1810 none outside that area followed its example. With the opening of the Virginian and Cumberland mines, ample supplies of bituminous coal became available, and even some of the best flint-glass factories burned it. Anthracite was also used in some places during the 1850's in spite of the fact that it required a stronger draft than bituminous. By 1860 only one furnace in New England continued to burn wood,[72] and a similar shift in fuel was fast taking place in all other glassmaking areas.

In some respects coal was less desirable than wood. Its use increased the cost of constructing furnaces because tunneled drafts and expensive understructures were necessary.[73] And when fans

[70] The numerous works in South Jersey, for example, burned chiefly wood until after 1855, when they began to use anthracite.—See Gaffield Collection, Glass Journal, I, 178, 214; II, 94–95; Weeks, *Report on Manufacture of Glass*, p. 97.

[71] Weeks reported in 1880 that "the fuel used in glass-making in the United States is chiefly coal, though at all works more or less wood is used for various purposes . . . [and] one or two [melting] furnaces . . . are reported as being wood furnaces."—*Ibid.*, p. 4.

[72] Jarves, *Reminiscences*, p. 114; Lee, *Sandwich Glass*, p. 62.

[73] Of course, entrepreneurs in this country did not have to develop the principles of such a furnace; they had only to copy European models, for coal furnaces had

propelled by steam were employed to build up intense heat,[74] the pots were cut and the life of the furnace shortened.[75] The New England Glass Company sought to avoid such difficulties and yet obtain a sufficient draft by building a huge chimney 235 feet tall. Another drawback of coal, especially bituminous, was that its smoke and fumes impeded work around the shop and colored the glass unless hooded pots were used in melting. The working hearths, flattening ovens, and annealing ovens and lehrs, consequently, burned coke, resin, or wood. All the disadvantages, however, were outweighed by the relative cheapness of coal and the possibility of obtaining a very high furnace temperature if adequate drafts were used.

An innovation which must be ranked in importance with pressing and the improvement of lime glass was the gas furnace. The first one of any importance in glassmaking was developed by Frederick von Siemens, brother of Werner von Siemens, who attained fame for his work in electricity; and it was installed in an English sheet-glass factory in 1861.[76] At first it did not operate with any great economy,[77] but by 1876 various improvements had induced at least 81 European houses to adopt it. Other furnaces which differed somewhat in principle had also been developed and introduced abroad on a wide scale.[78]

---

long been used in French and English glasshouses. See Scoville, "Technology and the French Glass Industry," *Jour. of Econ. Hist.*, I (1941), 158 ff.

[74] Gaffield Collection, Glass Journal, I, 17. All coal furnaces were not blast furnaces, and available data are insufficient for dating the first application of the blast principle in glassmaking and for gauging the extent of its popularity. Many factories, however, used a blast.

[75] *Ibid.*, I, 5–6. The greatest damage was done when new coal was put on the fire. Jarves' Cape Cod works and a few others adopted a special stoking device which pushed the fuel up through the glowing coals. — Davis, "American Glass Industry," p. 84; Gaffield Collection, Glass Journal, II, 22–23; J. Gillinder, in *One Hundred Years of American Commerce*, p. 279; Jarves, *Reminiscences*, p. 113.

[76] Gaffield Collection, Glass Journal, I, 85–86; Powell, *Principles of Glass-Making*, p. 110; *The Mechanics Magazine*, November 18, 1871, p. 395; Weeks, *Report on Manufacture of Glass*, p. 36. Somewhat later its principles found wide application in the iron and steel industry.

[77] Henry Chance wrote on December 24, 1867, that his firm had six Siemens furnaces at its Birmingham works but that there was no appreciable saving in fuel. "In all other respects the furnaces are satisfactory." — Gaffield Collection, Correspondence on Sunlight Experiments, I, 75.

[78] Powell, *Principles of Glass-Making*, p. 110. Of 600 glass furnaces in Germany

Only a few glassmakers in the United States followed the example of producers in England, France, and Germany.[79] Out of a total of 348 melting furnaces in 1880 only 21 burned gas, and this number included not only the Siemens but also the Gills, Nicholsons, and other American variations of European models.[80] The first Siemens furnaces in the United States were constructed at Pittsburgh in 1863 by Park, McCurdy, and Company (for refining copper) and by James B. Lyon and Company (for melting flint glass).[81] The Siemens model gained wide acceptance sooner in ironmaking and steelmaking than in glassmaking. Only three houses in the glass industry experimented with it before 1870,[82] and not one of these could proclaim it an unqualified success. Ten years later a few progressive firms in every branch save window glass had been won over by the merits of heating with gas.

The Siemens principle in 1861 embraced a regenerative furnace (which was, in fact, its distinctive feature) and a separate unit for converting solid to gaseous fuel. The regenerators of each furnace were two pairs of firebrick checker-works; through one pair the gas and air flowed into the furnace, and through the other they passed from the furnace after combustion into the chimney. A reversing valve was so arranged that the flow of gas and air be-

at the end of 1877, it was reported that 336 were heated with gas. — Gaffield Collection, Scraps, II, 185.

[79] Atwater, "Glass-Making in America," *Engineering Mag.*, IV (1893), 884. Gaffield, with reference to the Siemens furnace, observed on February 14, 1865: "We find that, as with all other useful inventions and discoveries, industry, time, and patience are necessary for their successful introduction." — Glass Journal, II, 65.

[80] Weeks, *Report on Manufacture of Glass*, Table VI. Not all of the 21 were in actual operation in 1880. Some were lying idle; others were under construction.

[81] In 1862 Thomas Gaffield had visited England to confer with C. W. Siemens about introducing the furnace in this country. C. W. Siemens, with his brother Frederick, held the basic patents issued in 1861. Gaffield, Tuttle, and Company was issued the first U. S. patent on March 1, 1864; but sometime between 1869 and 1871 its interests were sold to a group of construction engineers at Philadelphia. — Gaffield Collection, Copies of Letters, I, 149–150; Glass Journal, I, 41–42, 176, 187, 218, 264; II, 37–38.

[82] James B. Lyon and Company at Pittsburgh, the Lenox Plate Glass Works in Massachusetts, and Burgin and Sons at Philadelphia. — Atwater, "Glass-Making in America," *Engineering Mag.*, IV (1893), 884; V. S. Clark, *Manufactures in U. S.*, II, 72; Gaffield Collection, Correspondence on Sunlight Experiments, I, 91; Glass Journal, I, 275–276, 284–285; II, 28, 173; Glass-Making, Vol. VI; Scraps, I, 12–14, 28–30. Apparently, James B. Lyon and Company operated its furnace less than a year.

fore and after combustion could be switched from one regenerator to the other. This meant that before entering the furnace the gas and air were heated by contact with the brickwork, which had previously absorbed the waste heat of the exhaust gases. In other words, the regenerative chambers were designed to save the heat which otherwise would have been dissipated through the chimney and to utilize it to preheat the gas and air before they were ignited in the melting chamber.[83] That the regenerators permitted some economies in fuel was not seriously questioned after the first model was improved upon; but the Siemens system had at least three drawbacks: tar formed on the brickwork and had to be cleaned away and burned every so often; the gas did not give a uniform heat since its own temperature rose and gradually declined each time the reversing valve directed its course through the other regenerator;[84] and the checker-work chambers could not be added very easily to converted coal furnaces. The fact that entirely new melting units had to be built deterred many Americans from adopting the regenerative principle.

The other gas furnaces developed in Europe or modified in the United States were designed to avoid these disadvantages. Apparently not one of them used as little fuel as the Siemens, but their gas producers and flues could be installed with much less expense because only slight changes had to be made in the old furnaces.

Any sort of gas furnace, however, was definitely superior to furnaces fired directly with coal. Since gas makes little or no mess when burned, there was no smoke or pyrite fragments hurled from the fire pit to discolor the glass in the pots, and workers could easily keep the whole working area about the shop clean. The life of both pots and furnace walls was considerably extended by the substitution of gas for coal because the latter, especially in blast furnaces, yielded a very corrosive fire. Furthermore, the heat from gas was intense, relatively uniform, and easily regulated. The temperature could be quickly raised for melting and

[83] Gaffield Collection, Glass-Making, Vol. VI. It was said that the gas and air could be raised in this way to a temperature approaching 3,000°F.

[84] Tock, "Les différents systèmes de fours employés dans la fabrication des verres à vitres," *Revue Universelle des Mines*, XXXIX (1876), 579–580, 596–597.

refining the metal and then lowered for working it. Economy of fuel was perhaps the most important advantage claimed for gas furnaces. Not only did they consume smaller quantities of coal than the direct-firing type, but also their gas producers could burn the cheapest grade.[85] Consequently, the cost of coal, previously accounting for about one-third the variable costs of melting glass, fell by 40 to 60 per cent with the Siemens and by 30 to 50 per cent with the other gas models.[86]

Natural gas did not become an important factor in the industry until after 1880. The Rochester tumbler-works near Pittsburgh appears to have been the first to burn it. About 1875 this company sank a well 968 feet deep which yielded enough gas to heat its annealing ovens but not enough to keep one of its converted coal furnaces lighted except for a few months in 1875–1876. The opening of the great Pittsburgh and Middle Western gas fields several years later stimulated the expansion of glassmaking in a way unparalleled to that time.[87] Petroleum and its derivatives likewise played little or no part in the industry's growth before 1880, even though a few plants had already burned small quantities of benzine, naphtha, or crude oil in their auxiliary hearths.

A change far more widely evidenced than the shift from solid to gaseous fuel was the marked increase in furnace and pot capacity. During the first half of the nineteenth century most American houses had had a furnace of only six pots, each of which had held 600 or 700 pounds of metal; by the period 1861–1880 the ten-pot furnace was quite common, and the capacity of each pot was between 2,500 and 3,000 pounds.[88] The number of weekly meltings, moreover, had at least doubled.[89] These facts suggest that the

[85] Weeks, *Report on Manufacture of Glass*, p. 37. Hobbs, Brockunier, and Company at Wheeling, West Virginia, in 1879, was getting slack at 30 cents per ton, or the cost of carting it from the mines to the furnace. — Gaffield Collection, Glass Journal, III, 247.

[86] Tock, "Les différents systèmes de fours employés dans la fabrication des verres à vitres," *Revue Universelle des Mines*, XXXIX (1876), 556, 579; Colné, *Paris Exposition, 1878*, pp. 332, 353, 356; Gaffield Collection, Glass Journal, I, 275–276; Scraps, I, 91, 94, 160; Weeks, *Report on Manufacture of Glass*, p. 36 and note.

[87] See below, pp. 79–80.

[88] The data substantiating these generalizations are far too numerous to warrant specific reference here to their sources.

[89] A coal fire melted and refined glass in less time than wood, and gas was even faster than coal.

typical furnaces engaged in bottle and tableware production in 1880 probably turned out fourteen times as much glass each week as the small ones had melted near the beginning of the century.[90]

Tank furnaces did not really begin to supplant pot furnaces in this country until the 1890's.[91] The Anglo-American Glass Works at Poughkeepsie, New York, in 1879–1880 was the first in the United States to build and operate successfully a continuous tank. Only three or four so-called "day tanks" [92] had been in use for a few years before this in as many bottle- and tableware-shops. In 1880, throughout Europe, eighteen Siemens' gas furnaces were equipped with continuous melting tanks, and six others had day tanks. With very few exceptions American producers continued for another decade to ignore the marked superiority of continuous tank furnaces over those holding pots.[93]

There was very little improvement in the construction of annealing ovens, which manufacturers employed to reduce the strains and stresses occasioned in all ordinary glass products by unequal and uncontrolled cooling.[94] Such ovens ordinarily were

[90] The pot furnaces at the Rochester (Pa.) Tumbler Company — said to be the largest in the country — in 1877 would each hold 20 tons of metal at one time. — Gaffield Collection, Scraps, I, 98, 160. The increase in the capacity of window-glass furnaces had been appreciably less than fourteen-fold.

[91] The first continuous tank furnace for melting glass was patented in England by the Siemens brothers in 1870. Its basic principles were almost identical with those of the open hearth in steel.

[92] Day tanks melted and refined metal at night and were worked during the day. Their capacity (about 10 tons) was much less than that of continuous tanks. — Hower, "Some Contributions to the Glass Industry," *Western Pa. Hist. Mag.*, XVIII (1935), 36.

[93] Melting costs per ton of glass were halved; continuous melting, refining, and working were possible at the same time; tanks lasted longer than pots; the entire furnace was subjected to less wear and tear because of the constant temperature; fewer men were necessary to tend the furnace; and better hours and conditions of work could be arranged for the blowers and pressers. — Gaffield Collection, Glass-Making, Vol. VI; Rosenhain, *Glass Manufacture*, p. 72; Weeks, *Report on Manufacture of Glass*, p. 38. Flint glass, however, cannot be melted in tanks because direct contact with gas fire reduces lead and produces dark glass; and pots remain the only practical method of melting several different kinds or colors of glass in the same furnace. — Fettke, *Glass Manufacture of Pa.*, p. 119; Rosenhain, *Glass Manufacture*, p. 72. Even the tank furnace had an efficiency (heat utilized/heat applied) which rarely got above 25 per cent, seldom exceeded 15 per cent, and might have been as low as 5 per cent. Nonetheless, it was by far the most efficient glass furnace theretofore devised. — Phillips, *Glass*, p. 151.

[94] The United States issued patents on tempered or toughened glass on Decem-

square, squatty affairs which were first heated (most often with wood), then filled with glassware, and finally closed up to cool. Obviously, they wasted considerable floor space, time, and heat, but only a few American factories before 1880 had begun to experiment with a more efficient annealing device called the lehr. Lehrs were long, tunnel-like structures through which glass slowly moved in iron pans, hooked together in a sort of train. Since the practice was to keep the entrance of the lehr constantly heated and the other end at shop temperature, bottles and other items could be packed as soon as they emerged from it. That lehrs, in spite of their superiority as savers of time, space,[95] and fuel, did not replace the old-type oven in most shops until well after 1880 is just another indication of the backwardness of American producers in matters of technology.

---

ber 15, 1874. Such glass was made in the usual manner, then reheated, and suddenly plunged in an oleaginous bath. This procedure imprisoned the strains and stresses under great surface compression. Such glass was harder to fracture than ordinary annealed glass; but once the slightest fracture was made, the glass "exploded." For centuries glassmakers throughout Europe had observed such phenomena in Prince Rupert's drops, but none of them — before a Frenchman named De la Bastie in the 1870's — had seen an opportunity for commercial profit. One or two attempts to make "La Bastie" glass in this country were not very successful. — Gaffield Collection, Glass Journal, III, 71, 81, 121, 162, 163; Scraps, I, 54, 57, 101. The same basic principle of tempering or toughening has today been perfected and used in making such items as "safedge, fire-treated" tumblers (by the Libbey Glass Company) and thick plate-glass doors.

[95] Hence, they economized capital. Ovens ordinarily took days to cool, whereas glass was annealed while passing through lehrs within eight or ten hours.

CHAPTER II

THE HISTORY OF AMERICAN GLASSMAKING PRIOR TO 1880

(*Continued*)

From the technological developments outlined in the preceding chapter it is apparent that the five outstanding innovations were the perfection and mechanization of pressing, the improvement of lime glass, the shift from wood to coal as fuel, the beginning of rough and polished plate manufacture, and the introduction of producer-gas furnaces. The industrial effects of the last, however, did not become important until after 1880. Small-scale production continued to be characteristic of most plants in the bottle, window-glass, and tableware divisions.[1] Even though the industrial and technological patterns of growth were definitely interrelated, a complete understanding of each is still lacking. The scale of operations of individual works, for example, does not *wholly* account for the slow rate of change in technical procedures nor for the relatively backward state of glassmaking in America compared with certain European countries. Neither was the state of technology *entirely* responsible for the character of the industry. Other factors also influenced the rate of growth.

1. Labor and Labor Conditions: Comparative Wage Levels

Labor was significant because a high degree of skill was required in the work [2] and because wages constituted a substantial part of the total costs of production. According to the returns of the 1850 census, payments to labor amounted to about 43 per cent of the value of all glass made.[3] Calculations for 1850 and

[1] The few plate-glass factories built in 1869–1880 were very much larger than the average shop in these branches.

[2] Ordinary pressers and casters were semiskilled.

[3] Weeks, *Report on Manufacture of Glass*, Table VI. The importance of labor as an item in production costs varied somewhat according to the branch of the industry. Annual wage payments in 1880 were about 47 per cent of the total

1870 show this percentage to have been 45 per cent and 41 per cent, respectively; and the operating costs of the Boston and Sandwich Glass Company for four six-months' periods between 1826 and 1829 reveal that wages accounted for approximately 53 per cent of the total.[4] Most of the other data that have been collected suggest that at least two-fifths of the gross income of all shops (except those making plate glass) went as payments to workers.[5]

Nearly all the operations connected with gathering and blowing glass required extraordinary skill. For this reason it was long necessary for Americans to draw heavily upon European factories for their labor supply. Emissaries went to Great Britain to engage workers and shop foremen,[6] and manufacturers who sought to improve or enlarge their working force paid the passage for many Belgian, French, and German glassmakers.[7] Since blowers were extremely reluctant to impart their skill to others, and since the number of furnaces lighted in the United States increased rapidly during the century, entrepreneurs as late as 1880 were still seeking men from abroad.[8] They were not always fortunate in getting reliable workers who could resist the great demand for their services. Many an employer who had paid all traveling expenses found that he could not cope with the independence which

value of flint and lime glassware; 42 per cent of window glass; 40 per cent of green (container) glass; and 34 per cent of plate. Weeks, in Volume XX of the *Tenth Census*, reported that for one flint-glass shop the "percentage of wages to cost for the entire production" was 33 per cent in 1877–1880, and for another, 50 per cent in 1867–1870 and 65 per cent in 1880 (pp. 83, 91); for a green-bottle shop, 50 per cent in 1864–1880 (pp. 99, 100). For other shops, 33 per cent and 50 per cent were most frequently given.

[4] Lee, *Sandwich Glass*, pp. 58–60.

[5] Gaffield Collection, Scraps, II, 135, 156; Notes on Glass, IV, 52. The census returns for 1860 place the fraction at one-third. Unquestionably, the glassmakers who reported to the Tariff Commission in 1882 that labor was about 90 per cent of total blown glassware costs (Tariff Commission, *Report of 1882*, pp. 2,133, 2,512) were guilty of gross exaggeration.

[6] In spite of the fact that her laws, until 1825, prohibited the emigration of artisans. Apparently the men ran considerable risk and had to employ many ruses in order to get out. — Jarves, *Reminiscences*, pp. 58, 70; Knittle, *Early American Glass*, pp. 169, 181; Watkins, *Cambridge Glass*, p. 163; McKearin, *American Glass*, p. 101; Gaffield Collection, Glass Journal, II, 207.

[7] The Winslow (N. J.) Glass Company had obtained its entire working force from abroad around 1863. — Gaffield Collection, Glass Journal, I, 214.

[8] *Ibid.*, III, 242; Glass-Making, IV, 24; Scraps, II, 154; III, 60.

immigrant glassmen soon acquired nor keep them from going to work at higher wages elsewhere.[9]

Geographic mobility was not confined to immigrants. All glassmakers appear to have moved about from one shop to another to a greater extent than workers in other occupations.[10] Men from Maryland, New Jersey, New York, and Massachusetts frequently operated the western furnaces, and various managers and employees of the New England Glass Company, especially, helped establish the industry in other parts of the country. Employers sometimes traveled from one factory to another in an attempt to secure a sufficient working force for their own enterprise and left irate competitors and higher wage scales in their wake.[11]

That skilled glassmen were much better paid than common laborers or even most craftsmen in other industries has never been seriously questioned. Fortunately, Joseph D. Weeks, in Volume XX of the *Tenth Census*, has presented sufficient data to allow a comparison of the wages received by blowers, finishers, pressers, and flatteners in several shops located in different parts of the country with the daily earnings of common laborers in the same localities and for the same years [12] and of such artisans as painters, carpenters, blacksmiths, cabinet-makers and furniture carvers, machinists, shoe-leather cutters, temperers and forgers of cutlery, molders, and tobacco sorters. Skilled workmen in the glass industry around the decade 1870–1880 received at least two, and most often three, times the wages of ordinary laborers em-

[9] Gaffield Collection, Scraps, I, 48; Glass Journal, II, 2, 3, 64; Bining, "Glass Industry of Western Pa., 1797–1857," *Western Pa. Hist. Mag.*, XIX (1936), 258. Gaffield, speaking *ex parte*, observed that "English workmen are glad to come over to this country, but they act in a most shameful manner when here, where the laws regulating work and workmen are not so stringent as in England." — Glass Journal, II, 111. "They have had their passage money paid, and after arrival, some have become intemperate, discontented, spent their wages, run in debt, left their employers, gone to other factories, and some returned to England." — *Ibid.*, I, 261.

[10] Due to the general scarcity of labor and the abundant opportunities for gaining a livelihood, all workers in America were much more mobile geographically and occupationally than in Europe. — V. S. Clark, *Manufactures in U. S.*, I, 244. Glass blowers, however, were loath to change their occupation.

[11] Gaffield's manuscript journals cite many examples of such activity.

[12] When possible, the years 1880, 1870, 1860, and 1850 were selected for Boston, Philadelphia, Pittsburgh, St. Louis, Wheeling, and South Jersey. However, most of the comparisons were for only 1880 and 1870.

ployed in the same and other industries and from one-third to two-thirds more than many skilled craftsmen. One writer asserted in 1896 that blowers had larger annual incomes than many learned professors and divines.[13] Such wage differentials, moreover, appear to have been even greater before the Civil War than after.[14]

Gatherers were less skilled than finishers, blowers, or flatteners and received about one-third or one-half less in wages. Soon after pressing invaded the tableware division, other employees did not consider the men who operated the machines as glassworkers at all, and their wages reflected this difference in their status. From 1850 through 1863, for example, pressers at the New England Glass Company earned only one-half as much as blowers.[15] After 1865, however, this differential began to disappear until competent pressers of flint and lime glass around 1880 received as much as, and in some shops even more than, blowers.[16] Finally, it should be noted that the workmen who participated in casting plate glass in those enterprises set up for this purpose earned in 1880 less than half as much as skilled glassmen in other branches of the industry.

Geography is another factor to be considered in this matter of wages paid for skill. If they were employed in shops in Kentucky, Illinois, Indiana, Missouri, Ohio, and western Pennsylvania, pressers, finishers, gatherers, blowers, and flatteners in the same year received on the average about 25 to 30 per cent more than they would have obtained along the Atlantic Coast.[17] The differential was least in the lime- and flint-glass branch (15 per cent) and greatest in the window-glass (50 per cent for blowers). The situation in 1880 was not unusual.

Entrepreneurs considered the relatively high wage level in this

[13] Henderson, "Glass-making," *Popular Science Monthly*, XXXIV (1888–1889), 609. Some blowers of green bottles earned as much as $300 a month around 1896.

[14] See below, pp. 35–36.

[15] And no more than gatherers. — *Tenth Census*, XX, 81.

[16] *Ibid.* The New England Glass Company, however, still paid its blowers a third more than its pressers.

[17] Based upon wage data in *ibid.* Averages were struck for pressers, finishers, gatherers, and blowers in flint and lime; for blowers only in green bottles; and for boss-blowers, cylinder-blowers, gatherers, and flatteners in window glass.

country compared with European glass centers a much more serious problem than domestic regional differentials. They contended time and again that they had to pay two, three, and four times as much for skilled labor as their competitors did on the other side of the Atlantic.[18] Glassblowers and finishers would certainly fall into that class of skilled operatives and artisans which, according to Clark,[19] was at the greatest premium in America before 1860; but it appears probable that the owners of American shops exaggerated their disadvantage. A United States consul in Belgium reported in 1878 that Belgian glassworkers were earning from $56 to $65 a month; [20] and comparable workmen in Ohio and around Pittsburgh received in the same year from $80 to $100 a month. This means that American wages were about one-half greater than Belgian wages. The Tariff Commission was told in 1882 that the differential for blowers of window glass was about one-third.[21] At any rate, the amounts paid skilled glassworkers in the United States were high enough to constitute one of the important reasons why American producers operated under relatively disadvantageous conditions.[22]

It would be extremely helpful to know to just what extent the comparative wage situation around 1880 differed from what it had been before the Civil War, but unfortunately sufficient information regarding foreign wages is not available. There are enough data on domestic wages, however, to permit the construction of simple index numbers for the daily earnings of certain skilled glassworkers from 1841 through 1880, and these are presented in Chart I. For the earlier period — from 1800 to 1840 —

[18] See Davis, "American Glass Industry," pp. 113, 238; "Report on Glass Manufacturing in 1832" in Gaffield Collection, Glass-Making, Vol. III; Scraps, I, 67; III, 122; Atwater, "Glass-making in America," *Engineering Mag.*, IV (1893), 887; Jarves, *Reminiscences*, p. 93; J. Gillinder, in Gillinder and Evans, *Papers*, p. 33.

[19] *Manufactures in U. S.*, I, 390–391.

[20] Gaffield Collection, Scraps, II, 156.

[21] *Ibid.*, III, 122. Manufacturers of flint and lime glassware, however, reported to the same body that foreign wages were from one-half to one-fourth as high as the prevailing American level. Tariff Commission, *Report of 1882*, pp. 2,133, 2,513.

[22] Cf. Davis, "American Glass Industry," p. 141. Jarves wrote at the middle of the century: " . . . when the labor on glass constitutes its chief value, then glass can be imported cheaper than it can be manufactured in this country." — *Reminiscences*, p. 93.

the trend was probably upward, although it is impossible even to estimate the percentage increase. In 1865 glassworkers' wages were approximately 63 per cent higher than in 1860,[23] and the

CHART I

GENERAL WAGES, GLASSWORKERS' WAGES, AND WHOLESALE PRICES IN THE UNITED STATES, 1841–1880*

(Base = 1878–1880)

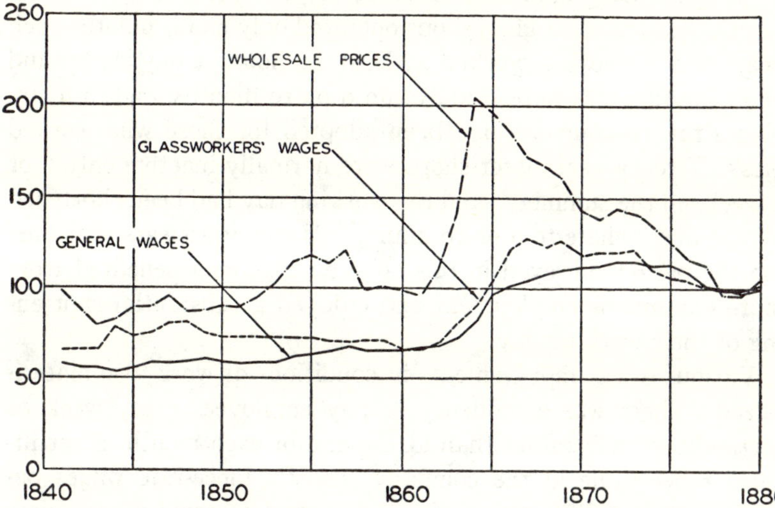

* The general wage series is that given in U. S. B. L. S., *Wages, Colonial Times to 1928*. In order to facilitate the comparison, the index numbers constructed with 1913 as the base have been divided by the average of the indices for 1878–1880. The glassworkers' index was compiled from wages paid pressers, finishers, gatherers, blowers, cutters, and flatteners from nine different factories (*Tenth Census*, XX) and from wages paid blowers of window glass and five different sizes of green bottles (Aldrich report on *Wholesale Prices, Wages, and Transportation*, IV, 1566, 1572). — The smallest number of relatives for any one year was 6 (for 1841–1844), and the largest was 38 (for several years following 1865). The wholesale price series is that given in Warren and Pearson, *Prices*, with the indices divided by the average for 1878–1880.

indexes show a decrease in real wages during the war period if one concedes that wholesale prices reflect (but imperfectly) the cost of living.[24] The ratio of glassworkers' wages to wholesale prices

[23] According to the figures used in Chart I, general wages rose by about 49 per cent.

[24] Admittedly, retail prices of commodities are more significant than wholesale

increased by 45 per cent from 1841 to 1880, whereas a similar ratio for general wages indicates a rise of about 62 per cent. This meant that during the four decades ending in 1880 skilled glassmen, on the whole, saw the differential in their earnings as compared with those of other workmen constantly lessen.

Money and real wages, however, are not the only factors involved in the well-being of any group; hours and other conditions of work are also significant. During the first decades of the nineteenth century, most glass shops operated only 6 or 7 months each year, and employees worked at least 12 hours a day.[25] Around 1860 the furnaces were usually run 9 or 10 months, and in many places the 10-hour day had been adopted for those who worked glass. Twenty years later shops were normally inactive only 6 or 8 weeks in the summer,[26] and the working day had been shortened almost everywhere to 9 or 10 hours.[27] Hence it appears that during the period surveyed in this chapter, glassmen benefited from more continuous employment and enjoyed a substantial shortening of their working day.

Various other improvements in conditions of work also materialized. There was a tendency to pay employees every week or fortnight in cash rather than at the end of each month in certificates redeemable at the company store.[28] Piece-rate wages became increasingly popular after 1860, and in some plants glass

---

prices in indicating changes in real wages, and retail prices ordinarily move more sluggishly than wholesale. But Mitchell found that it did "not matter greatly which series is used." — *Greenbacks*, pp. 337–338.

[25] Most of the following remarks regarding time worked are supported by information given in *Tenth Census*, XX, 78 ff.; Gaffield Collection, Glass Journal, I, 173; II, 43; III, 241; Scraps, I, 15; and in Weeks, *Report on Manufacture of Glass*, p. 90.

[26] Some, in the tableware division, maintained production the year around.

[27] The 7-day week did not become common until the twentieth century; during the nineteenth, 6 and sometimes 4 or 5 days constituted the working week. — Davis, "American Glass Industry," p. 229; Knittle, *Early American Glass*, p. 226; Lee, *Sandwich Glass*, p. 63; Watkins, *Cambridge Glass*, p. 26; Gaffield Collection, Glass Journal, I, 8–9.

[28] *Tenth Census*, XX, 78 ff. In 1880, 27 glassworks still maintained company stores as against 146 which reported no store. Many of the 27 stores were kept only as an accommodation with no obligation to trade there; but Weeks reported that "this question of 'truck' is [still] a burning one between employer and employed." — *Report on Manufacture of Glass*, p. 9.

was only melted and not blown or pressed at night.[29] Builders attained better ventilation and more healthful conditions in the workroom by placing chimneys over the furnace and by constructing more windows in the shop. Sometimes they built segments of the factory walls on pivots so that the walls could be swung open in warm weather.[30] Company-owned and employer-dominated villages gradually became much less common than during the first half of the century when most glass enterprises had bred practically self-contained communities.[31]

United action on the part of glassworkers occasioned few of these ameliorations, for strong unions did not become important until about 1880.[32] Local organizations existed shortly before and during the Civil War, but nearly always these were short-lived.[33] In either 1877 or 1878 a local gatherers' union in the Pittsburgh window-glass branch joined the Knights of Labor as Local Assembly 300, and by 1881 it was said to have enrolled in its membership nearly all blowers of window glass throughout the country. The American Flint Glass Workers' Union of North America resulted from the first successful national convention held by flint workers at Pittsburgh in 1878; and employees in shops making green and bottle glass also organized in the 1870's.

[29] Dyott, *Labor at the Glass Factory of Dyottville*, pp. 19–20; Gaffield Collection, Scraps, II, 118. Some of the flint-glass plants worked 24 hours a day, and their employees rotated in either 6- or 12-hour shifts. — Chipman, *Old Sandwich Glass*, pp. 38–39; Lee, *Sandwich Glass*, p. 63; *American Artisans*, April 17, 1867, p. 371.

[30] Especially for the bottle-houses in South Jersey.

[31] See Chipman, *Old Sandwich Glass*, pp. 23, 27–28; Gaffield Collection, Glass Journal, I, 19; II, 94–95; Scraps, I, 124, 162; II, 154, 193; McKearin, *American Glass*, pp. 100–101, 183–184, 185; Knittle, *Early American Glass*, pp. 93–94, 185; Lee, *Sandwich Glass*, pp. 25, 34, 65, 486; Watkins, *Cambridge Glass, 1818–1888*, p. 159; J. Gillinder, in *One Hundred Years of American Commerce*, p. 275; Thurston, *Pittsburgh*, p. 143. Thomas Dyott tried to establish at his Kensington works near Philadelphia a system of "moral and mental labor" rather similar to that of Robert Owens at New Lanark. — Dyott, *Labor at the Glass Factory of Dyottville*, *passim*.

[32] The great control which some of them (especially window glassworkers in 1890–1903) exercised upon wages, output, apprenticeship, and working conditions will be discussed in Chapter VIII.

[33] See Davis, "American Glass Industry," pp. 107, 175, 206–207, 226. There was, however, a sharp increase in the number of strikes during the 1860's and 1870's.

Bottle workers in the East and West, however, did not form a national union until 1890.[34]

Although formal associations were usually local or nonexistent earlier in the century, glassworkers formed a sort of brotherhood because of their craft. Entrepreneurs always faced a united front on the part of their employees when attempts were made to reduce wages, increase output, introduce new techniques, increase the number of helpers, or reorganize in any way the work about the furnace. Employers consequently claimed that they were not masters in their own shops and that the most dissolute[35] and independent workers determined the pattern of conduct for the others.[36]

Their antipathy to innovations was partially responsible for the little progress achieved in technological improvement. Deming Jarves wrote a friend:

> The glass blowers, on discovery that I had succeeded in pressing a piece of glass, were so enraged for fear their business would be ruined by the new discovery, that my life was threatened, and I was compelled to hide from them for six weeks before I dared venture in the street or in the glass house, and for more than six months there was danger of personal violence should I venture in the street after nightfall.[37]

Again, in the 1870's, workers in lamp-chimney shops refused to use a patented device for crimping the top of chimneys and went out on strike for almost two years.[38] The continuous melting tank likewise met with resistance in every branch of the industry because its introduction, necessitating considerable reorganization of work about the furnace, meant that glass had to be melted and blown at night as well as during the day if manufacturers were to realize fully its potential economies.[39]

[34] *Ibid.*, pp. 176, 207–208, 226; *Glass Industry*, p. 310.

[35] Intemperance seems to have been an unusually common failing of employees, and sometimes the men would not show up at the factory for days.

[36] Jarves, *Reminiscences*, pp. 24, 70; Gaffield Collection, Glass Journal, I, 165.

[37] From an extract of a letter displayed at the Philadelphia Centennial in 1876. — Gaffield Collection, Scraps, I, 95. This sort of reaction to innovations which imperiled their position, of course, was typical of craftsmen in most industries.

[38] U. S. Commissioner of Labor, *Regulation and Restriction of Output*, p. 643.

[39] Davis, "American Glass Industry," p. 202. The transition in fuel from coal to gas evoked little resistance, probably because glassworkers received immediate

2. Sources of Fuel and Materials: Their Effect Upon the Industry's Location

The availability of basic materials was another factor which affected the industry's development. Calculations based upon the census returns for 1860, 1870, and 1880 reveal that the cost of fuel and materials amounted to about one-third the value of all glass made;[40] and, although the evidence is not conclusive, the data found in the 1820 census suggest that materials accounted for something less than this percentage during the early part of the century. This may well have been true since the cost of fuel at that time was less significant than it was later. With the passage of years, the supply of cheap wood around each furnace dwindled, and increasing numbers of shops turned to coal which frequently had to be brought great distances.

The United States had most of the needed materials in such abundance that Jarves wrote around the middle of the century that "we can, in all the articles where the value of the materials predominates, compete successfully with importers of foreign glass."[41] The necessity of locating and developing domestic sources of supply, however, handicapped the first entrepreneurs. They could readily obtain fuel and potash from the vast forests, but they had to import fire clay for pots and furnace bricks from England, Holland, or Germany. American clay beds were first uncovered in New Jersey,[42] and later excellent deposits were found in western Pennsylvania and Missouri.[43] Since glassmakers did

---

benefits in the form of cleaner conditions about the shop and because they did not have to change their own manner of work.

[40] For 1880 the proportion of material costs to value of output was 50 per cent for plate glass, 34 per cent for glassware (tableware, chiefly), 37 per cent for window glass, and 43 per cent for green glass (containers, chiefly). Hence, in those branches where labor costs loomed large, the expense of raw materials was relatively small. See above, p. 30, n. 3.

[41] *Reminiscences*, p. 93. Clark wrote that "American glass makers were peculiarly favored in respect to raw materials." — *Manufactures in U. S.*, II, 498.

[42] The early Pittsburgh houses had to haul their supplies over the mountains in carts. — Bining, "Glass Industry of Western Pa., 1797–1857," *Western Pa. Hist. Mag.*, XIX (1936), 258; J. Gillinder, in *One Hundred Years of American Commerce*, pp. 275–276; Knittle, *Early American Glass*, pp. 211, 308. As later discoveries proved, this was like "carrying coals to Newcastle."

[43] Knittle, *Early American Glass*, p. 20; J. Gillinder, in *One Hundred Years of*

not know how to prepare domestic clay properly, they did not use it very much until after 1860;[44] and some were still tapping foreign sources as late as 1880.

Manufacturers also had to locate supplies of suitable sand because that yielded by most rivers and shores contained too many impurities and particles of iron for use in making clear glass.[45] The first important sources were found in New Jersey and Delaware, and their output was shipped to New England and as far west as Pittsburgh. Near the middle of the century, individuals were quarrying the excellent deposits at Juniata (Pennsylvania) and in the Berkshires. Eventually the United States was able to provide the best and purest sand anywhere available from many different sources.[46] Domestic coal was plentiful and cheap in the western area, and shops along the Atlantic coast could draw the quantities they needed by boat either from abroad or from the Cumberland region without too great a freight charge.[47] The Leblanc, and later the Solvay, process so cheapened soda that all glasshouses except those making flint ceased using potash almost altogether after 1850.[48] Throughout the period, therefore, the industry reaped substantial benefits from the increasing availability of materials and improvements in their quality.[49] Sand and clay, for example, were shipped to furnaces over shorter distances than

American Commerce, p. 283. The clays of Maryland, Ohio, and Indiana were also usable. — Weeks, *Report on Manufacture of Glass*, p. 29.

[44] Weeks, *Report on Manufacture of Glass*, pp. 39–40, 84; Bining, "Glass Industry of Western Pa., 1797–1857," *Western Pa. Hist. Mag.*, XIX (1936), p. 258; Jarves, *Reminiscences*, p. 113. Eastern manufacturers, especially, continued to import their clay, but this may well have been due to cheaper transport costs.

[45] Nevertheless, the first houses appear to have used it. — Knittle, *Early American Glass*, p. 7.

[46] E.g., in Berkshire County (Mass.), Juniata County (Pa.), Hancock County (W. Va.), Fox River (Ill.), and Crystal City (Mo.).

[47] Compared with western producers, however, those along the Atlantic were at a disadvantage, which in large measure explains the relative decline of the industry in New England. See Henderson, "The Glass Industry," *Popular Science Monthly*, XLII (1893), 595; Weeks, *Report on Manufacture of Glass*, p. 91; Lanier, "A Cloth of Glass," *Technology Rev.*, XLIV (1942), 334.

[48] Cooper, *Manual*, p. 115; Gaffield Collection, Scraps, I, 125. Soda took a shorter time to fuse, was available in a purer state, and was actually cheaper than potash. — Gaffield Collection, Glass Journal, I, 27; Correspondence on Glass, I, 59; Macbeth-Evans Glass Company, *Glass Making*, p. 54.

[49] As a result, manufacturers did not have to refine the various ingredients or frit the batch before melting; the amount of scum which rose to the surface in

in the early part of the century; and the construction of canals, turnpikes, and later railroads lowered inland freight rates per ton-mile on all goods.

Chart II portrays the price trends of materials used, along with index numbers for wholesale commodities and glassware. Prices of materials leaped more than twice as high as prices of general commodities in 1813–1814 because British naval action in Chesapeake and Delaware bays interfered with the shipment of Virginia coal to Philadelphia and because foreign coal prices in New York soared. From 1820 to 1855 the indices for items essential to glassmaking remained constantly higher than those for most goods sold at wholesale; thereafter, except for 1866, the two series converged. Aside from the two periods of war and inflation, the long-run trend of basic-material prices was definitely downward.

The locational pattern adopted by shops reflects one of the most important influences which fuel and materials had upon glassmaking.[50] Entrepreneurs had to weigh the advantages of being close to their materials against the savings which they could effect by locating near their own customers; and in the majority of cases, they decided to construct their works near important sources of fuel. If this was impossible, they selected sites along navigable rivers and harbors in order to minimize transport charges for both materials and finished product.

Fuel dominated their consideration because practically all of its great weight was lost in combustion. Manufacturers could escape, therefore, most of the freight costs occasioned by coal or wood if they located near ample sources of supply. Batch lost only about 20 per cent of its weight during melting;[51] hence, no matter whether glass was made near sand beds or near its own markets, most of the expenses involved in shipping sand, alkali, or lime could not be avoided.[52]

---

the pot diminished; and the glass increased in brilliance. See Weeks, *Report on Manufacture of Glass*, p. 34; McKearin, *American Glass*, p. 9.

[50] In this connection I have benefited greatly from discussion with my former colleague, Robert L. Bishop, who is studying certain aspects of the bottle industry.

[51] Gaffield Collection, Scraps, I, 125.

[52] If the price of fuel had been the same in every part of the country, the

Chart II

PRICES OF GLASSWARE, MATERIALS USED IN MAKING GLASS, AND GENERAL WHOLESALE COMMODITIES, 1801–1880*

(Base = 1856–1860)

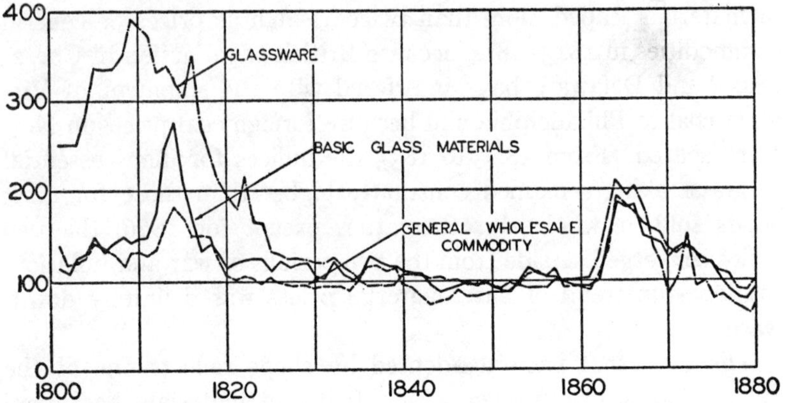

* The wholesale commodity indices are Warren and Pearson's (*Prices*, pp. 25–26), and it was not necessary to change them in any way because they averaged 100 for the base years used here.

The commodities used in the materials series are: potash at Philadelphia (1801–1860), potash at New York (1801–1860), Virginia coal at Philadelphia (1801–1860), foreign coal at New York (1801–1860), and anthracite at New York (1826–1860) — Cole, *Wholesale Commodity Prices, 1700–1861, Statistical Supplement;* red lead at Philadelphia (1801–1861) — Bezanson, Gray, and Hussey, *Wholesale Prices in Philadelphia, 1784–1861*, Part II, 121; anthracite pea (1843–1880), anthracite stove (1840–1880), bituminous (1857–1880), Rockland lime (1840–1880), and soda ash (1840–1880) — Aldrich Report, II, 176–179, 226–227, 261–262. Bezanson and Hussey have generously supplied unpublished prices for Philadelphia potash and red lead prices for 1861–1880. The weighting occasioned by using two series of potash prices (1801–1860), three of coal (for most years), and one of each of the other materials gives about the same result that selected weights would have yielded.

The commodities used in the glassware series are: window glass in New York (1801–1861) — taken from work sheets prepared by Warren and Pearson and borrowed from Dr. Cole's files at the Harvard Business School; window glass at Philadelphia (1801–1861) — Bezanson, Gray, and Hussey, *Wholesale Prices in Philadelphia, 1784–1861*, Part II, 87; window glass at New Orleans (1805–1864) — taken from work sheets prepared by George R. Taylor and borrowed from Dr. Cole's files; window glass at New York (1825–1863) — Juergens, "Wholesale Prices in New York City, 1825–1863," *Quarterly Publications of Amer. Statistical Asso.*, XII (1910–1911), 544; polished but unsilvered plate glass (1857–1880), third-grade window glass (1852–1880), first-grade window glass (1840–1880), bowls (1860–1880), goblets (1860–1880), pitchers (1860–1880), and glassware sets (1860–1880) — Aldrich Report, II, 233, 241, 242, 276–278. It is unfortunate that all of the series before 1857 are for window glass.

Sand, for example, would have to be shipped from its source to the consumer market (either as silica or as glass) in any case. Some producers, nevertheless, justified the location they chose for their plant by indicating the proximity of excellent sand beds. Fortunately for a few who took this erroneous view,[53] real economies were realized because the plant actually had a centralized position in regard to market areas.[54]

Most wood-burning furnaces were built near forests and, like sawmills, followed receding timber lines. The great advantage of Pittsburgh unquestionably lay in her excellent coal deposits; and throughout the century there was a marked tendency for the industry to shift to this and other coal-mining regions.[55] Table 2 suggests the extent of this westward trek. The data for 1820, 1840, and 1860 are not thoroughly reliable because the census returns were incomplete and probably inaccurate; yet they are helpful in indicating trends and proportions. The greatest change seems to have occurred between 1860 and 1880, when glassmaking declined sharply in the New England states and increased both absolutely and relatively in the western area.[56]

3. Demand for, and the Price of, Glassware

The growth of most industries reflects changes in the demand for their products, and glassmaking was no exception. During the first years of the nineteenth century, Americans needed window-

---

furnaces would have been built near major glass markets because the freight rates per ton-mile upon such packaged goods as glass were higher than upon such bulky items as sand.

[53] E.g., the plate-glassworks at Crystal City, Mo.

[54] It should be noted that the area east of the Mississippi River and north of the Mason-Dixon line not only provided most of the materials for glass (especially fuel) but also constituted the best market because of its relatively dense population.

[55] Depletion of local wood supplies was one of the most important reasons for the failure of shops before 1860.

[56] In 1880, Pennsylvania made about 44 per cent of all window glass, and New Jersey was a poor second with her 14 per cent; Pennsylvania also accounted for 51 per cent of all flint and lime glassware, while New York, which was second, made only about 12 per cent; New Jersey led all states in green glass (30 per cent) but was closely followed by Pennsylvania (29 per cent). The output of Allegheny County (the Pittsburgh area) alone, furthermore, was 65 per cent of all kinds of glass made in Pennsylvania and about 27 per cent of that produced in all states. The second ranking county (Philadelphia) accounted for only 7.67 per cent of the United States' output. — Weeks, *Report on Manufacture of Glass*, pp. 9–11.

TABLE 2

LOCATION OF AMERICAN GLASSHOUSES, 1820–1880

(Based on Census Data)

A. Number of Establishments

States (By Region)	Number of Establishments				Per cent of Total			
	1820	1840	1860	1880	1820	1840	1860	1880
New England (total)	(8)	(11)	(15)	(13)	(24)	(14)	(17)	(6)
Connecticut	2	2	2	1	6	2	2	0.5
Massachusetts	4	4	11	11	12	5	13	5.0
New Hampshire	2	3	2	1	6	4	2	0.5
Vermont	—	2	—	—	—	2	—	—
Middle Atlantic (total)	(14)	(38)	(43)	(76)	(42)	(47)	(49)	(36)
Eastern Pennsylvania	2	2	10	17	6	2	11	8
New Jersey	5	23	16	27	15	28	18	13
New York	7	13	17	32	21	16	19	15
South Atlantic (total)	(2)	(1)	—	(9)	(6)	(1)	—	(4)
Maryland	2	1	—	8	6	1	—	4.0
District of Columbia	—	—	—	1	—	—	—	0.5
West (total)	(9)	(31)	(30)	(113)	(27)	(38)	(34)	(54)
Illinois	—	—	—	7	—	—	—	3
Indiana	—	—	—	4	—	—	—	2
Kentucky	—	—	1	5	—	—	1	2
Missouri	—	—	2	6	—	—	2	3
Ohio	3	—	4	20	9	—	5	9
Western Pennsylvania	5	26	21	61	15	32	24	30
West Virginia	1	4	2	4	3	5	2	2
All Others	—	1	—	6	—	1	—	3
Total For All Areas	(33)	(81)	(88)	(211)[1]	(99)	(100)	(100)	(100)

B. Value of Output

States (By Region)	Value of Output			Per cent of Total		
	1840[2]	1860	1880	1840	1860	1880
New England (total)	($ 628,000)	($ 2,094,576)	($ 1,084,345)	(21)	(25)	(5)
Connecticut	32,000	58,000	160,000	1	0.7	0.8
Massachusetts	471,000	2,004,500	854,345	16	24.0	4.0
New Hampshire	70,000[3]	32,076	70,000	2	0.4	0.3
Vermont	55,000	—	—	2	—	—
Middle Atlantic (total)	(1,411,071)	(3,452,186)	(7,315,925)	(48)	(41)	(35)
Eastern Pennsylvania	95,000[3]	1,167,000	2,084,959	3	14	10
New Jersey	904,700	1,098,000	2,810,170	31	13	13
New York	411,371	1,187,186	2,420,796	14	14	12
South Atlantic (total)	(40,000)	—	(587,000)	(1)	—	(3)
Maryland	40,000	—	587,000	1	—	3
District of Columbia	—	—	—	—	—	—
West (total)	(866,222)	(2,954,543)	(12,167,301)	(29)	(34)	(57)
Illinois	—	—	901,343	—	—	4
Indiana	—	—	790,781	—	—	4
Kentucky	—	65,000	388,405	—	1	2
Missouri	—	282,000	919,827	—	3	4
Ohio	—	91,400	1,549,320	—	1	7
Western Pennsylvania	702,400	2,296,143	6,635,625	24	27	31
West Virginia	156,500	220,000	748,500	5	3	4
All Others	7,322	—	233,500	0.2	—	1
Total For All Areas	($ 2,945,293)	($ 8,501,305)	($21,154,571)[4]	(99)	(100)	(100)

[1] Only 169 of these, however, actually made glass in 1880; 42 were idle or being built.
[2] Includes the output of glass-cutting establishments although the number of establishments listed in Part A is for glassworks only.
[3] Includes one interpolated value for a shop which did not report its output.
[4] Naturally, this total includes the output of only those shops which were active.

panes for their houses more than they needed any other kind of glassware; therefore, the major portion of the output was sheet or crown. Later on, fine grades of flint glass found a ready market, and bottles were produced in increasing quantities to supply druggists, concocters of patent medicines, distillers, brewers, and household food packers. The demand for plate glass did not become sufficiently great to warrant domestic production on a large scale until the latter half of the century.[57]

Glass panes were exceedingly rare in houses built during the seventeenth century. At first they were small, diamond-shaped pieces which were fitted into casement windows with lead. After 1700 the double-hung frame grew in popularity, and windows came to be regarded from an architectural as well as from a purely utilitarian standpoint. The swing-out variety, patterned after the French and sometimes reaching the floor, enjoyed great esteem in the 1850's, but the ordinary two-sash window remained the most common. Regardless of the type of frame employed, there was a persistent tendency to use larger and larger panes.[58] A shop in New Hampshire around 1814, for example, finding that its 8″ by 10″ squares were larger than the public wanted, had to discontinue this size.[59] Thirty years later, in the Boston area, twelve panes of 11″ by 16″ or 12″ by 18″ were commonly used in each window; then the sash with three long lights (for example, 26″ by 10″) made its appearance, only to be superseded in new constructions by two-light sashes. Gaffield remarked that in 1867 only a few people had yet installed sashes with single panes (which measured from 30″ by 30″ to 42″ by 40″).[60] Eventually this single-light sash became the most prevalent.

Some of the wealthiest people in Boston had begun to use polished plate instead of sheet glass in their front windows before 1850, and thereafter this practice spread. More and more city

[57] Clark mentions the "existence of an almost unlimited market for it" in the Middle and Far West during the 1880–1893 prosperity. — *Manufactures in U. S.*, II, 502.

[58] This was the chief reason why sheet or cylinder glass finally displaced crown altogether. See above, pp. 15–16.

[59] Burbank, "Keene Glass Works," *Antiques*, IX (1930), 312.

[60] Gaffield Collection, Notes on Glass, IV, 151 ff.

stores used plate glass for display purposes, and during the decades 1840–1870 the ordinary size of each pane increased from 26″ by 16″ to 84″ by 60″.[61]

Bottles and flasks, like windowpanes, enjoyed a ready market throughout the century. At first, they were used chiefly for whiskey and other spirits, but the patent medicine industry soon began to require large quantities also. The preservation of foodstuffs by housewives (to whom the advent of the Mason screw-top jar in 1858 was a boon) increased the demand still more. Commercial food packers, however, did not use glass containers to any great extent before 1880.

The tableware division probably reflected changes in consumers' buying habits more than any other branch of the industry. Until 1864 flint, and thereafter lime, glass was blown or pressed into all sorts of items for household use, and beautiful designs were cut or etched by acid on many pieces. If one shop developed a pattern which met with popular approval, others immediately duplicated it. After the Civil War, manufacturers fashioned complete sets of tableware in the same pattern, and colored glass returned to favor. Since pressing and the improvement in lime glass had placed tableware within the reach of the masses, its market in 1880 was many times larger than it had been a half-century earlier.

One major division in the flint-glass branch arose in response to technological changes outside the industry. This was the manufacture of chimneys for oil lamps. Although a few chimneys had been blown and shaped by hand earlier, the discovery of petroleum and the perfection of the kerosene lamp shortly before the Civil War eventually led to a tremendous increase in the demand. The peak in production occurred in the decade following 1890, when the annual output averaged about 22,500 tons and actually reached 26,554 tons in 1899.[62] Meanwhile, the wholesale price per dozen chimneys had fallen from $1.75 in 1854 (and $2.25 in 1864) to as low as 30 cents in 1888.[63]

[61] *Ibid.*, IV, 153–154. Gaffield concluded that these two trends resulted in a great increase in the demand for polished plate.

[62] Gillinder and Evans, *Papers*.

[63] Davis, "American Glass Industry," p. 238, n. 2.

Since the amount of all kinds of glass taken off the market in any one year partially depended upon selling price, it would be interesting to know whether items other than lamp chimneys also became cheaper as industrial output expanded. Bottles appear to have declined in price by about two-thirds from 1804 to 1832; flint-glass tumblers by something like three-fifths; and common glassware by one-half.[64] Pressed lime tumblers in 1877 were bringing only 19 cents per dozen in the East, whereas in 1808 those blown of flint glass around Pittsburgh had sold for $2.00 per dozen.[65] These fragmentary data, admittedly, are unsatisfactory for tracing price trends for the entire period. Fortunately, it was possible to construct simple index numbers from a few continuous price series compiled by others, and these are shown in Chart II. Until 1857 window-glass prices in several different places alone were available; consequently, the indices at best give only the probable trend of prices for all kinds of glass prior to the Civil War. That the long-run trend was markedly downward, however, is not open to serious question.

The domestic market expanded not only because of the relative decline in the price of glass compared with most other commodities but also because of the increasing population, the rising standard of living, and a remarkable improvement in transportation facilities. During the eighteenth century most shops had to sell their output in neighboring regions unless they were happily located along natural water routes. In fact, the high cost of carting windowpanes and bottles over the Alleghenies between 1796 and 1800 was one reason why men constructed furnaces around Pittsburgh at that time. By 1832, however, both eastern and western shops were selling their products from Maine to New Orleans. First canals and later railroads drastically reduced inland freight costs.[66] A hundredweight of glass shipped between

[64] *Ibid.*, p. 122; Dyott, *Labor at the Glass Factory of Dyottville*, p. 53; "Report on Glass Manufacturing in 1832," in Gaffield Collection, Glass-Making, Vol. III; White, "Early Pittsburgh Glass Houses," *Antiques*, X (1931), 366; Knittle, *Early American Glass*, pp. 214, 314.

[65] Knittle, *Early American Glass*, p. 314; Gaffield Collection, Glass Journal, III, 174; *National Glass Budget*, XXIV, February 27, 1909.

[66] The rate per ton-mile between the Atlantic Coast and the interior for most

Pittsburgh and Philadelphia, for example, had to pay about $10 in 1800; after the state's canal system was completed, the cost fell to around $1.25. By 1853, owing to the competition of railroads, it had declined to 40 or 50 cents.[67]

4. Government Policy, Wars, and Depressions

Government agencies did not participate directly in the development of glassmaking to any great extent after 1800. During the eighteenth century, however, colonial and state legislatures had encouraged various entrepreneurs by permitting them to raise money through lotteries, extending them loans with or without interest, exempting their capital from taxation for a limited time (usually five years), and waiving obligations of military and jury service for their employees. Less frequently, shops had been accorded exclusive production privileges;[68] and Massachusetts seems to have been the only state to give employers bounties on production.[69] Shortly after 1800 nearly all of these encouragements were discontinued, and glassmakers could only hope at the very most to obtain exemption from local taxation and military service.[70]

In 1790 Congress refused the request of John Frederick Amelung for a substantial loan so that he could continue making glass in Maryland,[71] and throughout the nineteenth century, as in other

---

goods fell by six-sevenths or seven-eighths from before the completion of the Erie Canal to 1850–1860. — V. S. Clark, *Manufactures in U. S.*, I, 350.

[67] Bining, "Glass Industry of Western Pa., 1797–1857," *Western Pa. Hist. Mag.*, XIX (1936), 267.

[68] E.g., Massachusetts granted such privileges to a group of men in 1752 and to another in 1783, 1787, and 1793. — Gaffield Collection, Glass Journal, III, 212–214; Notes on Glass, V, 78–79; Jarves, *Reminiscences*, p. 56; J. Gillinder, in *One Hundred Years of American Commerce*, p. 276; Knittle, *Early American Glass*, pp. 105, 181. Connecticut did likewise in 1747, 1779, and 1782. — Knittle, *Early American Glass*, p. 192; Gaffield Collection, Scraps, I, 147.

[69] By an act passed on June 15, 1793, the Boston Crown Glass Company was to receive 6*d*. per crown (for the first 10,000 made in any year) for three years. — Gaffield Collection, Notes on Glass, V, 79.

[70] *Ibid.*, V, 88–89, 99–100. It is interesting to note that many decades later, when natural gas was discovered throughout the Middle Northwest, local communities vied with one another in offering factory sites and free or cheap gas to those who would build furnaces in their neighborhood.

[71] Gaffield Collection, Scraps, II, 166; Knittle, *Early American Glass*, pp. 172 ff.; McKearin, *American Glass*, p. 103.

industries, the federal government followed such precedents. It did, however, encourage domestic production by subjecting glass imports to very high duties.

Pearce Davis, in his intensive study of the relation of the industry to the tariff, found that the *ad valorem* rates [72] on most kinds of glass increased from 10 to around 25 per cent between 1789 and 1812. They were doubled when war broke out. In the years 1816–1824 they were lowered almost to their pre-war level, but from 1824 to 1832 they were raised to between 50 and 100 per cent. The Compromise Tariff of 1833 provided that the rates be gradually reduced to around 20 per cent by 1842, when they were again increased to such an extent that most specific duties were equivalent to *ad valorem* rates of over 100 per cent. Congress lowered them in 1846 and again in 1857 to an average of 30 and 24 per cent, respectively. During the Civil War and until 1872 they varied from 30 to 65 per cent, depending upon the kind of glass. In 1872 only slight revisions downward were effected. Even though such figures as these admittedly are not altogether reliable (for there is no adequate method for averaging the protection afforded by specific duties),[73] they nevertheless support two generalizations. First, glassmaking enjoyed protection from foreign producers throughout the period; and, second, the amount of protection increased from 1800 to 1833, fell from then until the Civil War (except for 1842–1846), and remained substantially high thereafter.

Partially as a result of the tariff, domestic shops were able to supply around 80 per cent of the glass bought in America between 1831 and 1880 [74] compared with probably less than 60 per cent

[72] Davis cautiously attempted to estimate the *ad valorem* equivalents of the specific duties levied during the nineteenth century, and his estimates have been utilized in this summary.

[73] See, e.g., Ellsworth, *International Economics*, pp. 354 ff.

[74] The formula, (domestic output minus domestic exports) divided by (domestic output minus domestic exports plus foreign imports), yields the following:

1831	82 per cent	1860	79 per cent
1840	83	1870	81
1850	81	1880	80

Many of the data were taken from Davis, "American Glass Industry," *passim*, and some from Weeks, *Report on Manufacture of Glass*, p. 2. The percentages for certain branches of glassmaking varied much more than the averages for the

before 1820.[75] Chart III shows the balance of trade in all kinds of glass from 1826 through 1880. With each increase in duties, imports declined; with each decrease, imports increased. Hostilities,

CHART III

AMERICA'S BALANCE OF TRADE IN GLASS, 1826–1880*
(In thousands of dollars)

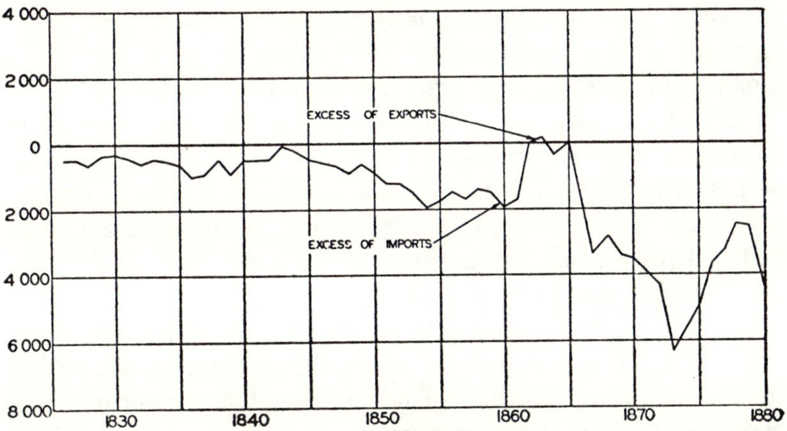

* Source: Davis, "American Glass Industry," Appendices A and B. In 1873–1880 an average of 4,552,500 pounds of bottles filled with liquids and other stuff were imported each year, but these are not included with the data used in this chart. The figure for 1843 is for nine months only, due to a shift from the fiscal to the calendar year.

new tariff rates, and especially premiums on gold[76] caused imports to decline drastically during the Civil War and exports to spurt upward.

In all branches of the industry save pressed ware, the majority of American manufacturers had no competitive advantage over

entire industry. The output of domestic window glass, for example, amounted to 94 per cent of domestic consumption in 1831; with reduced protection in 1860, this figure dropped to 65 per cent; in 1880, it was around 78 per cent.—Davis, "American Glass Industry," pp. 141, 191, and n. 1.

[75] Davis estimates that in 1820 "the domestic production of flint and common glass constituted considerably less than 50 per cent of the total U. S. consumption of these articles." For window glass, the comparable figure was judged to be 87 per cent.—*Ibid.*, pp. 24, 66.

[76] As the premium on gold in terms of greenbacks gradually declined to zero from the latter months of 1864 to the last few days in 1878, imports of foreign glass became two and three times greater than ever before.

foreigners. By maintaining higher prices than abroad, the tariff encouraged domestic production (which in itself may not necessarily be a good thing), but in no instance can one unquestionably judge it responsible for technological improvement. The development of pressing, following 1820, seems to have occurred in spite of protection, and one might well argue that after 1860 the tariff removed a powerful incentive to industrial and technological betterment.[77] The manufacture of plate glass may have been an exception, for there was greater justification on purely economic grounds for granting temporary protection to this branch of the industry than there was in the case of window glass.[78] After the duties rose from a "revenue" to a "protective" level, the domestic output of rough plate increased from insignificance in 1860 to 97 per cent of total consumption in 1890, and the domestic output of polished plate from nothing in 1870 to 82 per cent of domestic purchases in 1890.[79]

As was typical of industry in general, wars and depressions influenced glassmaking greatly. From the embargo of 1807 until the end of the war with England in 1814, the number of glasshouses multiplied. Entrepreneurs incorporated fourteen in New York State alone, and production in the Pittsburgh area almost quadrupled in value from 1810 to 1815. Between 1808 and 1814 at least forty-four glass furnaces were lighted throughout the country for the first time.[80] This growth was mainly attributable to a marked decrease in glass imports as shown in Table 3. The annual imports are given as percentages of the average number of black quart bottles and square feet of window glass bought from foreigners each year in the period 1806–1808. The effects of the Nonintercourse Act (March, 1809) and of the war were most pronounced.

The end of the war brought a flood of imports which, together

[77] At least, this is Davis' conclusion.

[78] Since plate glass necessitated production on a very large scale, the "infant industry" argument could be reasonably applied.

[79] *Ibid.*, p. 253 and Table XXIII. According to the 1880 census, the United States made 74 per cent of the rough plate it used and only 32 per cent of the polished (including silvered plate).

[80] Davis, "American Glass Industry," p. 14; Gaffield Collection, Notes on Glass, V, 98; McKearin, *American Glass*, p. 133.

with the crisis of 1819, was responsible for acute distress. It has been estimated that at least half of the glasshouses built between 1808 and 1815 failed during the first five years of peace following the Treaty of Ghent;[81] and around Pittsburgh the number of men

TABLE 3

IMPORTS OF BLACK QUART BOTTLES AND WINDOW GLASS, 1806–1820*
(Percentages of the Average for 1806–1808)

	Bottles	Window Glass		Bottles	Window Glass
	Per Cent	Per Cent		Per Cent	Per Cent
1806–1808	100	100	1815	5	9
1809	15	15	1816	220	79
1810	39	93	1817	100	51
1811	119	102	1818	114	50
1812	18	26	1819	78	44
1813	20	22	1820	55	22
1814	11	10			

* Based upon figures given by Davis, "American Glass Industry," pp. 55, 59.

employed in the industry fell by 76 per cent and the value of their output by 85 per cent.[82] Many of the entrepreneurs who submitted returns for the 1820 census made such statements as "business dull" and "supply overrunning demand in consequence of large foreign importations." During the recovery following 1820–1821 and until 1837, glassmaking prospered. At least 64 new works were established. The crisis, which brought an end to this expansion, witnessed the complete failure of some twenty or thirty shops and a temporary suspension of activity on the part of many more.

The industry grew steadily during the 1840's and 1850's despite the lowered tariffs of 1846 and 1857; not even the distress of 1857 interfered very much. Additional expansion was fostered by the Civil War and the accompanying depreciation of the greenback because they enabled Americans to improve their position in both domestic and foreign markets.[83] Glassmaking was unscathed

[81] McKearin, *American Glass*, p. 133.

[82] Davis, "American Glass Industry," p. 16, n. 3; see also, Bining, "Glass Industry of Western Pa., 1797–1857," *Western Pa. Hist. Mag.*, XIX (1936), 261–262.

[83] Until 1860 glass imports had been from five to ten times greater than ex-

by the conflict because the industry was almost entirely centralized in northern and western areas which escaped the ravages of war. The number of plants at work near Pittsburgh was 40 per cent greater in 1865 than eight years earlier, and output increased elsewhere. The cessation of hostilities and the crisis of 1873 occasioned only temporary setbacks. In 1880 there were almost twice as many factories making glass as there had been in 1860, and the value of their combined output was 2.4 times greater.[84]

5. Business Organization, Extent of Competition, and Entrepreneurship

Nearly all the companies organized in the United States before 1800 had been simple proprietorships or partnerships, and these two forms of business organization remained popular for the next hundred years. New York was the first state to pass a general incorporation law on March 22, 1811. Her example was followed by most northern and western states between 1816 and 1860. Partly as a result of these acts [85] and partly as a result of the new impulse given manufacturing by the embargo, nonintercourse, and war in 1807–1814, there was a marked increase in the number of glasshouses which sought corporate charters. Fifteen were incorporated in New York between 1805 and 1814; six in Massachusetts between 1812 and 1818.[86] Thereafter, fewer petitions were filed and some were turned down. The owner of extensive bottleworks near Philadelphia unsuccessfully sought permission to incorporate in 1834,[87] and a year earlier the New York Senate com-

---

ports. In 1862–1865 they were only 7 per cent more, and in 1863 exports actually surpassed imports.

[84] *Eighth Census* and *Tenth Census*.

[85] "At first there was a disposition on the part of company promoters to pass by the general statute in order to procure charters by special acts." — V. S. Clark, *Manufactures in U. S.*, I, 266.

[86] Gaffield Collection, Notes on Glass, V, 80–81, 98. Between 1809 and 1815 at least 22 glass enterprises were incorporated in the United States. — Davis, "American Glass Industry," p. 15.

[87] This was "Dr." Thomas Dyott, who was convicted in 1838 of fraudulent insolvency in connection with the operations of his Manual Labor Bank in Philadelphia. — Van Rensselaer, *Early American Bottles and Flasks*, p. 162; Knittle, *Early American Glass*, p. 149. Evidently the real reason for Dyott's attempt to incorporate the glass factory was his desire to disentangle this from his banking enterprise and not the altruistic reasons he advanced in several open letters appended

mittee and legislature had refused a similar request advanced by other promoters on the grounds that the amount of capital required for the proposed factory was not too great for a single proprietor or partnership to raise.

With only a few exceptions (like the New England Glass Company), most corporations in the industry merely cloaked the operations of one or two individuals. No elaborate corporate organization or specialization of functions appeared, and the number of stockholders in most firms remained quite small.[88] The majority of glassmakers who incorporated before 1880 must have been motivated not so much by a wish to raise large amounts of capital as by a desire to escape unlimited personal liability in case of business failure.[89]

No matter what form their organization took, a great many companies were short-lived. A furnace built in 1849 at Lancaster, New York, for example, operated under five different firm names within thirty-two years.[90] The Saxon Sheet Glass Company in Boston began work in 1865 only to fail within six months. It was reorganized and failed again in as short a time; a third group of men, kindling its furnace once more, admitted defeat a few weeks later.[91] The numerous but small shops in New Jersey were constantly failing and passing into new hands; one in Bridgeton underwent reorganization six times between 1836 and 1880.[92]

---

to his book.—*Labor at the Glass Factory of Dyottville*, pp. 55 ff. Pennsylvania was one of the last industrial states to pass a general statute for incorporation.—V. S. Clark, *Manufactures in U. S.*, I, 267.

[88] The advantage offered by the corporation in raising large amounts of capital, however, was utilized by the plate-glass factories built in the 1870's.

[89] The Massachusetts law did not limit the liabilities of stockholders in manufacturing corporations until February 23, 1830, and then each one was severally responsible for all debts until the whole amount of capital stock had been paid in.—Commonwealth of Massachusetts, "Act Defining the General Powers and Duties of Manufacturing Corporations," passed on February 23, 1830, and printed in full in the "Act of Incorporation of the New England Glass Company" (Boston, 1842), found in Gaffield Collection, Glass-Making, Vol. VI. British commissioners to the New York Exposition in 1854 were impressed by the advantages which Americans derived from the low cost of incorporation and the limited liability enjoyed by stockholders.—V. S. Clark, *Manufactures in U. S.*, I, 266–267.

[90] White, "Lancaster Glass Works," *Antiques*, XII (1933), 300.

[91] Gaffield Collection, Glass Journal, II, 90, 110, 113, 114, 120, 121; Lanier, "A Cloth of Glass," *Technology Rev.*, XLIV (1942), 334.

[92] Nichols, *Bridgeton, New Jersey*, p. 24.

One reason why bankruptcies were so frequent was the failure to adopt some reliable system of accounting. If books were kept at all, entries were not faithfully made nor was any one procedure followed throughout. Some of the books were more like diaries than business records,[93] and usually no allowance was made for depreciation and other capital charges. The individual or family who ran the shop as a proprietorship or partnership either was too careless or probably saw no reason for carefully calculating profits and losses.[94] Even most incorporated firms had to render an account of their affairs only to small groups of stockholders who were often employees as well. Such corporations as the New England Glass Company and the Boston and Sandwich Glass Company, with relatively numerous shareholders drawn from the business world and with their complete set of accounting papers, were exceptions rather than the rule. After the Civil War noticeable improvement was made in bookkeeping methods, and many more firms kept adequate records than formerly.[95]

Very few fortunes were made in glass before 1860.[96] An annual profit of $1,000 in bottlemaking was said to have afforded evidence of success;[97] and a leading window-glass manufacturer of New Jersey [98] stated in 1865 that substantial profits had never been made in glass. Those who were well off, he continued, had

[93] See, e.g., those of "Baron" Stiegel preserved by the Pennsylvania Historical Society and those of O'Hara and Craig in the library of the Historical Society of Western Pennsylvania.

[94] Many small businessmen may be classed in the same category today. They do not include as "costs" of doing business their "imputed" interest on capital supplied, "imputed" wages for their own labor, or "imputed" rents on their own land.

[95] As late as 1916, however, the U. S. Department of Commerce found that 109 out of 213 glass establishments submitting reports made no allowance for depreciation charges. — *Glass Industry*, p. 44.

[96] De Pauw of the New Albany plate-glass factory testified in 1882 that his company's losses up to January, 1879, amounted to $600,000 and that "no plate-glass has been made in America without loss to the maker prior to 1879." — Tariff Commission, *Report of 1882*, pp. 937, 1,530.

[97] Knittle, *Early American Glass*, pp. 356–358. Samuel Slater, in 1831, reported that most cotton manufacturers made very small profits after deducting interest on their investments and their own wages. — V. S. Clark, *Manufactures in U. S.*, I, 372.

[98] John B. Hay of the Winslow (N. J.) Glass Company. — Gaffield Collection, Glass Journal, II, 83. Gaffield remarked in about the same year that "window glassmakers have a poor chance of success or money making in our country."

made their fortunes through investments outside the industry. According to Deming Jarves, who was himself a successful manufacturer, of the 42 concerns which made flint glass in the Atlantic states before 1852, 28 were complete failures, 10 were still operating, and 4 had ceased activity. The owners of only two of the four had been fortunate enough to have "retired with property." [99] The Sandwich and the New England companies were perhaps the most profitable of the longer-lived firms. The former was able to pay yearly dividends averaging about 6 per cent from 1825 to 1888; and the latter from 8 to 20 per cent for most years between 1837 and 1873.[100] In addition, each company had reinvested from $300,000 to $400,000 of its earnings in plant and equipment.

The desire to widen profit margins did not result in horizontal or vertical expansions to any great extent before 1880. The founders of the New England Glass Company incorporated the New England Crown Glass Company in 1824 and the New England Bottle Glass Company in 1826, but this was unusual. About the only area where several manufacturers appear to have operated more than one plant was in and around Pittsburgh. In 1857 at least five firms there each owned three or more glasshouses.[101] Neither was there a persistent tendency to expand operations vertically.[102] Coal and basic materials [103] were usually purchased from outsiders, and by 1880 small companies were buying their molds from others.[104] The largest tableware factories, however, still maintained mold shops, and nearly all insisted upon making their own melting pots.[105]

Since there were no monopolistic "giants" or combinations but

[99] *Reminiscences*, p. 65.

[100] Gaffield Collection, Glass Journal, II, 265; Lee, *Sandwich Glass*, p. 57; *National Glass Budget*, XXVIII, December 12, 1914.

[101] Thurston, *Pittsburgh*, p. 142. Twenty-five years later, 59 companies operated the 75 factories in the region. — Gaffield Collection, Scraps, II, 196.

[102] Soon after the New England Glass Company began its successful career in flint glass, it decided to make its own red lead, part of which it also sold to other manufacturers.

[103] Occasionally a glasshouse would own its own sand pits.

[104] Knittle, *Early American Glass*, pp. 32–33. There were five metal-mold works at Pittsburgh in 1860 with a combined capital of $250,000 and with a total of 334 male employees. — *Ibid.*, p. 35.

[105] Weeks remarked in 1884 that "this opposition to outside potmakers is not so great as it was a few years since." — *Report on Manufacture of Glass*, p. 40.

only large numbers of separate firms in the window-glass, bottle, and glassware branches, there was much competition.[106] Practical glassmen were constantly entering the industry because the amount of capital required for building and operating bottle, window-glass, and small tableware plants was not prohibitive. Manufacturers along the Atlantic coast who faced high fuel costs increasingly complained that producers in western coal regions were ruining them, and everywhere they decried the steady increase in the number of furnaces. The improvement in lime glass increased competition in tableware to such an extent that flint-glass factories either failed or had to cease using lead in their batch.[107] Furthermore, charges of "selling below cost" were frequently hurled back and forth. Since most glassmakers did not use an adequate cost-accounting system and ordinarily overlooked all capital charges, such claims may not have been unwarranted. At any rate, the least capable entrepreneurs were continually being weeded out. Another significant result of this domestic competition, especially after the Civil War, was a tendency toward specialization in production. Firms increasingly limited their output to a particular line such as containers, window glass, or tableware; and some factories even began to develop certain types of bottles and jars as their main interest.

Foreign competition was also keen. After the War of 1812, manufacturers in England were accused of dumping all kinds of glassware in the United States in an attempt to ruin the many new shops set up following the embargo of 1807, and the British government continued until 1845 to subsidize its nationals with bounties on all exports of flint glass. Even had there been no dumping, most American producers along the seacoast could not have met the prices of imported ware because production costs in all except the pressed-glass field [108] were much greater in America

[106] As is usually the case in industry, competition was not "perfect," and no firm could have sold unlimited quantities of glass simply by lowering its prices below those of other producers.

[107] This, together with the relatively high cost of fuel in Massachusetts, was primarily responsible for the failure of the New England and the Sandwich companies in the 1880's.

[108] In this connection, it is significant that nearly all the glass exported before 1880 was pressed ware.

than abroad. Overseas freight charges and losses due to breakage in transit were not sufficient to compensate for this disadvantage. For example, only one light out of 864 pieces of plate glass imported in 1863 by a Boston firm from Birmingham, England, was broken, and European houses were able to ship their products to America in returning grain and cotton freighters at only nominal freight rates. A Belgian told Gaffield in 1862 that he knew of one manufacturer who actually had been paid 50 cents per ton by a skipper for the privilege of carrying a cargo of glass to New York so that his ship would not have to buy ballast only to bear the expense of getting rid of it upon arrival.[109] Following the Civil War the practice spread of dumping unpacked bottles as ballast into the holds of vessels returning to the United States.[110] Manufacturers for inland markets did not suffer as much from imports as did those located along the seacoast because heavy overland freight charges gave them a cost advantage.

High tariff barriers eventually gave American producers substantial protection from foreigners,[111] but they still had to contend with stiff competition among themselves. As a result, window-glass manufacturers offered little organized opposition to the unionization of their workmen, and they later found in the strong labor union an effective means for controlling total output.[112] Earlier, between 1860 and 1865, several manufacturers along the Atlantic Coast had met at least once each year to set selling prices, fix wages, and agree upon discounts and packaging costs; [113] but, since no permanent organization resulted, these first attempts to limit competition were not very successful. Around 1875 the owners of window-glass factories west of the Alleghenies got together and formed an organization, which on July 9 and 10, 1879, gave way to the American Window Glass Manufacturers' Association. This included both eastern and western producers and set out to control competition, curb overproduction, and maintain profits.

[109] Gaffield Collection, Glass Journal, I, 115.

[110] The American tariff was amended in 1883 to counteract this. — V. S. Clark, *Manufactures in U. S.*, II, 499. [111] See above, p. 50. [112] See below, pp. 217 ff.

[113] Gaffield Collection, Glass Journal, I, 84 ff., 217, 222–223, 263; II, 75, 83, 87; Prices Current of Window Glass, *passim*. The largest number of firms represented at any one meeting was eighteen on December 1, 1863.

For more than twenty years following 1880, it and the workers' union completely dominated the window-glass branch.[114] Flint-glass manufacturers also had met a few times during the Civil War,[115] but the National Flint and Lime Glass Association was not formed until May, 1880, when the Western Association (dating from 1874) and the Eastern Association (from 1879) merged. A strong Chimney Manufacturers' Association was also organized in 1880. Hence, combinations of employers strong enough to control domestic competition in their respective fields did not become influential factors in the industry's expansion until after 1880.[116]

The extent to which producers themselves conditioned the development of glassmaking should also be considered. The personal element in the economic or industrial equation is hard to evaluate because it is so elusive and not subject to easy generalization. The entrepreneur lends himself to the abstraction demanded by much of today's economic theory less readily than do labor, capital, and natural resources.[117] His thought processes and most of his actions are not deterministic because his functions are essentially those of a human being and not of any mechanistic object. It is dangerous, therefore, to postulate that from a purely economic standpoint the entrepreneur always, or even most of the time, acts rationally. This means that his behavior is never truly predictable nor entirely understandable in the way "scientific" phenomena are. Nonetheless, he plays a significant role in the evolution of most industries, and some attempt must be made to describe his character and background.

[114] See below, pp. 217 ff.

[115] Gaffield Collection, Glass Journal, I, 284–285; II, 27–28; Prices Current of Window. Glass, p. 99. Twenty-eight separate firms signed an agreement in 1864 to discontinue the sale of goods on time, to make all prices net cash, and to make their customers pay the national excise.

[116] No producers' association comparable to that in window glass ever developed in the green glass (container) field.—Davis, "American Glass Industry," p. 212.

[117] "The philosophical problem of [entrepreneurial] action is that of freedom, or especially rational freedom. A little reflection should make it clear why this topic has been so controversial. As far as scientific thinking is concerned, there can be no 'theory' of such activity, and it must be treated as residual, as what transcends interpretation in positive-scientific terms."—Knight, "Profits and Entrepreneurial Functions," *Tasks of Econ. Hist.*, II (1942), 130.

During the eighteenth and first part of the nineteenth centuries many persons in America who had never had any experience in glassmaking attempted to build furnaces and to direct the activities of skilled blowers.[118] Several people, with abundant civic pride, frequently got together and built glassworks, only to fail immediately because of inadequate capital or a lack of knowledge.[119] Dissatisfied workmen, who had quit one shop, found little difficulty in organizing co-operative ventures among themselves or in fomenting promotional schemes among other groups, because glassmaking itself was a fascinating business and because prospective profits could be made to appear very large on paper.[120] In a new country, and especially in a new industry, it was to be expected that most of the first entrepreneurs would be unfamiliar with the technical and organizational requirements for financial success. Later, as glassmaking became more widespread, executive ability developed within the industry.[121]

At times this bordered on inbreeding and created an unhealthy state, for glassmakers in all countries seem to have had a strong tendency to become slaves of their craft and its techniques.[122] Various compositions of different kinds of glass long remained cherished secrets which were handed down from father to son; [123]

[118] Only a few examples can be given here. Caspar Wistar was originally a manufacturer of brass buttons; Thomas Dyott concocted patent medicines; "Baron" Stiegel owned iron furnaces; James O'Hara provisioned the army under contract; Albert Gallatin's background caused him to be appointed secretary of the United States Treasury under Jefferson; John Mather made powder; and Edmund Monroe was a Boston banker.

[119] Burbank, "Keene Glass Works," *Antiques*, IX (1930), 312; Snow, "Franklin Glass Factory," *Antiques*, XII (1933), 134; Knittle, *Early American Glass*, pp. 168–170, 196–197, 240, 242–243, 383; Barber, *American Glassware*, pp. 55–59.

[120] Three different workers estimated for Gaffield, who formed the Boston Crystal Glass Company in the 1860's, that the annual profit one could expect from a window-glass plant during a fifty-week year would be $100,000, $109,000, or $136,850.— Gaffield Collection, Glass Journal, II, 136, 138–139, 147–148; Lanier, "A Cloth of Glass," *Technology Rev.*, XLI (1942), 317. Gaffield, nonetheless, discovered that in 1865–1866 his shop sustained a net operating loss of $123,689 to which would have to be added a substantial capital loss in case the buildings and tools (valued at $170,000) could not be sold.— Gaffield Collection, Glass Journal, II, 139–140.

[121] Likewise, capital for new shops. See Bining, "Glass Industry of Western Pa., 1797–1857," *Western Pa. Hist. Mag.*, XIX (1936), 259.

[122] It may well be argued that this tendency was common in most crafts.

[123] See Gaffield Collection, Correspondence on Glass, I, 49; Glass Journal, I,

consequently, natural improvement through industrial give-and-take was limited. Men who had grown up in the industry during the first half of the nineteenth century were inclined to make conservative or even reactionary business heads during the second half.[124] Several firms, moreover, became family affairs, with each succeeding generation loath to depart from the way of its predecessor. Only a few seem to have encouraged experimentation,[125] and those in the tableware division competed more actively with one another in trying to develop or duplicate popular glass patterns than they did in trying to improve technology by hiring chemists or technicians.[126] Most shops, it is true, did not have sufficient resources to finance research or promote innovation even if their owners had so desired.

Fortunately, after 1865 entrepreneurial talent from without once again began slowly to infiltrate the old ranks, and this time

---

48–49. A glassmaker at Millville, N. J., wrote in the *Scientific American* (August 21, 1865): "The principal reason why matters pertaining to the [glass] art are not more publicly discussed, I presume, is the desire to keep the formula secret. This, I believe, is the principal reason why so few real improvements occur. . . ." — Gaffield Collection, Scraps, I, 17. Today, by physical and chemical analysis, the components (and their proportions) of any piece of glass can be readily ascertained.

[124] See Jarves, *Reminiscences*, p. 64. A lack of entrepreneurial aggressiveness, generally, was an important factor in impeding technological and industrial growth. Of course, there were exceptions, and among the outstanding leaders must be ranked Deming Jarves, James B. Lyon, the Leightons, the Bakewells and Pages, Christian Dorflinger, the Gillinders, and many others. At least Jarves and Lyon, among the most aggressive, had not grown up in the industry. All of these men, moreover, were engaged in making tableware, and this was the division which made the most creditable showing of any in the industry before 1880.

[125] Deming Jarves, for example, secured the services of a certain Rice Harris from England around 1838 and paid him $5,000 to come over for six months and teach his workmen at Sandwich how to make opalescent glass. — Chipman, *Old Sandwich Glass*, p. 120; Lee, *Sandwich Glass*, p. 123; *National Glass Budget*, XXVIII, December 7, 1912. He also encouraged with liberal rewards for achievement those employees who played around with mixtures and colors. With his approval, two men set up and operated miniature furnaces in the basement of their homes for purposes of such experimentation. — Lee, *Sandwich Glass*, pp. 34–35; Chipman, *Old Sandwich Glass*, p. 53. The other owners of the Boston and Sandwich Glass Company, however, seem to have resented these activities and considered them wasteful. Accordingly, in 1858, Jarves withdrew and organized the Cape Cod Glass Company. Cf. Lee, *Sandwich Glass*, pp. 48 ff.

[126] Many employed professional designers. Gaffield stated that "the educated designer is the man who is always in demand about a factory, for by the results of his lucubrations, are the factories maintained." — Scraps, III, 5.

it often proved a quickening influence.[127] The greatest problem confronting the industry during its formative period had been the acquisition of technical information. Practical glassmen familiar with European practices, therefore, had been needed to build, staff, and direct the early shops. After 1820, as men skilled in production methods became increasingly available as workers and foremen, aggressive business ability supplanted an intimate knowledge of basic techniques as the primary prerequisite of a successful manufacturer.[128] Several co-operative ventures undertaken by skilled workmen collapsed as much from poor business ability as from inadequate financial backing;[129] and failures were frequent among shops set up and financed by individual glassmen who had become prosperous while in another's employ. The belief that only an entrepreneur who had had practical experience as a glassmaker could be successful nonetheless prevailed. This in part deterred many outsiders from entering the industry where profits

[127] Captain John B. Ford is an outstanding example. Successively and successfully saddle- and harness-maker, drygoods and grocery merchant, flourmiller, maker of flour cabinets for household kitchens and of feed-cutting boxes for farmers, iron manufacturer, shipbuilder, and operator of river steamboats, Ford became interested in a bottle factory in 1865 and soon thereafter, when he was 56 years of age, began to build his first plate-glass plant at New Albany, Ind. This he had to sell in 1872, but he built other works at Louisville, Ky., and at Jeffersonville, Ind. When he was 69, he was practically penniless; but with a fresh start made possible by one or two successful real-estate deals, he started the Creighton, Pa., plate-glass factory in 1880. Later the Pittsburgh Plate Glass Company was organized to take over this highly successful plant, and Ford helped build two branch factories at Tarentum and Ford City.

Unquestionably, Ford, through his perseverance and entrepreneurial enthusiasm, was largely responsible for the successful establishment of the American plate-glass industry. When he was 80 years of age and worth well over a million dollars, he started the Michigan Alkali Company at Wyandotte. See Pound, *Salt of the Earth.*

[128] The organization and supervision of the work around the furnace still required an intimate knowledge of glassmaking processes, but a foreman or plant manager could be hired for this purpose. During the greater part of the nineteenth century, there was little reason other than the small scale upon which most plants operated for burdening the entrepreneur with shop management; yet this was the common practice.

[129] Gaffield Collection, Scraps, I, 7; Jarves, *Reminiscences*, pp. 60 ff.; Knittle, *Early American Glass*, pp. 300–303, 372, 373–374. Some co-operative organizations prospered, but in such cases one or two of the men usually bought out the others within a few years.—Knittle, *Early American Glass*, pp. 325–326; Gaffield Collection, Scraps, II, 118, 156; Macbeth-Evans Glass Company, *Glass Making*, pp. 30–31.

were low and the fire hazard great. American capital had too many opportunities for remunerative investment in other fields to be attracted into glassmaking before mechanization enhanced its profitableness in the twentieth century.

6. Conclusion

In spite of entrepreneurial mediocrity, business crises, and other hindrances, however, the glass industry grew in stature. Between 1820 and 1880 the number of active shops increased five fold, the number of workmen employed twenty-five fold, the physical annual product about seventy fold, and the amount of capital invested, almost one hundred and seventy fold.[130] The present and foregoing chapters have suggested several factors which contributed to this growth. Among the more general ones were the increasing wealth and population of the country as a whole, improvements in transportation, and the discovery of abundant raw materials of good quality. High import duties, the embargo of 1807 and nonintercourse, the war with England in 1812, and the Civil War with its exchange depreciation tended to reserve the expanding domestic market for American producers. Such major technological changes as side-lever pressing, lime glass, and improvements in furnace construction resulted in lowered production costs and increased consumption. The shift from small wood furnaces to larger and more costly ones fired with coal, the increased investment in expensive molds and pressing machinery, the construction of more permanent factories, and the building of several plate-glass plants with their heavy outlay in equipment, all aid in accounting for the seven fold increase from 1820 to 1880 in the amount of capital employed per worker.

There were also a number of handicaps which help to explain why the industry did not grow at a faster rate and why its small-

[130] The 1820 census, upon which this statement is based in part, does not give complete returns for even those shops listed. Davis' estimates for the numbers employed and the amount of glass made in that year ("American Glass Industry," pp. 65, 106) have been used, and the capital for three shops has been interpolated. The *value* of the annual output increased only twenty-seven fold from 1820 to 1880, but the price of glassware declined by 62 per cent (see Chart II, above). The ratio of wholesale commodity prices in 1820 and in 1880 suggests that the real value of capital invested increased one hundred and eighty fold.

scale character remained essentially unchanged. A shortage of labor with sufficient skill and technical knowledge, the accompanying high wage scales, and the reluctance of glassworkers to adopt new methods of production exercised a retardative influence upon the expansion of individual shops. High tariffs tended to maintain the status quo. By reducing competition, they prolonged the life of many inefficient firms and may have removed an important incentive to industrial and technological improvement.[131] Few innovations made before 1880 were of a revolutionary character, and most of the technical processes did not necessitate production on a large scale.[132] With some justification it could perhaps be argued that the industry had already improved its handicraft methods as much as possible; and, until the big hurdle of adapting the processes to machine principles had been cleared, little further advance could be expected.[133]

In 1880 glassmaking was approaching the threshold of a technological revolution. The basic features of precision machinery were known, tank furnaces were available, and temperature control — so important in working glass — was heightened by the substitution of gaseous for solid fuel. When inventors then learned how to build machines which could handle metal in the same manner as skilled blowers, epoch-making advances occurred. Before this happened, it was reasonable only to expect most producers to make the best of what had already been accomplished. One of the arguments advanced thus far, however, is that

[131] This point is effectively made by Davis, "American Glass Industry," *passim*. He concludes that the development of pressing and lime glass — America's main contributions to glass technology before 1880 — occurred in spite of protection.

[132] Operation upon a larger scale than was then common would have required only a duplication of the production unit and would not have added many economies from a greater division of labor or from a fuller utilization of available equipment. In order to justify expansion, therefore, entrepreneurs had to weigh the advantages to be derived from buying materials and selling glassware in greater quantities against market limitations as expressed by an insufficiently elastic demand for their own output.

[133] For technological reasons glass has to be formed as soon as it is made and before it cools. This means that all stages from the mixing of batch to the annealing of finished products are intricately interrelated; therefore, drastic improvement in either melting or forming operations would strain production continuity. For metals and most things made with fire, the situation is different. After metals are smelted and refined, they can be converted into finished goods at a later time and at another place.

the majority of American manufacturers in 1880 actually were behind the times in matters of technology. Especially after 1860, their shops had less efficient furnaces and used older processes than those in England, Belgium, and Germany. What factors occasioned a reversal in this state of affairs in the decades between 1880 and 1920 constitutes one of the main questions which subsequent chapters attempt to answer.

CHAPTER III

INDUSTRIAL GROWTH, 1880–1920

In the census year 1919 American producers of glass, operating 317 firms, were responsible for an output valued at $261,884,080 and for almost $101,000,000 paid out as wages and salaries to some 83,600 persons. Forty years earlier 169 active establishments, with the help of some 24,000 wage earners, had made glass worth $21,154,571. This meant that, after adjustments are made for changes in the value of the dollar, the average productivity per worker roughly doubled and the volume of glass melted and shaped into useful form probably increased anywhere from six fold to ten fold.[1] The value of all goods manufactured in the United States shows an increase of around five fold for the same period.[2] It appears, therefore, that the output of glass expanded somewhat more rapidly than the output of manufactured goods generally.[3] With this growth of the in-

[1] If the value of output in 1919 is adjusted by the index of glassware prices (see below, p. 251), the physical increase appears to have been around ten fold; if the general wholesale commodity price index is used in making the adjustment, the apparent increase becomes approximately six fold. Neither adjustment would be faultless. The glassware price indices do not include different kinds of bottles and are unduly affected by the sharp drop in plate glass prices following 1880; the general wholesale commodity price series cannot be considered as representative of glass prices because glass was subjected to more cost-reducing improvements than were most commodities entering into the general index.

[2] It is admittedly hazardous to draw such a generalization from the census returns because the figure for 1919 ($62,418,078,773) includes only factory industries whereas the figure for 1879 ($5,369,579,191) includes factory, hand, and neighborhood industries. The general wholesale commodity price series compiled by Warren and Pearson (*Prices*) provides a means for compensating variations in the general price level. Had indices for only manufactured goods been available for the entire period, they would have been more reliable for present purposes than the Warren and Pearson figures. Since the manufactured goods price index compiled by the Bureau of Labor Statistics (U. S. B. L. S., *Wholesale Prices*, Bull. 415, p. 31) increased 2.478 fold from 1890 to 1919 and the Warren and Pearson wholesale commodity price index increased 2.463 fold between the same years, it is unlikely that any great distortion of the data resulted from the method adopted.

[3] Correcting changes in the value of output by making allowances for variations in the value of the dollar provides only the roughest sort of production index. Carl Snyder, using unweighted physical outputs for 87 items as data, has

dustry, the size of the average firm (measured by the number of persons employed, capitalization, and value of output) also increased as shown in Table 4. Almost twice as many persons were employed per firm in 1919 as in 1879. After making allowances

TABLE 4

GROWTH IN SIZE OF THE AVERAGE GLASS FIRM, 1879–1921 *

Census Year	Number of Active Firms	Capital Employed per Firm	Number of Employees per Firm	Value of Annual Output per Firm
1879	169	$111,270	143	$125,175
1889	294	139,343	156	139,629
1899	355	173,025	155	159,267
1904	399	224,033	168	199,519
1909	363	356,166	200	253,706
1914	348	442,316	226	353,693
1919	317	680,380	264	826,133
1921	329	**	182	648,849

* Based on data presented in *Glass Industry*, p. 25; in Weeks, *Report on Manufacture of Glass*; and in the various censuses.

** Data not given in the Census.

for changes in the value of the dollar and in the price of glass, it appears that each firm, on the average, used about three times as much capital and made between four and five times as much glass per year in 1919 as it had done forty years earlier.

Unfortunately, sufficient data are not available for breaking down the glass industry into its four major branches and analyzing in similar fashion the growth in size of the average firm in each. From the various census returns, however, it is possible to obtain averages for the value of the annual output produced by firms making (1) polished plate glass, (2) window glass, (3) con-

suggested that total production increased around 2.7 fold between 1879 and 1919 (*Business Cycles*, p. 239); and Warren and Pearson's indexes reveal an increase of approximately 3.5 fold (*Volume of Production in U. S.*, pp. 6–7). Since both of these series include items other than just those manufactured, the production of manufactured goods probably increased more than the two series indicate. Warren M. Person's index of manufactures increased about 6 fold whereas his index of total production for the United States quadrupled for the same period (*Forecasting Business Cycles*, p. 170). The physical output of polished glass expanded almost 55 fold from 1879 to 1919; window glass, only 4 fold; glass containers, around 12 fold; and pressed and blown glassware, probably from 3 fold to 6 fold. See below, pp. 253 ff. and notes.

tainers, and (4) pressed and blown glassware. These averages are given in Table 5. The plate-glass factories were the largest in 1879, and thereafter they expanded their scale of operations on the average more rapidly than did firms in the other branches. Window-glass plants and those producing bottles and jars experienced a greater increase in the value of their annual output between 1879 and 1919 than did factories specializing in other kinds of pressed and blown ware.

Since it is difficult to appreciate the rate of growth experienced by an isolated industry, Table 6 presents statistics for fourteen other manufacturing fields in addition to data on glassmaking. The primary consideration guiding the selection of these industries was the desire to obtain an adequate variety. The list includes some which made capital goods and some which produced only consumers' goods; some which were characterized by firms operating on a very large scale and others whose output was produced in either small or moderately-sized plants; some which were already mechanized before 1879, others which experienced important technological improvements between 1880 and 1920, and some which continued to rely heavily upon handicraft production; some which loomed largest in the American economy and some of relatively less importance; and, finally, some which, along with glass, were "fire" industries and others which were not. The increase in size of individual establishments in each industry is measured by arithmetic averages of the amount of capital employed, the number of persons engaged, the cost of materials used, and the value of the annual output in the two census years 1879 and 1919, with the data given as percentages of the figures for 1899. In order to facilitate comparisons of the actual size of each industry (as distinct from the average size of each firm), the number of establishments is given for each of the three years and also the actual averages for the amount of capital employed, the number of persons engaged, the cost of materials, and the value of output per firm in each industry for 1899.

Compared with the manufacture of glass, seven industries [4]

[4] They were (1) leather boots and shoes, (2) chemicals and coal-tar products,

Table 5

GROWTH IN SIZE OF THE AVERAGE FIRM IN THE FOUR MAJOR BRANCHES OF GLASSMAKING AS MEASURED BY INCREASES IN THEIR ANNUAL OUTPUT, 1879–1919 *

Census Year	Plate Glass **			Window Glass			Glass Containers ***		Pressed and Blown Glassware ***	
	Number of Firms	Annual Output per Firm (dollars)	Annual Output per Firm (sq. ft.)	Number of Firms	Annual Output per Firm (dollars)	Annual Output per Firm (sq. ft.)	Number of Firms	Annual Output per Firm (dollars)	Number of Firms	Annual Output per Firm (dollars)
1879	5	$ 158,800	208,400	49	$103,006	1,902,796	42	$135,010	73	$131,076
1889	16	260,780	568,757	84	107,843	2,243,381	69	123,499	125	148,810
1899	16	322,412	1,055,224	100	108,794	2,170,640	147	147,461	84	203,287
1904	17	469,309	1,605,479	103	112,727	2,355,495	158	212,854	103	213,167
1909	..	....	....	...	....	....	166	216,978	114	240,337
1914	19	777,568	3,178,080	64	273,374	6,265,608	150	346,393	107	282,984
1919	17	1,961,633	3,342,573	79	520,262	4,669,775	145	652,899	130	544,222

* Based on data found in the various census reports; in Tariff Commission, *Glass Industry as Affected by War*, p. 28; and in Tariff Commission, *Flat Glass*, pp. 40, 42.

** Includes only polished plate glass, which was the chief product of these factories.

*** The census returns for 1879 and 1889 classified firms (other than those producing plate and window glass) as either "green and black" glass factories or as "glassware" factories. Since the first group made predominantly bottles and jars and the second produced mainly other kinds of glassware, there is some justification for considering them as "container" and "pressed and blown glassware" factories, respectively.

TABLE 6

INCREASE IN THE SIZE OF THE AVERAGE FIRM IN FIFTEEN AMERICAN INDUSTRIES, 1879–1919
(Based on Census Data)

	Actual Number of Establishments			Capital Employed			Numbers Engaged in Industry			Cost of Materials Used			Value of Output		
	1879	1899	1919	1899 (in thous. of $)	1879 (% of 1899)	1919 (% of 1899)	(aver. no.) 1899	1879 (% of 1899)	1919 (% of 1899)	1899 (in thous. of $)	1879 (% of 1899)	1919 (% of 1899)	1899 (in thous. of $)	1879 (% of 1899)	1919 (% of 1899)
Boots and Shoes, Leather	1,956	1,600	1,449	$ 64	35	627	94	61	174	$106	49	466	$163	52	489
Chemicals and Coal-Tar Products	595	459	781	194	25	435	46	35	246	75	55	479	137	48	536
Clothing, Men's	6,166	5,729	5,258	21	62	500	23	113	174	25	84	460	48	70	460
Clothing, Women's	562	2,701	7,711	18	81	283	33	136	79	31	111	284	59	97	266
Cotton Goods	1,005	1,055	1,288	443	49	325	292	63	118	167	68	594	322	65	512
Furniture	6,008	7,972	3,154	15	54	893	13	77	392	8	81	1038	19	74	953
Iron and Steel Blast Furnaces	490	278	195	573	37	718	148	57	163	473	25	674	744	25	548
Iron and Steel Mills	515	519	500	878	28	605	367	52	230	753	34	446	1,151	35	492
Lumber and Timber Products	25,708	33,035	26,119	19	38	274	9	67	233	10	59	180	17	53	312
Marble and Stone Works	2,846	6,070	4,240	11	52	245	10	80	100	5	89	240	14	79	214
Paper and Wood Pulp	742	763	729	220	30	565	69	51	248	92	51	697	167	46	647
Pottery, Terra Cotta, Fire Clay, Brick, Tile	6,383	6,423	2,754	23	24	665	17	71	253	4	56	800	15	44	687
Soap and Candles	629	558	356	68	34	894	22	36	377	59	53	1146	95	44	946
Woolens and Worsteds	2,066	1,221	852	210	27	465	106	48	195	121	49	645	196	48	638
Glass	211	355	371	173	54	336	155	74	145	47	81	521	159	63	444

experienced a more rapid increase in the total amount of capital invested; six [5] found the total number of persons they employed expanding more rapidly; four [6] were confronted with a relatively greater rise in total expenditures for basic materials between 1879 and 1919; and only four [7] had a greater percentage increase in the value of their annual product. Judged on the basis of these four criteria of growth, only the industries producing chemicals and coal-tar products, women's clothing, milled iron and steel, and paper and wood pulp seem to have expanded more rapidly than did glassmaking during the forty years under consideration.[8] In regard to comparative increases in the scale of operations of the *average firm* in each industry, glassmaking trailed behind the majority. Only two [9] of the fourteen industries had smaller percentage increases than did glass in the average amount of capital employed per plant in 1919 as compared with 1879; and four [10] had smaller percentage increases than did the manufacture of glass in (1) the average number of persons employed per plant, (2) the average amount expended on basic materials, or (3) the average value of their annual output.

(3) women's clothing, (4) iron and steel mills, (5) paper and wood pulp, (6) pottery and allied branches, and (7) soap and candles.

[5] They were (1) chemicals and coal-tar products, (2) women's clothing, (3) iron and steel mills, (4) lumber, (5) paper and wood pulp, and (6) soap and candles.

[6] They were (1) chemicals and coal-tar products, (2) women's clothing, (3) cotton goods, and (4) paper. Only in women's clothing, however, was the increase much greater than in the production of glass. The relative increase in the cost of materials used by the iron and steel mills was almost as great as the percentage increase in material costs in glassmaking.

[7] They were (1) chemicals and coal-tar products, (2) women's clothing, (3) iron and steel mills, and (4) paper and wood pulp.

[8] In terms of physical expansion the soap industry may also have expanded more rapidly than glass. The cost of materials used and also the price of soap may not have risen as much as the materials purchased by glassmakers and glassware prices during the war years. This possibility serves to indicate the limitations of the four criteria of growth here employed.

[9] They were (1) women's clothing, and (2) marble and stone works.

[10] For the average number of persons employed, they were (1) men's clothing, (2) women's clothing, (3) cotton goods, and (4) marble and stone works; for the average annual expenditure on basic materials, they were (1) men's clothing, (2) women's clothing, (3) lumber and timber products, and (4) marble and stone works; and for the average value of the annual output, they were (1) men's clothing, (2) women's clothing, (3) lumber and timber products, and (4) marble and stone works.

In other words, firms in ten of the fourteen industries listed in Table 6 appear to have expanded their scale of operations more rapidly in the period 1879–1919 than did glass factories on the average.[11] A partial explanation of this is to be found in the fact that in 1879 glasshouses were already producing on a relatively larger scale than were most establishments in the majority of these industries; hence the same or even a greater absolute expansion would constitute a smaller rate of growth in the size of glasshouses than it would for firms in the other industries. Only iron and steel mills, iron and steel blast furnaces, and plants making cotton goods were as large on the average in 1879 as glassworks. On the other hand, during the subsequent forty years the average firm in some of the other industries expanded so much more rapidly than did establishments making glass that they actually became larger than the average glass factory.[12]

The arithmetic average, it should be pointed out, has definite limitations as a device for describing the nature and rate of growth in any industry, especially during a period when phenomenal changes are occurring. This is because the innovations and expansions of scale of the few firms in the forefront are only gradually adopted by the remainder. Hence strong tendencies for expansion in scale of operations are minimized by including in the average all the imitator and laggard plants. What is happening to the establishments in the vanguard may be of equal or even greater significance to the economic historian than changes in the "average." This was certainly true of the glass industry. The Pittsburgh Plate Glass Company in 1920 was more than twenty-five times larger than the most important plate-glass factory had been in 1880; [13] both the Libbey-Owens Sheet Glass Company and the American Window Glass Company in 1920

[11] Women's clothing, men's clothing, lumber, and marble and stone works were the only ones which, apparently, experienced a smaller percentage increase in average size than did those plants producing glass.

[12] In 1919, apparently, the average firm in chemicals and coal-tar products, cotton goods, iron and steel blast furnaces, iron and steel mills, paper and wood pulp, soap and candles, woolens and worsteds, and possibly in leather boots and shoes was larger than that in glass.

[13] See above, pp. 11–12, and below, p. 239; Pittsburgh Plate Glass Company, *Glass*, pp. 37–38.

were around twenty or thirty times larger than the largest window-glass factory had been forty years earlier; [14] and approximately the same comparison could be drawn between the Owens Bottle Company in 1920 and any of the outstanding bottle firms in 1880.[15] The contrast between the size of the leading firms making tableware before 1880 and around 1920 was less marked than the foregoing contrasts drawn for the other major branches of the industry. The Libbey Glass Company in 1920, for example, may not have been operating on a scale much more than six or eight times greater than that of the largest tableware factory during the middle decades of the nineteenth century.[16]

Production data for 1904 and 1919, given in Table 7, shed additional light on the tendency of glass factories to expand. In 1904 five factories, constituting only 1.3 per cent of the total, each produced an annual output valued at $1,000,000 or more and together were responsible for 11.3 per cent of the total value of all kinds of glass made in the United States. In 1919 each of sixty-nine factories, constituting 18.6 per cent of the total number making glass, had an annual product valued at $1,000,000 or over and collectively accounted for 60.8 per cent of the country's glass output. This increase in the relative importance of the very largest to the total number of firms was much more pronounced in the glass industry than in all manufacturing industries considered collectively.

The physical appearance of glassmaking factories also changed markedly. Most of the plants engaged in production in 1880 were located either in small communities or in the country. Their drafty wooden and corrugated iron structures represented little "sunk" investment, and hence they could be moved rather easily

[14] See above, p. 12, and below, pp. 127 ff., 218 ff.

[15] See above, p. 12, and below, pp. 109 ff.

[16] See above, p. 13, and below, pp. 138 ff. Unfortunately, in view of the inadequacy of available data, it seems inadvisable to be more exact than this in drawing comparisons. The census and other government publications were careful in presenting their information lest the reader be able to link up the statistics to particular firms. Since exact figures on the size of leading firms in 1880 and of the American Window Glass Company and the Pittsburgh Plate Glass Company in 1920 have not been found, the estimated increases in size given in the text are to a considerable extent impressionistic in character with a conscious bent toward conservatism.

(*Courtesy of James Gillinder*)

FIG. 3A. A TYPICAL BOTTLE FACTORY *Circa* 1860 — THE SAMUELS BOTTLE WORKS AT PHILADELPHIA

Table 7

COMPARISON OF GLASS FIRMS WITH ALL MANUFACTURING ESTABLISHMENTS, CLASSIFIED ACCORDING TO VALUE OF THEIR ANNUAL OUTPUT, 1904, 1919 *

Value of Products per Establishment	1904				1919			
	Establishments		Combined Output		Establishments		Combined Output	
	Number	Per cent of Total	Value	Per cent of Total	Number	Per cent of Total	Value	Per cent of Total
Glass Firms								
$100,000 to $1,000,000	230	57.6	$ 62,274,058	78.2	256	69.0	$ 100,456,890	38.4
Over $1,000,000	5	1.3	8,992,772	11.3	69	18.6	159,317,630	60.8
Firms in All Industries								
$100,000 to $1,000,000	22,246	10.3	$6,109,012,538	41.3	48,855	16.9	$15,433,003,954	24.8
Over $1,000,000	1,900	0.9	5,628,456,171	38.0	10,414	3.6	42,301,103,617	67.8

* Sources: *Thirteenth Census*, VIII, 91; X, 878; *Fourteenth Census*, VIII, 180; X, 833.

to new areas where natural gas had just been discovered in abundance or where more plentiful and cheaper supplies of coal were available. Even as late as 1916 the government agents who investigated the glass industry for the purpose of ascertaining its costs of production reported that "glasshouses are generally antiquated. They are usually flimsy structures put up in the formative period of the industry and have outlived their usefulness. Very few plants are constructed on the plan of a modern scientifically laid out factory."[17] By the time of this report, however, the tendency to construct clusters of substantial brick buildings, spreading over several acres, had already become pronounced. In 1920 the more important plants in each branch of glassmaking differed little in external appearance from glass factories of the present day or from any other mechanized factories. They gave the impression of what the term "factory" signifies to any high-school student today. This meant that they had anchored to one particular spot. The age-old wandering of glasshouses was disappearing.

Much of this change in physical appearance and equipment had been necessitated by developments in the technology employed in melting and fashioning glass. The relatively small and inexpensive furnace[18] which melted glass in pots rapidly gave way to the much larger, longer lasting, and more costly tank furnace in all branches except those producing plate and lead glass.[19] In 1880 there was only one continuous tank furnace in an American glasshouse, and not more than four smaller intermittent ones for melting and refining glass at night and for working glass during the day.[20] A continuous tank successfully melted window glass for the first time in 1888; twelve years later enough were being worked to produce about one-half the nation's output of windowpanes.[21] The data in Table 8 indicate how rapidly tanks

[17] *Glass Industry*, p. 317.

[18] See above, pp. 27–28.

[19] For a discussion of the relative merits of the two types of furnace, see above, p. 28 and n. 93.

[20] See above, p. 28.

[21] Libbey-Owens-Ford Glass Company, Colburn Records, Vol. B, p. 49; *Twelfth Census*, Part III, p. 966.

grew in popularity for the industry as a whole, but even these figures are apt to leave the reader with too conservative an impression. Tank furnaces could melt much more glass per year than pot furnaces. Hence the amount of glass melted and refined in tanks constituted an even greater proportion of the total quantity produced in 1919 than the relative increase in the number of continuous and intermittent tanks would indicate.[22] The vast majority of the pot furnaces in that year were employed in factories making plate, obscured, and wire glass and either blown or pressed lead glassware. Nearly all containers, lime [23] glassware, and window glass were fashioned from metal melted in tanks.[24]

TABLE 8

NUMBER OF MELTING FURNACES IN ALL ACTIVE GLASS FACTORIES, 1879–1919 *

	1879**	1899	1904	1909	1914	1919
Pot Furnaces Being Worked	280	391	349	370	328	289
Pot Furnaces Lying Idle	63	60	49	59	58	41
Continuous Tank Furnaces Being Worked	1	192	340	369	454	598
Continuous Tank Furnaces Lying Idle	—	14	26	48	48	47
Intermittent Tank Furnaces Being Worked	4	130	94	73	59	116
Intermittent Tank Furnaces Lying Idle	—	17	13	13	16	9

* Based on data found in Weeks, *Report on Manufacture of Glass*, pp. 13–15; *Census of Manufactures*, 1914, p. 834; *Fourteenth Census*, X, 837.

** All furnaces of active establishments are considered as having been worked; the furnaces lying idle belonged to inactive factories or were in process of construction.

[22] Since figures on the average annual output of either tank or pot furnaces are not available, there is no reliable way of measuring the relative importance of either in terms of total output.

[23] For the differences between lime and lead glass, see above, pp. 21–22.

[24] The tank furnace was not invented in America and, as a matter of fact, was only slowly introduced into the country. See above, p. 28. After manufacturers realized its superior efficiency over the pot furnace, however, they outran producers in other parts of the world in its utilization.

Another reason why there was a tendency after 1880 to construct permanent buildings and hence to remain at one location was the fact that technological changes in several branches of the industry revolutionized the manner of fashioning glass. And the new processes necessitated heavy capital investments.[25] In 1880 machinery of any sort was used only in casting or rolling plate glass and in pressing lime and lead glassware. All other objects were either blown or shaped by simple tools in the hands of highly skilled glassmakers. In the three decades following 1890, epochal inventions paved the way for forming bottles and jars either partially or completely with machines; for blowing or drawing sheets of window glass without utilizing the services of skilled blowers; for pressing or blowing tumblers mechanically; for blowing lamp chimneys, tubing, and incandescent light bulbs with either automatic or semiautomatic devices; and for pressing bowls, plates, and other similar pieces of tableware without the direct intervention of human hands. Whereas in 1890 individuals fashioned all bottles and jars produced in American factories, it has been estimated that by 1917 about 50 per cent (and by 1922–1923 approximately 80 per cent) of the total output resulted from machine processes.[26] No window glass was made by any method other than hand blowing as late as 1900; by 1919 approximately 75 per cent of the country's output was either blown by cylinder machines or pulled out in a continuous sheet by drawing machines.[27] Skilled laborers blew and shaped all lamp chimneys before 1895 but made only about 40 or 50 per cent of

[25] The amount of capital employed by those factories producing window glass by hand in 1916 averaged $118,117; the capital invested in plants making the same commodity by machines averaged $260,779. Mechanized bottle factories in the same year had $1,245,357 capital invested on the average; semiautomatic machine bottle factories had $318,292; and hand bottle factories, only $136,101. Establishments producing jars with the aid of machinery had capital investments averaging $504,919. Hand-blown tableware shops averaged $205,360, and quasi- or fully mechanized plants producing blown and pressed tableware had almost twice this amount of capital invested. Factories making glass lighting ware employed machines and supported capital investments averaging $730,454. — *Glass Industry*, p. 77.

[26] Davis, "American Glass Industry," pp. 315, 331.

[27] Tariff Commission, *Flat Glass*, p. 39. By 1925 only 1 per cent of the nation's output was blown by skilled labor. — *Productivity of Labor in the Glass Industry*, p. 161. For a description of these machines, see below, pp. 170 ff., 191, 329 ff.

(*Courtesy of Arthur E. Fowle*)

FIG. 3B. THE LIBBEY-OWENS SHEET GLASS FACTORY *Circa* 1920 NEAR CHARLESTON, WEST VIRGINIA

the total in this manner in 1920.[28] Similar workers fashioned the entire output of incandescent light bulbs before 1895; in 1920 they made only about 40 per cent, and semiautomatic machines accounted for the remainder.[29] Edward Danner in 1917 patented the first successful automatic machine for drawing and blowing glass tubing, and hand drawing had completely passed from the scene nine years later.[30] Most of the simple, inexpensive items for table use such as bowls and tumblers were also made in 1920 on semiautomatic or completely automatic machines.

Plate glass and glassware containing lead were the only important items which escaped revolutionary technical changes during the period under study. Plate glass continued to be rolled out on large metal tables and to be subjected to separate and discontinuous grinding and polishing processes. Minor improvements in the equipment used, however, made necessary a much larger capital investment than that required before 1880 and led to increased output per worker and to a sharp decline in unit values.[31] Skilled blowers and other handicraftsmen were — and still are — largely responsible for the objects made of lead glass.

The amount of horsepower employed reflected the increasing mechanization of the industry as a whole between 1880 and 1920. As Table 9 indicates, this amount increased thirty-six fold over the forty years, whereas the consumption of power for industry generally showed only an eight-fold increase.

Still a third factor helps account for the increasing permanence of factory buildings in the glass industry. During the latter part of the nineteenth century natural gas, the fuel *par excellence* for glass furnaces, was discovered at many different places in the middle northwestern states. Whenever a new well was brought

[28] *Productivity of Labor in the Glass Industry*, p. 114. The only reason why some was made by hand was the unwillingness of manufacturers to invest additional capital in a decadent industry.

[29] *Ibid.*, p. 126. Only 0.1 per cent of the output for that year was made on completely automatic machines.

[30] *Ibid.*, p. 143.

[31] The price of two sizes of polished plate in 1915 was about one-third their 1880 price. — *Wholesale Prices, Wages, and Transportation*, II, 241–242; U. S. B. L. S., *Wholesale Prices*, Bull. 320, p. 203.

in, all sorts of industries dependent upon fire hastily set up plants in the immediate area. Glass was no exception. Few wells, however, produced gas abundantly beyond a few years or sometimes even months. Hence, as long as new sources of supply were constantly appearing, glasshouses abandoned old sites and opened up again at new locations.[32] The natural-gas boom in Indiana, Ohio, and Illinois collapsed several years before the First World War, but improvements in producer-gas and oil-burning furnaces permitted the economical operation of many factories which depletion of natural-gas sources left stranded. Consequently, it was no longer imperative for entrepreneurs to retreat to new gas fields in the Southwest and West Virginia in order to continue their competitive struggle for markets.

TABLE 9

AMOUNT OF POWER USED IN AMERICAN INDUSTRY GENERALLY AND IN THE MANUFACTURE OF GLASS, 1879–1919 *

Census Year	Primary Horsepower Employed in All Industries		Primary Horsepower Employed in the Glass Industry	
	Amount	Percentage Increase	Amount	Percentage Increase
1879	3,410,837	100	5,672	100
1889	5,938,635	174	28,241	498
1899	10,097,893	296	52,943	934
1904	13,487,707	395	91,476	1613
1909	18,675,376	547	123,132	2172
1914	22,437,072	658	163,139	2877
1919	29,504,792	865	207,430	3659

* Based on data provided by *Fourteenth Census*, VIII, 122; X, 831.

The metamorphosis of the glass industry between 1880 and 1920 had other important aspects than those which have already been described. The composition of its labor supply and the nature of its labor market were transformed. In 1880 the production of all glass except plate and pressed ware required the services of highly skilled workmen. These men, even before they

[32] Thirty-one factories, it is said, were built at Findlay, Ohio, between June, 1886, and June, 1887, after natural gas was discovered there. The supply diminished so quickly that five years later not more than one-tenth of the factories within the producing area of Indiana and northwestern Ohio were still using natural gas. — V. S. Clark, *Manufactures in U. S.*, II, 517.

organized strong national craft unions, formed something of a fraternity, whose members enjoyed substantially higher standards of living than most workers in other industries.[33] Child labor constituted a fifth of the working force.[34] During the last two decades of the nineteenth century powerful labor unions obtained jurisdiction over nearly all glassworkers and were able to exercise compelling influence over such matters as apprenticeship, wages, output, working conditions, and length of production season.[35] By 1920 their voice in industrial affairs had dwindled to a mere echo of the challenges and ultimatums hurled at manufacturers in former years. The introduction of machinery had hastened the disappearance of child labor and had forced blowers and other skilled workers to seek employment in other industries, to congregate in those less important branches of glassmaking not then invaded by machines, or to sacrifice their skill and become machine tenders. The relatively high wages formerly paid declined, and glassworkers ceased to enjoy a favored position in the general labor market.

Also, the competitive position of American producers in the world market greatly improved. Chart IV reveals the trade balance in glass of the United States with all other countries. Until 1915 a greater value of glass was imported than was exported each year; during the war years, the reverse was true. The maintenance of high protective duties on most kinds of imported glass until the passage of the Underwood Tariff Act in 1913 [36] and the dislocation of European production and overseas shipping wrought by the war were, of course, partially responsible for the later trends in export and import values. But another basic factor was the lessening of the cost disadvantage under which American glassmakers had been laboring for decades. No one has seriously questioned the statement made by various leaders in the industry that from 1880 to 1920 the United States led the

[33] See pp. 32 ff., above, and pp. 200 ff., below.

[34] Out of a total of 24,177 persons employed in making glass in 1879, 5,566 were boys under seventeen years of age and 92 were girls under sixteen. — Weeks, *Report on Manufacture of Glass*, p. 15.

[35] See below, pp. 200 ff.

[36] See below, pp. 245–246.

rest of the world in developing and improving glass furnaces, lehrs, and forming machines. The advantages of mass production caused several American entrepreneurs to expand their operational scale beyond the point attained by leading European manu-

CHART IV

AMERICA'S BALANCE OF TRADE IN GLASS, 1880–1920 *

(In thousands of dollars)

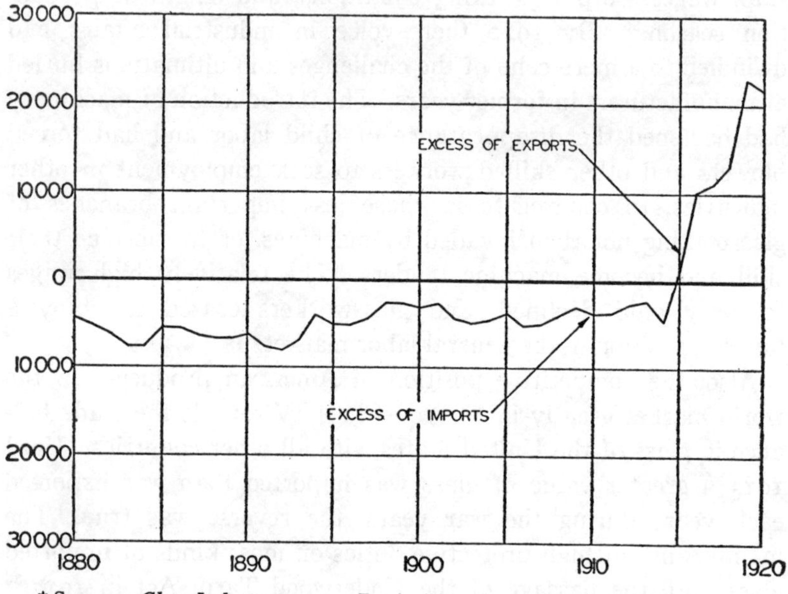

* Sources: *Glass Industry*, p. 334; Davis, "American Glass Industry," Appendices A and B.

facturers. Before 1880 the situation had been the reverse. Europeans had been the pacemakers, and American glassmen had trailed behind. Whereas the latter occasionally had visited factories in Europe to obtain workers and information on the latest methods of melting and working glass, in 1920 Europeans were striving to bring their plants up to the level of operating efficiency of American factories.

Throughout the forty-year period the price of glass declined relative to the general price level.[37] And Americans bought in-

[37] See below, p. 251.

creasing quantities of glass containers, plate glass, window glass, lighting equipment, and even cheap tableware. The quality of products also greatly improved during this time.

The sweeping changes in the complexion of glass manufacturing which have been mentioned in this chapter offer the economic historian an opportunity to essay a case study in the factors responsible for industrial growth. At certain times and under certain circumstances the growth of any industry may appear to have been largely a product of nonpersonal and environmental factors; at other times the personal or entrepreneurial effort may seem to have been the dominating, activating force. As a working hypothesis, however, one should always consider that both personal and nonpersonal elements have participated (although in varying degrees of importance) in bringing to pass industrial expansion. This is true of the development of glassmaking from 1880 to 1920, even though the personal element appears to have played an especially important role.

A group of men, endowed with diverse talents and resources, were drawn together at Toledo and for a quarter century or more devoted their time and energies to perfecting and introducing to the trade glassmaking machines of various types and to establishing model, large-scale factories in several branches of the industry. Their efforts were liberally rewarded with large pecuniary profits and with world-wide recognition as the outstanding entrepreneurs engaged in making glass. How they erected an industrial empire upon an offshoot of a small tableware factory forms the subject matter of Chapters IV and V. Since the success of their efforts stemmed primarily from certain technological innovations for which they were chiefly responsible, Chapter VI describes the origin, nature, and importance of the processes and machines they developed. Other inventors and entrepreneurs, stimulated by the success achieved at Toledo, launched experimental projects of their own. An era of general technological ferment ensued, described in Chapter VII, which had decided repercussions upon the continued activity of the Toledo Group. Subsequent chapters treat such general environmental factors as the composition of the labor market, the extent of competition, the evolution of consumer buying habits, the trend of glass prices, the pattern of factory location, and the protection provided by tariffs. The concluding chapter attempts to examine and evaluate the influence exerted by E. D. Libbey, M. J. Owens, and other individuals.

CHAPTER IV

THE ORGANIZATION AND HISTORY OF THE GLASS FIRMS FOUNDED BY THE TOLEDO ENTREPRENEURS, 1888–1920

The first decades of the twentieth century brought an end to the dearth of aggressive entrepreneurial leadership which had persisted in the American glass industry during most of the nineteenth century. A relatively small number of men whose activities centered around Toledo, Ohio, played a powerful role in the transitional state; and the industrial and technological developments in which they actively participated were responsible to a great extent for the sweeping changes in glassmaking which have been summarily described in Chapter III. Under the leadership of Edward Drummond Libbey and Michael Joseph Owens, the men at Toledo busied themselves with inventing, introducing, and actually using many mechanical contrivances for pressing, blowing, and drawing glass. The results of their efforts eventually rendered obsolete most of the industry's technology, which had remained for the most part at the handicraft level throughout the world for many centuries. Large-scale production, consequently, became a characteristic feature of the industry; and both corporate and private fortunes were amassed. In order to understand and appreciate fully the revolutionary tendencies emerging in glassmaking after 1900, it will be helpful — if not essential — to describe in some detail exactly what transpired at Toledo from 1888 through 1920 and to suggest what appear to have been the chief factors involved.

At the outset, the industrial setup at Toledo was simple, but it became increasingly complex as new technical processes were perfected and as many separate companies were concomitantly incorporated. Some of the companies devoted their chief attention to developing and promoting glassmaking machinery and to licensing outsiders in its use; others engaged directly in manufacturing. The control of all, in one way or another, remained in the hands of the Toledo Group. The Libbey Glass Company, organized in 1888, preceded by seven years the incorporation of its offshoot, the Toledo Glass Company, from which most of the others stemmed. After 1900 the two original companies pursued almost completely independent courses, and a rift tended to develop between the two groups of entrepreneurs. The Libbey Glass Company and its directors had little or nothing to do with the lamp-chimney, bottle, and sheet-glass enterprises which the Toledo Glass Company organized. Instead, they became interested in experimental work on automatic tube drawing and sponsored the formation of other companies which developed and licensed paste-mold bulb machines. A system of cross-licensing and the personality and activities of Libbey himself, who maintained his interest in both entrepreneurial groups, however, managed to preserve a semblance of unity between what may be conveniently called the "Libbey Glass Faction" and the "Toledo Glass Faction."

The crosscurrents of technological interests and interconnecting controls between the two divergent groups and also between the various enterprises which they promoted diminish somewhat the value of any chronological account of the separate companies' growth. Nevertheless, a chronological history of the composite industrial development emanating from Toledo would prove so bewilderingly complex that it would be of even less value. The procedure adopted in this chapter, therefore, roughly conforms to the more simplified company approach. Since E. D. Libbey had operated the New England Glass Company at Cambridge, Massachusetts, during its closing years before he moved its equipment to Toledo, a historical sketch of that factory (1814–1888) serves as an introduction to the second section of the chapter, which treats the organization and early activities of the Libbey Glass Company (1888–1895).

Promotional work in the tumbler and lamp-chimney fields and experiments with automatic bottle-blowing machinery marked the years from 1895 (when the Toledo Glass Company was formed) until 1903 (when the same group of men incorporated the Owens

Bottle Company). Section three treats this eight-year period, and section four is concerned with the expansion in the Owens Bottle Company until 1921. The following chapter describes the purpose and history of the Owens-European Bottle Machine Company from 1905 through 1920, the early activities and successes of the Libbey-Owens Sheet Glass Company, and the aims and accomplishments of the Westlake companies. Its two concluding sections once again direct the reader's attention to the parent concerns — Libbey Glass and Toledo Glass — and summarize the innovational, manufacturing, and financial achievements attained after their first years.

1. The New England Background of the Libbey Glass Company (1814–1888)

A small group of New Englanders, on February 14, 1814, incorporated the Boston Porcelain and Glass Manufacturing Company and then proceeded to construct a melting furnace in East Cambridge.[1] When their venture proved unsuccessful shortly after the Treaty of Ghent had terminated the war with England, another group of men leased the plant in 1815. They too failed, and the original owners disposed of the property at public sale in 1817 to Deming Jarves and several associates. These men organized the New England Glass Company.[2] Hence, one of the most successful glass manufactories of the nineteenth century had its origin in that period of business confusion and bankruptcy which followed the cessation of hostilities with Great Britain.

From the very first the New England Glass Company specialized in beautifully blown and richly cut flint glass and also was among the first two or three concerns to make pressed ware in large quantities.[3] Since its products enjoyed great popularity, the company prospered. From 1837 to 1873 it declared annual dividends varying from 8 to 20 per cent[4] and in addition reinvested several hundred thousand dollars of earnings in plant and equipment. A few scattered figures which trace in bold outline

[1] McKearin, *American Glass*, p. 592.

[2] Its corporate charter was issued on February 18, 1818.

[3] See above, p. 18.

[4] Gaffield Collection, Glass Journal, II, 265.

the extent and rate of its growth are available, and these are given in Table 10. The company, expanding more than ten fold between 1818 and 1855, apparently attained its peak of prosperity around the middle of the century. Following 1865, however, two increasingly important factors sapped its strength.

TABLE 10

GROWTH OF THE NEW ENGLAND GLASS COMPANY, 1818–1885 *

Year	*Capitalization*	*Numbers Employed*	*Value of Yearly Output*
1818	$ 40,000	40	$ 40,000
1820	80,000	96	65,000
1823	......	140	150,000
1849	320,000	...	......
1851	400,000	450	......
1855	......	500	500,000
1865	500,000	"over" 500	500,000
1874	500,000	250	......
1875	500,000	"over" 400	......
1876	500,000	...	232,304
1878	300,000	...	......
1882	......	"over" 100	......
1885	......	285	......

* Sources: Gaffield Collection, Glass Journal, II, 265; III, 9, 12–13, 39; Notes on Glass, IV, 50; Scraps, I, 123, 161; Jarves, *Reminiscences*, pp. 61–62; Watkins, *Cambridge Glass, 1818–1888*, pp. 14–15, 16, 18, 28.

As was pointed out in an earlier chapter,[5] the fuel most commonly employed in melting furnaces at that time was coal. And coal cost much more in the New England area than it did around Pittsburgh. Consequently, all glass factories along the Atlantic coast had to cope with rather keen competition from western producers; and Massachusetts, which had formerly been one of the most important glassmaking centers, was less successful in its fight than some of the other Atlantic states. Her industry after 1860 stagnated, both absolutely and relative to other producing regions.[6] The second factor directly weakened the heart

[5] See above pp. 23 ff.

[6] See above, Table 2, and below, Table 22. Today (1947) Massachusetts can

of the New England Glass Company. In 1864 William Leighton, Sr., one of its former superintendents, began a series of experiments with lime glass while in the employ of a plant at Wheeling, West Virginia. So fortunate were his endeavors that lime glass turned out to be almost as beautiful as the much costlier flint glass.[7] As a result, most manufacturers began to use lime in place of flint glass whenever they made tableware and other items for household use. Since the New England Glass Company had acquired a national reputation largely because of its excellent flint ware, its owners were reluctant to lower their standards of quality despite the rapid shift in consumer preference. The depression following 1873 made their plight still more difficult. They finally conceded defeat in the competitive struggle evoked by cheap western coal and the advent of lime glass. They banked their fires and on April 12, 1877 ran an offer in a Boston newspaper to sell or lease the works.[8] William L. Libbey, their own "agent" or general manager, acquired the lease the first months of the following year.[9]

Libbey had received his training in glassmaking at the Mount Washington Glass Works in South Boston.[10] In 1860 he and one Captain Timothy Howe took over the plant there, and six years later Libbey became its sole owner. After transferring the works to New Bedford in 1866, he continued production for about four years and then sold out to others in either 1870 or 1872 in order to become general manager at the New England Glass factory.[11] E. D. Libbey, William Libbey's son, also became associated (in 1874) with the Cambridge plant as one of its

count only one small shop at New Bedford, which makes small quantities of hand-blown ware.

[7] See above, p. 22.

[8] Gaffield Collection, Glass Journal, III, 180; Scraps, I, 123.

[9] A newspaper clipping dated March 29, 1878, states that Libbey had recently taken the lease. — Gaffield Collection, Scraps, I, 161. But his son, writing in September, 1903, said that he had continued as manager until 1880, when he bought out the company. — Owens-Illinois Glass Company, Libbey Letters, II, 52.

[10] This enterprise, like the New England Glass Company, had also been started by Deming Jarves. — Lee, *Sandwich Glass*, p. 36; MacManus, *Glass Manufacture*, pp. 48–49.

[11] McKearin, *American Glass*, p. 142. The Mount Washington Glass Works later became part of the Pairpoint Manufacturing Company and today makes a fine quality product as the Gunderson Glass Works.

clerks and remained there after his father took up the lease in 1878. The firm name was changed two years later to the New England Glass Company, W. L. Libbey and Son, Proprietors; and when his father died in August, 1883, E. D. Libbey continued operations.

He did not inherit a thriving business. At the time of the *Tenth Census* only one of the four melting furnaces had been lighted,[12] and the working force had dwindled to less than a fourth of what it had been in 1855 or 1865. A significant revival, however, was in prospect. The proceeds from his father's insurance policies permitted E. D. Libbey to meet many pressing financial obligations and also to undertake the production of "Amberina." From the sale of this patented product he reaped profits which later "served as the principal foundation of his successful career." [13]

Amberina was melted from a flint-glass batch into which a little gold dust had been sprinkled. After it was blown, such glass had a pale amber color of yellowish tone. Subjecting each piece to various degrees of heat at the working furnace caused its color to shade from light amber to the deepest red.[14] Amberina ware captivated the buying public at once, and sales mounted from $45,000 in 1883 to $80,000 the following year. Its popularity led one New York glass importer to contract with a foreign manufacturer to copy its appearance by plating plain flint glass with a thin layer of the much more expensive gold flint. The dealer then sold the "counterfeit" pieces at lower prices than real Amberina. And the Mount Washington Glass Works went so far as to bring out an identical pattern under the name "Rose Amber." Since this was clearly an infringement of the patent which he

[12] Weeks, *Directory, U. S. Glass Works*, p. 15. This directory was drawn up preparatory to Weeks's investigation of the industry for the census report, and one copy forms part of the Gaffield Collection. See also Weeks, *Report on Manufacture of Glass*, p. 91.

[13] Owens-Illinois Glass Company, "Lives of E. D. Libbey and M. J. Owens," p. 2.

[14] Gold ruby glass (flint glass containing gold) was not new, and glassmen had long learned to reheat it in order to develop sufficiently its ruby color. The distinctive process in making Amberina consisted of reheating ruby glass in such a way as to obtain variations of amber color.

had obtained for his new product on July 24, 1883, Libbey brought suit against Mount Washington in June, 1885.[15]

The public's taste in tableware, however, was fickle, and Amberina's popularity gradually declined. New colors and styles — such as Pomona, Peachblow, and Agata — were also introduced at Cambridge, but these, too, enjoyed only temporary success.[16] At best, they gave the New England Glass Company brief respite in its competitive struggle with western producers making cheap lime glass and enjoying low fuel costs. In 1888 Libbey faced an annual deficit of about $40,000,[17] and in the same year his flint workmen went out on strike. The strike was a recurrence of labor difficulties which had first erupted in October, 1885. In order to appreciate the nature of the controversy it will be necessary to recall that wage levels in glass shops along the Atlantic coast were lower than those prevailing in western areas.[18] In spite of their competitive disadvantage in the cost of coal and other materials flint manufacturers in the East had been able to continue operations by paying their employees less per hour than glassmen on the other side of the Alleghenies regularly received and also by exacting a larger hourly output. As long as the workers in the two regions did not belong to the same labor union, it was rather easy to maintain such wage and output differentials. In 1878, however, the American Flint Glass Workers' Union had combined theretofore independent organizations. Its western members, in October, 1885, persuaded flint glassworkers in the New England and Central Atlantic states to strike for wages and a "fair hour's output" which would equalize laboring conditions between the East and the West. The strike, lasting only about two weeks, ended after both workers and employers made concessions.[19] The western group, in control of the union's policies, was not satisfied with the settlement and instigated

[15] *E. D. Libbey* v. *Mt. Washington Glass Company*. The hearings and testimony presented by many outstanding figures in the tableware industry, a copy of which is preserved in the Gaffield Collection, have provided most of the information presented in this paragraph.

[16] MacManus, *Glass Manufacture*, p. 50.

[17] *Ibid.*, p. 54.

[18] See above, p. 33.

[19] Gaffield Collection, Scraps, III, 8–10.

another strike in 1888. Libbey's plant was among the ones affected. E. D. Libbey appealed to his workers' loyalty and reason by pointing out that he would be forced to close down permanently if they insisted upon obtaining the same pay and maximum output quotas that more fortunate producers around Pittsburgh had allowed. Since the men did not harken to his warning but obeyed the exhortations of the union's western representatives, Libbey drew all fires at the New England Glass Company and sought a new beginning in the West.[20]

2. The First Years of the Libbey Glass Company at Toledo (1888–1895)

After visiting several possible locations in the natural-gas belt of the North Central West, Libbey finally selected a site at Toledo. His decision seems to have been partially influenced by the fact that a solicitous real-estate agent had recently introduced him to Florence Scott, his future wife and the charming daughter of one of the city's leading citizens.[21] In addition, many local businessmen agreed to raise jointly $4,000 to pay for four acres of factory land which he would select. Libbey filed his petition to incorporate the W. L. Libbey and Son Glass Company on April 14, 1888; and the ensuing charter, authorizing a capitalization of $100,000, permitted the new enterprise "to manufacture and sell rich cut and artistic glass, and to do all things incident thereto for profit." [22] By June 21, 1888, all its stock had been subscribed and paid in full.[23] The directors then voted to pay Libbey $100,000 in cash for the assets of the old New England Glass Company which he had transferred from Cambridge, for the factory in process of construction at Toledo, and

[20] It is significant that the same strike sounded the knell of the Boston and Sandwich Glass Company, another illustrious Massachusetts house. The New England Glass Company formally surrendered its charter on September 18, 1890.—Watkins, *Cambridge Glass, 1818-1888*, p. 38.

[21] According to John H. McNerney of the Owens-Illinois Glass Company, who had received the story from the real-estate agent himself.

[22] Libbey Glass Company, Libbey Glass Company Minutes, I, 1. Joseph W. Robinson of Toledo, who has preserved the early minutes of this company, obligingly permitted me to use all pertinent material in his possession.

[23] Libbey on February 6, 1889, held 1,598 of the 2,000 shares. Each share had a par value of $50.—*Ibid.*, I, 24.

for the real estate which had been donated to him.[24] Before the year ended they lighted a new thirteen-pot furnace (the capacity of which was slightly over sixteen tons); and 115 men who had followed their employer from Cambridge started work. Additional blowers and other skilled workmen were obtained from the glassmaking area around Wheeling, West Virginia.

The new enterprise did not fare very well at first.[25] Many employees became homesick, and about thirty-five actually returned to Boston; the plant manager was found incapable of controlling the men under him; the furnace was improperly built and never worked successfully; and the customary batch formulas did not yield as high-grade glass as they had done at the old Cambridge works. By borrowing heavily at a local bank, however, Libbey managed to meet each payroll during these critical years. Labor difficulties at Corning, New York, provided him with an unexpected opportunity in 1890 and practically saved his venture from early failure. The opportunity involved the big business in incandescent lamp bulbs which the Corning Glass Works had built up with Edison General Electric. The owners of Corning had long insisted upon operating an open shop, but their employees struck in 1890 for union recognition. Libbey's company meanwhile had already begun to make some bulbs for a few independent companies in the electric-lighting field. When the strikers at Corning approached one of Libbey's representatives with the suggestion that his company supply Edison and that they would furnish the necessary blowers, Libbey agreed. In order not to interfere with the regular line of production at Toledo, the manager of Libbey's bulb department rented an idle window-glass plant at neighboring Findlay, where his men made large quantities of bulbs during eighteen months following January 1, 1891. At the end of that time the bulb workers were transferred to the Toledo plant because a new ten-pot furnace had recently been completed there.

[24] *Ibid.*, I, 20–21.

[25] A great part of the information which follows is based upon the remarkable memory of E. J. Barry, who had accompanied Libbey to Toledo and who later became plant superintendent at the Libbey works.

That the Findlay venture marked the beginning of a new era for the Libbey enterprise cannot be seriously questioned. Prior to 1891 Libbey Glass had made and cut artistic flint glass of the finest quality almost exclusively; thereafter, bulbs and tubes for incandescent lamps constituted a second specialty. Its total sales of lamp parts from 1904 through 1917, for example, amounted to $14,406,977 whereas all other glassware sold during the same period brought in $10,312,772.[26] More immediately, according to the balance sheet drawn up for January 1, 1890, the company had had a deficit of $3,000; by August 11, 1891, and January 1, 1892, this had turned into a surplus of $50,000 and $75,000, respectively.[27] Exactly one year later, the surplus totaled $109,500; [28] a 10 per cent cash dividend amounting to $10,000 had been paid in the meantime, and a stock dividend of 15 per cent had been distributed.[29] The directors, in a meeting held on February 3, 1892, gave further evidence that the company had just passed through its first major crisis successfully. Upon the suggestion of E. D. Libbey, they awarded $10,000 in fully paid capital stock [30] to Solon O. Richardson, Jr., and increased his annual salary by $3,000 "in compensation for extra services in the extraordinarily successful operation of the plant at Findlay." [31] At the same meeting the directors decided to change the name of their firm to the Libbey Glass Company and acceded to Libbey's request that they erect and operate an exhibition factory at the forthcoming Columbian Exposition at Chicago.

From all indications, Libbey's personal enthusiasm concerning participation in the World's Fair was not shared by his business associates. Nonetheless, they organized a separate corporation

[26] Libbey Glass Company, Documents relating to licenses, etc., Safe No. 2, Special Examination and Report by Ernst and Ernst, Auditors.

[27] Libbey Glass Company, Libbey Glass Company Minutes, I, 26, 34.

[28] *Ibid.*, I, 59.

[29] *Ibid.*, I, 35–36. This is the first record I have found of any dividend paid by the company.

[30] Together with the 15 per cent stock dividend this raised the company's capitalization to $125,000.

[31] *Ibid.*, I, 35–36. While in Libbey's employ at Cambridge, Richardson had experimented with bulb-blowing and evidently had been responsible for stirring up interest among the other directors in the Corning strikers' proposition.

known as the Libbey Glass Company of Illinois, with a capital of $75,000, of which two-thirds was furnished by the parent company and subscribed for in the name of its individual stockholders on a pro rata basis.[32] The financial outlay at Chicago eventually approached $250,000, and Libbey himself shouldered most of the risk. Since he owned on February 7, 1894, almost 68 per cent of the stock in the Libbey Glass Company at Toledo, his share in its $50,000 liability for the Chicago subsidiary amounted to around $34,000. Most of the remaining $200,000 expended was borrowed upon Libbey's personal signature.

The opening months of the exposition proved quite discouraging. Few people visited the model glass factory, even though admittance at first was free; and sales of glassware were discouragingly small. Libbey was forced to go before the exposition's finance committee for funds in order to keep his men at work. After the harvesting season in the middle west ended, however, attendance at the fair increased markedly. Upon the suggestion of one of his female employees, who considered the typical American disdainful of witnessing any exhibit that was free, Libbey began to exact an admission charge of 10 cents per person. Almost immediately crowds of people overran his building, and he joyfully discharged the hundred or so persons whom he had been paying to mill around the exhibit as decoys. Attendance grew so large that the building proved inadequate; consequently, the price of admission was raised first to 25 cents and finally to 50 cents.[33] Each visitor received a little spun-glass bow on the end of a stickpin; [34] and the sale of paperweights, cut-glass tableware,

[32] Much of what follows is based upon the early records and minute books of the Libbey Glass Company. Several individuals who knew E. D. Libbey and M. J. Owens intimately have supplied additional information.

[33] The fee could be used as partial payment on any one piece of glassware which the customer bought.

[34] Spun glass was no recent discovery. The practice of drawing it out into such fine threads that it could be woven into tissues and feathers for baby caps had been discussed by Réaumur in 1713. — "Ductilité de diverses Matières," *Histoire de l'Académie Royale des Sciences*, pp. 206–210. One Signor Olivi of Venice is supposed to have made very beautiful and flexible glass cloth around 1837; M. Dubus Bonnel at Lisle at about the same time wove a cloth "which is perfectly flexible, and may be applied to the hangings of rooms and other purposes;" and Williams and Sowerby in London exhibited a few years later how glass could be spun with

and other objects soared. The Spanish Princess Eulalie during her trip to the fair was enchanted to see how the men actually melted and blew glass, and was especially captivated by a spun-glass dress which they had made previously for Georgia Cayvan, a well-known American actress of that day. Such interest flattered Libbey, who was also quick to see its advertising value. He accordingly offered to make a duplicate for the Princess, and she reciprocated by bestowing the honor on his company of being named the "Glass Cutter-in-Chief to Her Royal Highness, the Infanta Eulalie."

When the exposition closed, the Libbey Glass Company of Illinois was able to return the $50,000 advanced by the parent firm at Toledo, to repay all the other sums it had borrowed, to turn over to the fair's committee $122,583 for its concession rights, and to distribute among its owners approximately $100,000 as profit.[35] The glass exhibit, however, yielded even more significant benefits in the long run. Because of the favorable publicity received at the fair, the Libbey trademark became known throughout the country as a guarantee of excellence in glassware; and Michael J. Owens, the plant manager at the exhibit, had found sufficient leisure during the first months at Chicago to toy with some ideas concerning the possibility of blowing glass mechanically.

3. The Formation and Early History of the Toledo Glass Company (1895–1903)

As soon as he and the rest of the staff returned to Toledo, Owens began to work on a machine which he hoped would successfully blow glass for incandescent lamps. On September 17, 1894, he assigned two patents which he had just received on "ap-

steam power and then woven on a loom. — Gaffield Collection, Notes on Glass, III, 111-114. Austrian glass shops exhibited many different articles of clothing made of spun glass at the Parisian Exposition in 1878. — Weeks, *Report on Manufacture of Glass*, p. 75. And Gaffield of Boston recorded in his private journal on December 17, 1881, that he had recently seen some spun-glass pieces, but that he seriously doubted whether suitable clothing of glass would ever be made in large quantities. — Glass Journal, IV, 11.

[35] According to an oral report given William H. Boshart by Owens himself.

(*Courtesy of Owens-Illinois*)

FIG. 4. MICHAEL J. OWENS, 1859–1923

paratus for mechanically operating paste-glass molds" [36] to the Libbey Glass Company for use in making light bulbs only.[37] During the following year both he and Libbey obtained additional patents on various improvements and assigned the bulb rights to the same company in return for $200.[38] Hence, the Libbey Glass Company obtained exclusive world rights to use and sub-license all present and future inventions of these two men in so far as they pertained to manufacturing bulbs.[39]

The machine, however, had additional possibilities which Libbey thought could best be realized by forming a separate promotional and developmental company. He accordingly filed an application with the State Department of Ohio on December 17, 1895, to incorporate the Toledo Glass Company. Its capital stock, consisting of 1,500 shares with a par value of $100 each, was then opened to subscription,[40] and the stockholders held their first meeting on February 15, 1896. At that time E. D. Libbey and M. J. Owens granted the new enterprise all the rights in their present inventions which they had not already assigned to the Libbey Glass Company. They further agreed to turn over to Toledo Glass all the patents they might obtain during the next seventeen years or as long as they continued to be in its employ. In return, they each received $37,500 in its fully paid capital stock.

The Toledo Glass Company then proceeded to build a plant at Toledo and equip it with a fourteen-pot furnace and several machines which Owens had already adapted for making tumblers.[41] Production there had scarcely begun, however, when the

[36] See above, p. 16, n. 47, for a description of the "paste-mold."

[37] These patents were numbered 482,526 and 489,543. — Libbey Glass Company, Documents relating to licenses, etc., Safe No. 3.

[38] On February 5, 1896. — Owens-Illinois Glass Company, Owens' Private Files. These patents bore the numbers 534,840 (dated February 26, 1895) and 548,587, 548,588, 548,600, and 548,605 (dated October 22, 1895).

[39] Libbey-Owens-Ford Glass Company, Toledo Glass Company Minutes, I, 37–40. The minutes of the directors' and stockholders' meetings contributed most of the information presented in this section.

[40] All was subscribed by August 3, 1896.

[41] These machines were essentially the same as those used in blowing bulbs. The major difference lay in the shape of the molds used. Owens' first patent specifically on the tumbler machine was numbered 524,840, and was issued on February 26, 1895.

Rochester Tumbler Company, an important Pennsylvania firm, became interested in what was transpiring at Toledo and made overtures to obtain exclusive machine rights. According to agreements between the two companies, concluded on September 1, 1897, the Rochester Tumbler Company received exclusive, nonassignable rights to use the machines in the manufacture of tumblers, finger bowls, lemonade glasses, and stemware throughout the United States. In return, it was to pay Toledo Glass $100,000 over a period of ten years.[42] The latter company retained the privilege of producing as many tumblers and the like as it could with its original furnace and equipment. Rochester Tumbler, moreover, consented to market the combined output of both plants under its own name and at prices which it thought desirable.[43] These provisions were modified on July 1, 1900, as follows: the licensee agreed to pay $35,000 in cash over a period of three years for the furnace, equipment, and tumbler rights theretofore reserved by the licensor; it obtained reaffirmation of its exclusive tumbler rights for the United States; and it had all limitations upon their assignability removed. Less than a year later, on January 9, 1901, the Toledo Glass Company, the Rochester Tumbler Company, and the National Glass Company [44] entered into a new contract providing that the tumbler license be assigned to National Glass and that it assume the financial payments which Rochester Tumbler still owed Toledo Glass. During the latter part of 1904 the remaining outstanding payments, amounting to $30,000, were settled for $20,000 cash. Hence, the Toledo Glass Company ultimately received from either National or Rochester approximately $125,000.[45]

After having thus disposed of its machine's bulb rights and tumbler rights for the United States to the Libbey Glass Company and the Rochester Tumbler Company, respectively, the

[42] The Toledo Glass Company contracted to furnish six machines free of charge and any additional ones at cost.

[43] Its commission on sales of Toledo Glass's output was fixed at 6 per cent.

[44] The National Glass Company was a combination of the Rochester and several other companies in the tableware field. See below, pp. 237–238.

[45] Toledo Glass made discount allowances for all advance payments of the yearly installments met by Rochester Tumbler in conformity with the 1897 purchase agreement. See, *ibid.*, I, 234–236.

Toledo Glass Company had proceeded to organize the American Lamp Chimney Company in 1898.[46] The new corporation obtained the nonexclusive machine rights for lamp chimneys, the plant and tools of the parent organization, and $112,500 cash for working capital. In return, it gave Toledo Glass $300,000 in its fully paid common stock and $125,000 in its fully paid preferred.[47] Almost immediately George A. Macbeth and Thomas Evans, heads of two other concerns which had long made chimneys by hand processes, became gravely concerned lest the new enterprise steal the major portion of their business. Through an intermediary they were able to reach an understanding with Libbey personally, whereby they would incorporate the Macbeth-Evans Glass Company under Pennsylvania laws with a capitalization of $2,000,000.[48] The Toledo Glass Company was to get all the preferred stock of the new corporation in exchange for an exclusive machine license covering lamp chimneys, and the American Lamp Chimney Company was to turn over its plant, equipment, and nonexclusive license for $330,000 of the common stock.[49] The Macbeth-Evans Glass Company duly received its charter in the summer of 1899.[50] The stockholders immediately elected E. D. Libbey vice-president and director and appointed M. J. Owens superintendent at the Toledo branch factory.[51] The American Lamp Chimney Company surrendered

[46] The Toledo and American Lamp Chimney companies signed a nonexclusive license agreement on January 9, 1899.

[47] Toledo Glass sold all the preferred for $112,500 before January 1, 1899; set aside $162,000 in common to be paid as dividends to its own shareholders; gave E. D. Libbey $100,300 of common for a $78,000 debt which it owed him; and offered 377 shares ($100 par value) for sale to the public at $78 each. On February 6, 1900, the company still had 339 of the 377 shares of common on its hands.

[48] One-tenth of this amount was to be in 7 per cent cumulative preferred ($50 par value) and the remainder to be in common shares of $50 par value.

[49] In addition, it was to receive a fair value for its goods on hand and accounts receivable.

[50] Application to incorporate under the name Illuminating Glass Company was first filed on April 17, 1899. The stockholders, however, voted on May 25 to change the name to Macbeth-Evans Glass Company and received a new charter on June 19. The original incorporators and directors, all figureheads, resigned on July 11, and the real owners then took over. — Libbey-Owens-Ford Glass Company, Licenses and Correspondence of Toledo Glass Company, Vol. I.

[51] Shortly after the Macbeth-Evans Glass Company began operations, it had six active factories comprising those formerly worked by the separate companies

all its assets on July 11 and wound up its affairs by declaring a final dividend on March 21, 1900.

Because of the complicated way in which it disposed of its exclusive machine rights for lamp chimneys, it is difficult to determine exactly how much the Toledo Glass Company ultimately obtained for them. The American Lamp Chimney Company appears to have been a wholly owned subsidiary. If it was, then Toledo Glass or its stockholders eventually received about $500,000 in Macbeth-Evans common and preferred stock in exchange for both the exclusive license and the physical assets of the plant at Toledo.[52]

The Toledo Glass Company was not nearly so successful in marketing its foreign tumbler and lamp-chimney rights as it was in selling the domestic rights.[53] On July 13, 1897, the Diamond Glass Company, Ltd., of Montreal, bought the Canadian rights for $15,000, which figure included the cost of making, delivering, and installing two machines. Toledo Glass voted in December of the same year to try to sell all European tumbler rights for $800,000 net and, if necessary, to build a demonstration factory in Europe. Libbey himself journeyed to England in 1898 with power of attorney to effect the sale under the best possible terms. On March 5, 1900, he reached an agreement with certain English interests whereby they would form the Automatic Glass Blowing Patents Syndicate, Ltd., with a nominal capital of £85,000.[54] The syndicate was to pay £1,500 in cash and £54,500 in its ordinary and preference shares for Toledo Glass's exclu-

---

which participated in the merger. Two were located in Pittsburgh, one at Charleroi (Pa.), another at Elwood (Ind.), still another at Marion (Ind.), and, finally, the one at Toledo. In all, their furnaces held 180 pots. A seventh factory, at Pittsburgh, was lying idle. See Macbeth-Evans Glass Company, *Glass Making*, p. 41; and Owens-Illinois Glass Company, American Lamp Chimney Company Letters, p. 147.

[52] According to the agreement reached by Libbey, Evans, and Macbeth on May 11, 1899, $30,000 of the $330,000 common stock paid the American Lamp Chimney Company was to be turned over to one Sidney Mason, the intermediary who had effected the consolidation. Evans and Macbeth also were to give Mason $30,000 and $40,000, respectively, from their holdings of Macbeth-Evans common.

[53] The Libbey Glass Company held the exclusive bulb rights for all countries.

[54] £44,500 of this was to be 6 per cent cumulative preference shares, of which Toledo Glass was to receive £35,000 and an intermediary the remaining £9,500.

sive European rights.[55] Before the year ended, however, the syndicate was in serious financial difficulty, and Libbey had to return to England to protect the rights and interests of his company. In November, 1901, the directors of Toledo Glass consented to surrender all the stock and claims they held against the syndicate for £850 cash. They never received this relatively small sum, although it was actually paid, because the English solicitors who collected it while acting for Toledo Glass were forced into bankruptcy, disbarred, and eventually imprisoned. Hence, Libbey and his associates obtained nothing from the sale of the European tumbler rights on their machine.

4. The Owens Bottle Company (1903–1920) [56]

The Toledo Glass Company meanwhile had diverted its chief attention from introducing tumbler and lamp-chimney machines to developing completely automatic methods for blowing bottles. According to M. J. Owens' testimony given under oath in 1919, he had conceived the basic ideas for his bottle machine in 1898.[57] On October 27 of the following year he and a competent draftsman contracted to construct experimental models.[58] Their efforts met with such success by 1903 that the Toledo Glass Company decided it was time to incorporate a separate firm for the purpose of licensing domestic producers. The Owens Bottle Machine Company accordingly received its charter under the laws of New Jersey on September 3, 1903.[59] With an authorized

[55] £38,500 shares was to go to Toledo Glass; and £16,000 shares and £1,500 cash were to be paid one Samuel Snell, who had acted as intermediary.

[56] Unless otherwise specified, the information for this section was supplied by the minutes of the meetings held by the Toledo Glass Company's stockholders and directors (4 volumes), the Owens Bottle Company's stockholders and directors (8 volumes), and the Owens Bottle Company's Executive Committee (2 volumes).

[57] Hearings and proceedings of the *Owens Bottle Company* v. *the Libbey Glass Company* in Libbey Glass Company, Documents relating to licenses, etc. Safe No. 1.

[58] Libbey-Owens-Ford Glass Company, Toledo Glass Company Minutes, I, 202.

[59] The directors filed an application to incorporate under Ohio laws on December 16, 1907, and then transferred the assets of the old company to the new. In February, 1908, the New Jersey corporation surrendered its charter. In 1919 the stockholders voted to change the name of their company to the Owens Bottle Company, and ten years later they agreed to merge with Illinois Glass. This was how the Owens-Illinois Glass Company came into existence. Hereafter, the original company will be referred to as "Owens Bottle" rather than "Owens Bottle Machine."

capitalization of $3,000,000 (five-sixths of which was in common stock and one-sixth in 7 per cent cumulative preferred), it enjoyed the right to manufacture and deal in glass, to make and sell all kinds of glass machinery and tools, to obtain patents and grant licenses thereof, and to hold the stock of other companies. Toledo Glass then issued it an exclusive bottle license for the United States and its territories and also handed over its experimental plant which had been built on Libbey Street in Toledo. The newly elected directors of Owens Bottle paid Toledo Glass $2,800,000 in cash and securities for these properties.[60]

On September 1, 1904, the new enterprise had $400,000 preferred and $2,500,000 common stock outstanding. Since it had obtained very little operating capital from the disposal of these shares,[61] it found itself in desperate need of additional funds a few months later. Its preferred at that time was selling below par on the open market,[62] so when Toledo Glass offered to purchase

[60] According to the agreement dated October 16, 1903. Only $276,000 of this amount was to be in cash. $124,500 was to be in Owens' fully paid preferred and $2,399,500 in its fully paid common. The Toledo Glass Company did not retain all of this. It turned over (1) $100,000 in cash and $400,000 in Owens common to the Glass Machinery Company, which had been experimenting with various processes and machines at Pittsburgh, for the assignment of certain of its patents (generally known as the "Proeger Patents"); (2) $11,250 in cash and $25,000 in Owens common to a John J. Power for his patents; and (3) $12,500 in cash to a Herman Hilde for the exclusive right to his inventions. Toledo Glass also agreed to set aside $500,000 of Owens' common stock to be used, if necessary, for paying bonuses to subscribers of the preferred. (Less than three-fifths of this amount, however, was ever used for this purpose.) In 1917 Owens Bottle paid Toledo Glass $40,000 cash for the right to manufacture all other kinds of glass on the machines in addition to bottles and jars. — Libbey-Owens-Ford Glass Company, Toledo Glass Company Minutes, IV, 181, 188, 207. Hence, the Toledo Glass Company received and retained $192,250 cash, $124,500 preferred, and $1,674,500 common from the sale of all its domestic bottle machine rights.

[61] It is impossible to ascertain exactly how much. Presumably it used all the preferred to cover its obligations to Toledo Glass because it handed over directly 1,245 certificates ($100 par value) and sold the remaining 2,755 shares to raise nearly all the $276,000 cash paid. Only 1,005 out of 25,000 common shares ($100 par value) were left after the settlement with Toledo Glass, and most of these were probably distributed as bonus to purchasers of preferred. Certainly the promoters never intended to use the common stock to raise working capital. See Owens-Illinois Glass Company, Annual Reports of Owens Bottle Company, I, 1–7. Everything suggests that they underestimated the amount of capital that would be needed to put the machine on the market.

[62] The first dividends on preferred were paid on January 10, 1907, and all accumulated dividends were paid in full on October 1 of the same year.

500 of the unissued shares at face value, it immediately accepted the offer. Its directors authorized the sale of 140 additional shares in 1905–1906 in order to meet promotional expenses. This financial embarrassment centered around the inability of the Owens Bottle Company to find suitable licensees as easily and as quickly as had been anticipated.

At that time skilled laborers blew most bottles in the United States, and semiautomatic machines fed by hand made only wide-mouth jars.[63] When established producers heard that the promoters of the Owens machine claimed that it could both gather and blow glass automatically, they were incredulous. Although the demonstration amazed them all,[64] few were anxious to secure a license. Trade conditions were not altogether favorable in 1903–1904, and the expense of installing the machines appeared prohibitive. New furnaces, lehrs, and other accessory equipment would be necessary, and in most cases these in turn would occasion new factory construction.[65] Consequently, the Owens Bottle officials soon realized that they would have to help create their own licensees.

In September, 1903 — the same month Owens Bottle was organized — E. D. Libbey opened negotiations with a group of Pittsburgh beer-bottle manufacturers, who wanted to form an association to operate the machine under exclusive license.[66] When he saw that the men were not likely to reach a satisfac-

[63] See below, pp. 178 ff.

[64] On October 3, 1903, E. D. Libbey wrote another official in the company, "I had some big brewers here yesterday from Pittsburgh, and the same old story — everyone is thunderstruck with the machine." — Owens-Illinois Glass Company, Libbey Letters, II, 61.

[65] *Ibid.*, "History," p. 12. It is significant that each of the first three licensees built entirely new plants. The Northwestern Ohio Bottle Company and the Baldwin-Travis Glass Company (later the Thatcher Manufacturing Company) constructed works at West Toledo and Kane (Pa.), respectively; and, although the Ohio Bottle Company acquired several factories from the concerns which merged to form it, Owens machines were first installed in its new plant completed at Newark in 1905. — *National Glass Budget*, XX, March 25, 1905. The machine itself was costly. The one with six-arms sold for around $7,000–$7,500 in 1907–1912, and the ten-arm machine for about $12,000–$13,500 in 1910–1913. — Owens-Illinois Glass Company, Owens Bottle Company Minutes, I–III, *passim*.

[66] See his letter, dated September 25, 1903, to W. S. Walbridge. — Owens-Illinois Glass Company, Libbey Letters, II, 44.

tory understanding among themselves, he secretly approached Edward H. Everett, one of their number. Libbey and Everett, who owned a bottle factory at Newark, Ohio, together decided upon another plan.[67] As a result of their agreement, Everett's company merged with some others which kept three plants running at Massillon and Wooster, Ohio, and formed the Ohio Bottle Company capitalized at $4,000,000.[68] The Owens Bottle Company issued it an exclusive license for beer, porter, ale, and soda-water bottles on November 1, 1904.[69] On the same day another enterprise, the Northwestern Ohio Bottle Company, received exclusive rights for wine, brandy, and a few special "branded" bottles. Libbey and the other major stockholders in Owens Bottle had just incorporated this company, whose authorized capital was $100,000.[70] Neither Ohio nor Northwestern, however, was Owens' first licensee, because the Baldwin-Travis Glass Company had secured the right to manufacture milk bottles on September 16, 1904. Less than a year later it assigned its license to the Thatcher Manufacturing Company, which had never made glass but which had manufactured a rather complete line of dairymen's supplies.[71]

[67] In April and May, 1904, they exchanged several letters under fictitious names ("Raymond" and "Potter") and with no identifying marks on the envelopes or stationery. Libbey's expressed desire to keep the whole affair secret until all details were definitely arranged, however, was not realized. The bottle-makers at Pittsburgh somehow learned of the intrigue shortly before May 12, 1904, and in no uncertain terms claimed that they had been betrayed. — *Ibid.*, III, 1–13.

[68] Owens-Illinois Glass Company, Walbridge Letters, I, 151; Libbey-Owens-Ford Glass Company, Colburn Records, IV, 2. The Ohio Bottle Company, the Busch Bottle Company (with plant at Belleville, Ill.), and the Streator (Ohio) Bottle and Glass Company consolidated in August-September, 1905, to form the American Bottle Company, capitalized at $10,000,000. — *Commercial and Financial Chronicle*, LXXXI (1905), 1,101–1,102.

[69] The Ohio Bottle Company assigned this license to the American Bottle Company on September 7, 1905. The new enterprise consolidated most of its production equipment at Streator and Newark, where it installed many Owens automatic machines. It worked its other plants a few years as hand shops and then shut them down.

[70] Owens-Illinois Glass Company, Libbey Letters, II, 362. The Owens Company did not pledge or subscribe to any of Northwestern's stock in its own name. In the Spring of 1905 some of the same men bought a plant at Newark, Ohio, and incorporated the Newark Machine Bottle Company. It marketed its entire output through the Ohio Bottle Company, from whom it had received a sublicense. — *National Glass Budget*, XXI, July 15, 1905.

[71] It had also been an important distributor of milk bottles. — *National Glass*

It is significant that as late as January 1, 1909, only these three companies, organized for the express purpose of using the automatic machine, had received licenses. Ohio and Northwestern had been promoted entirely or partially by men interested in the Owens Bottle Company. And Baldwin-Travis, formed by "outsiders" desirous of obtaining the milk-bottle rights, sold out to Thatcher, who previously had not made glass. This was not an altogether gratifying record in view of the fact that there were over one hundred firms then engaged in manufacturing bottles and jars by hand.[72] Producers and workmen alike simply refused at first to accept the Owens machine in spite of its obvious merits.[73] Options on fruit-jar rights, for example, had been successively held by James A. Chambers in 1906 and by Frank C. Ball in 1907,[74] but both men had allowed their options to expire. For the Owens Bottle Company, 1909 was a banner year. Partly as a result of some recent improvements in the construction and operation of the machine and partly because of the success achieved by the original licensees, many established firms sought licenses in that year.[75] Five were issued.[76] By 1914 fourteen were

Budget, XX, June 11, 1904. It seems that a Francis E. Baldwin, who was both president and treasurer of Thatcher, had intended that Baldwin-Travis (capitalized at $50,000) would make the bottles which Thatcher would sell. As the *National Glass Budget* predicted on November 19, 1904, the two companies soon combined under the Thatcher name. H. E. Travis, with whom Baldwin had associated to obtain the Owens license, was a practical glass manufacturer who had been superintendent at the Fidelity Glass Company in 1903 and who later established his own plant and made milk bottles by hand processes.

[72] Owens-Illinois Glass Company, "History," p. 12. [73] See below, pp. 204 ff.

[74] The Ball Brothers Glass Manufacturing Company at Muncie, Indiana, obtained the fruit-jar license from an assignee in November, 1909, and by the end of 1910 had paid more than four times the amount stipulated in Ball's original option covering the same rights.

[75] The Owens board of directors was informed on November 19 that thirteen producers wanted to install its machines and that a fourteenth would be incorporated if there was any chance that it could get a license. — Owens-Illinois Glass Company, Owens Bottle Company Minutes, I, 162.

[76] To Louis Hollweg on January 2 for fruit jars, to the Hazel-Atlas Glass Company on May 20 for most kinds of packers' ware, to the H. J. Heinz Company on July 19 for various bottles to be used only in merchandizing its own food products, to the Whitney Glass Works in December for druggists' ware, and to the Owens West Virginia Bottle Company on December 27 for certain kinds of beverage bottles. Hollweg assigned his rights in November to the Greenfield (Ind.) Fruit Jar and Bottle Company, from whom Ball Brothers bought them a week or so later. Owens West Virginia was a new corporation.

outstanding, covering nearly every important kind of bottle and jar.[77]

From the outset Owens Bottle adopted the policy of limiting each licensee to particular lines of containers such as milk bottles, brandy bottles, whiskey bottles, domestic fruit jars, and certain kinds of packers' ware. Usually the lines were those which had previously been the licensee's specialty. Five manufacturers each enjoyed exclusive rights on their line, and two others shared similar privileges for whiskey bottles.[78] Whitney at Glassboro, New Jersey, secured sole rights for oval-shaped ammonia bottles, but its privileges relating to general "prescription" ware were nonexclusive. Following 1910 the Owens Bottle officials stopped issuing broadly exclusive contracts. This change in policy, however, almost came too late to do any good. After deciding to manufacture bottles themselves on a large scale, they found that entry into the field had already been seriously limited. In 1913 and 1914 they begged their exclusive licensees — American Bottle, Hazel-Atlas, Illinois Glass, Thatcher, and Charles Boldt — to sublicense them so they could make various bottles on their own machine.[79]

They also became convinced that exclusive licenses were not the best means for maximizing royalty earnings. They repeatedly tried to get their licensees to expand output and to sublicense other producers, pointing out that such action not only would increase the income of both companies but would also re-

[77] Licenses were issued to Illinois Glass on June 11, 1910, January 18, 1911, and May 22, 1914; to Charles Boldt on June 11, 1910; to Macbeth-Evans on July 15, 1910; to D. C. Jenkins Glass in 1912; to the Owens Eastern Bottle Company on June 11, 1912; and to Maryland Glass in December, 1913. Of these, Owens Eastern was the only new company. Macbeth-Evans' rights were not very important because they covered glass (other than bulbs) used in artificial lighting.

[78] Thatcher for milk bottles (after 1908); Northwestern for wines and brandies (after 1904); Ohio for beers, porters, ales, and soft drinks (after 1904); Ball Brothers for fruit jars (after 1910); Hazel-Atlas for many items in the line of packers' ware (after 1909); and Charles Boldt and Illinois Glass for whiskies (after 1910).

[79] See, especially, Owens-Illinois Glass Company, Owens Bottle Company Minutes, III, 70–72; IV, 80 ff. E. D. Libbey reported after one of the conferences that "the representatives of the Illinois Company . . . repeatedly stated that they preferred to sell out their entire business rather than surrender [their exclusive rights in] the liquor bracket."

move an important incentive for others to invent bottle machines. On the whole, their pleas went unheeded.[80] Furthermore, each time Owens Bottle wished to issue a new license, those who had already received sole rights for particular bottles felt compelled to contest the validity of the proposed action.[81]

The exclusive licensees, of course, paid more for their rights than did the others. Thatcher (for milk bottles) and Louis Hollweg (for fruit jars) each paid $50,000; Hazel-Atlas turned over $500,000 of its capital stock; and the others had to accept large minimum production quotas. With two exceptions, all manufacturers paid royalties on their output.[82] The rates originally were supposed to be one-half the estimated reductions in labor costs brought about by the machine, but the royalties stipulated in the licenses issued between 1909 and 1914 were 50 per cent or 66.7 per cent lower than those stated in the 1904 contracts.[83] These reductions had become increasingly necessary as bottle blowers in hand shops accepted substantial wage cuts and as various semiautomatics gradually were installed in other plants.[84]

If Owens Bottle licensed more than one firm in any line of bottles, it usually set maximum production quotas for each firm. On April 29, 1918, its executive committee decided to allow additional companies to employ its machines in the future only for

[80] Thatcher, for example, was not supplying all the milk bottles bought in the United States and could "not hope to do so." Yet, it refused to sublicense such important nonmachine producers as Travis, who was able to compete with Thatcher because his plant was located near cheap supplies of natural gas. In return for reducing Thatcher's royalty from 40 cents to 10–15 cents per gross on all additional machines installed, Owens Bottle finally obtained permission to make from 50,000 to 150,000 gross of milk bottles each year in its Clarksburg plant for Thatcher's account.

[81] Illinois Glass, for example, contested Owens Bottle's right to allow Maryland Glass to make Bromo-Seltzer bottles.

[82] Hazel-Atlas paid practically no royalties but shared its profits with Owens Bottle through dividends on the 5,000 shares of stock. Ball Brothers, the only other exception, had commuted its royalties for $500,000 cash in 1910.

[83] In 1904 the royalty rates were 40 cents per gross (for milks) and 50–54 cents per gross (for wines, whiskies, and the like); in 1909–1914 they were 10-15 cents per gross (for milks), 15–25 cents per gross (for prescription ware), and 15–25 cents per hundredweight (for whiskies, prescription, and packers' ware).

[84] See below, pp. 205 ff.

making their usual output.[85] Certainly at no time did it show any desire to stir up competition among its own licensees. All nonexclusive contracts made prior to 1912 contained provisions for the control of prices. Usually Owens Bottle reserved the right to fix selling prices if the licensees could not arrive at a satisfactory understanding among themselves. An article published in the *National Glass Budget*,[86] however, caused the directors to fear lest their price-control stipulations be judged in violation of the Sherman Antitrust Act; and on December 19, 1911, they voted to revoke all price-fixing clauses in their contracts. The following month they agreed upon additional modifications. They vested the ownership of their machines in each company using them (instead of merely leasing the machines as they had always done previously) [87] and removed all limitations theretofore imposed upon their disposal. In other respects, Owens Bottle did not change substantially its licensing policy. Each contract covered the life of the basic patents and exacted an explicit recognition of their validity. Owens Bottle retained the right to apply for patents on all improvements made by licensees or their employees on any part of the equipment it had supplied. Although it did not forbid them to follow other production methods in their plant, licensees could not use the Owens furnace, revolv-

[85] This applied chiefly to the Graham machine, which Owens Bottle had acquired by purchasing the Graham Glass Company in 1916. No new licenses for the Owens machine were issued after 1918.

[86] Owens-Illinois Glass Company, Owens Bottle Company Minutes, II, 127, 130. The article in the *Budget* (XXVII, December 2, 1911, p. 10) was a reprint of the "Information Circular No. 15" recently issued by Patent Attorney George H. Benjamin, of New York City. Benjamin pointed out that the Attorney General, in filing suit against the tobacco and Standard Oil trusts, was not attempting to invalidate the protection accorded inventors by existing patent laws. The institution of previous antitrust cases had simply been predicated upon the idea that "after a manufacturer working under a patent had sold the article manufactured, his right to exercise any further control over it (either as regards its use or price for which it might be sold) ceased . . . . To keep within the Sherman law, therefore, manufacturers should remember . . . that they can make no agreement with purchasers, agents or distributors, by reason of which the use or the re-selling price of the article is fixed and determined . . . . Whether the position taken by the Government, is or is not correct, must be finally decided by the Supreme Court."

[87] Each licensee had paid for its machines at the outset (cost plus 10 per cent) and had then held back a small portion of its royalties each year until it had been fully reimbursed.

ing basin, fire polisher, or lehr except in conjunction with its automatic machine.

The primary reason for incorporating the Owens Bottle Company in 1903 had been to create an agency for licensing established producers. Its promoters themselves knew nothing about the bottle business and certainly never intended to make glass containers in commercial quantities. Yet, according to its profit and loss statement for the fifteen months ending on December 31, 1918, its net income from bottle sales definitely surpassed its earnings from royalties and licenses. Owens Bottle and its subsidiaries had become by 1920 the largest manufacturer of bottles in the United States. How had this change come about?

From the outset it had made in its small demonstration plant on Libbey Street some bottles which it had marketed through the Ohio Bottle Company or its successor.[88] Next, it initiated in April, 1906, regular production at this plant to supply the needs of local catchup firms, and its bottle profits for fourteen months prior to October, 1907, totaled $12,687. The third step was taken in 1908. E. D. Libbey informed the stockholders in his Fifth Annual Report (September 30, 1908) that their directors had bought up all the capital stock in the Northwestern Ohio Company on the first day of January [89] and that they had decided to complement their licensing of "outsiders" with the construction and operation of additional plants. In the Sixth Annual Report he pointed out that the company's basic patents would eventually expire and that its royalties would taper off. It therefore seemed desirable, he said, to create a more permanent source of revenue and at the same time establish model factories which could be used by all licensees as yardsticks to gauge their own operating efficiency. Two years later he announced that the company planned to manufacture bottles on an even larger scale than had been envisaged theretofore.

[88] On January 18, 1906, this licensee told Owens Bottle to discontinue making bottles for its account. — Owens-Illinois Glass Company, Owens Bottle Company Minutes, I, 178.

[89] Owens-Illinois Glass Company, Annual Reports of Owens Bottle Company. The purchase agreement, concluded the previous month, entailed payments amounting to $134,219. Northwestern had a two-machine plant in West Toledo.

The next move directed toward fulfilling the new production program, following the purchase of Northwestern, was the incorporation of the Owens West Virginia Bottle Company (with an authorized capital of $100,000) during the last two months of 1909. The directors immediately began to construct a modern three-furnace, six-machine plant at Fairmont, West Virginia.[90] Both Northwestern and Owens West Virginia consolidated with the parent concern in the last quarter of 1911. The State Department of West Virginia chartered the Owens Eastern Bottle Company (with a capitalization of $1,000,000) the following year. Owens Bottle originally subscribed to 51 per cent of its stock, and four other companies [91] — three of which had inadequate resources for installing automatic machinery in their own plants — agreed to take the rest. A large factory to make "prescription" ware was then built at Clarksburg. Operations there during the first two years, however, were not very profitable so the other owners sold out to Owens Bottle on January 11, 1915. Then the two companies amalgamated. Owens Bottle likewise absorbed the Whitney Glass Works at Glassboro, New Jersey, on July 1, 1918, after having first obtained control in 1915.[92] It also purchased an idle factory at Greenfield, Indiana, from Ball Brothers on February 9, 1917, at a cost of $62,500 and installed three machines there. Finally, it constructed a very large and highly mechanized plant at Charleston, West Virginia, in 1917–1918.

[90] By September 30, 1910, the factory was completed. Six years later it had six machines.

[91] The Whitney Glass Works (5 per cent), the Fidelity Glass Company (14.66 per cent), the Cumberland Glass Manufacturing Company (14.66 per cent), and the Bellaire Bottle Company (14.66 per cent).

[92] It had licensed Whitney in December, 1909. One year later Owens Bottle loaned it $50,000 on its 6 per cent second-mortgage bonds ($34,000 in cash and $16,000 for accrued royalties). George D. Whitney, the president, died on February 24, 1915, and the administrators of his estate asked Owens Bottle to take up Whitney's personal notes in order to prevent the immediate and ill-advised sale of 1,954 shares of Whitney stock which the creditors held as collateral. Owens Bottle complied, advanced an additional $50,000, and received an option to buy the 1,954 shares for $136,780. On June 22, 1915, the directors voted to exercise this option. Three years later Owens Bottle bought all the other outstanding shares and dissolved the Whitney Glass Works on July 1, 1918.

In 1916, as a major stockholder, it had caused the Whitney directors to build a new six-machine factory. The old seven-machine plant was abandoned in 1919.

Owens Bottle's expansion on the horizontal plane was not, however, limited to such complete absorptions as these. Two other companies were fully acquired, and control over two more was obtained, but all four retained their own identity until after 1920. The American Bottle Company, licensed in 1904, was one of the largest bottle manufacturers in the country. Its net earnings for the twelve months ending on August 31, 1914, amounted to $1,201,968, and the value of its assets at that time was around $10,000,000. Its plants at Streator, Illinois, and Newark, Ohio, equipped with forty-six machines, were said in March, 1916, to have had an annual capacity of 324,000,000 bottles and to have employed at least 1,200 people regularly.[93] In 1913 it had made twice as many containers as Owens Bottle; in 1914 it had made about one-third more than Owens Bottle. A majority of its owners agreed on November 16, 1915, to sell their shares at par to E. D. Libbey, who immediately assigned the option to his Owens Company. Within twelve months Owens Bottle purchased nearly all [94] the capital stock and thereby doubled its own production capacity.

At the same time Libbey and his associates acquired for $650,000 [95] the Graham Glass Company of Indiana (which had controlling interest in the Graham Glass Company of Oklahoma) primarily because Graham had developed a promising bottle machine [96] and because its business with Coca-Cola was the largest and most profitable account in the soft-drink field. Graham had two factories equipped with its own machines at Evansville and Loogootee, Indiana, and its Oklahoma subsidiary had a small plant at Okmulgee.[97] On December 15, 1919, the directors of

[93] Owens-Illinois Glass Company, Owens Bottle Company Minutes, V, 132. At the same time Owens Bottle was operating three factories with a total of 20 machines, 700 employees, and a capacity of 230,400,000 bottles a year.

[94] It acquired all but $300 preferred and $800 common out of $8,419,500 shares outstanding. Owens Bottle decided to exchange $5,000,000 of its preferred for an equal amount of American stock and to sell $2,000,000 of its own preferred and enough common (at $225 per share) to complete the purchase.

[95] The agreement, concluded on June 28, 1916, stipulated that $185,000 was to be paid in cash, $315,000 in Owens Bottle preferred (3,000 shares at $105 each), and $150,000 in Owens Bottle common (2,000 shares at $75 each).

[96] See below, pp. 162 ff.

[97] In 1917 it also built a furnace at Checotah, Oklahoma.

Owens Bottle and of the Charles Boldt Glass Company, a licensee, jointly agreed to reorganize the latter as the Charles Boldt Glass Manufacturing Company.[98] Since 51 per cent of its new capital stock went to Owens Bottle,[99] the Toledo Group gained control of two more factories (one located at Cincinnati, Ohio, and the other at Huntington, West Virginia).

The reader may find Table 11 helpful in summarizing and visualizing fully the horizontal expansion of the Owens Bottle Company from 1903 through 1920. The number of its plants increased from one small demonstration shop in 1903 to fifteen mechanized factories by the end of 1920. In the latter year, five were being operated under the name of two wholly owned subsidiaries and four under the name of two controlled companies. In addition, Owens Bottle had something less than one-third interest in the Hazel-Atlas Glass Company,[100] the second largest bottle producer in the United States; a small amount of common stock in the Thatcher Manufacturing Company; and over half the bonds issued by the Fairmount (Indiana) Glass Works. Even this expansion, however, fell short of the grandiose plans entertained in 1917, when E. D. Libbey had tried unsuccessfully to effect the consolidation of Owens Bottle, Ball Brothers, Hazel-Atlas, and Illinois Glass.[101]

The vertical growth of the Owens Bottle Company was never commensurate with its spectacular expansion on the horizontal plane. On February 22, 1906, its directors bought the Woodruff

[98] Between 1910 and 1916 Owens Bottle had acquired 1,666 shares of Charles Boldt common and had tried to buy enough more to give it controlling interest. Upon Charles Boldt's refusal to co-operate in 1916, Owens Bottle surrendered its holdings for $1,000,000. — Owens-Illinois Glass Company, Owens Bottle Company Minutes, V, 283–284.

[99] For the $1,683,000 common received in 1919, Owens Bottle was to pay $183,000 cash, to exchange 30,000 shares of its own common stock, and to cancel a $30,000 note which it held.

[100] That is, 21,314 out of the total 74,486 common shares. In return for a license in 1909, Owens Bottle had received 5,000 shares; in September, 1911, it had subscribed to 1,440 more (paying therefor $180,000); and apparently it had acquired the remaining shares through stock dividends. In 1926 Hazel-Atlas repurchased all these for something like $3,600,000. — T. N. E. C., *Hearings*, Part II, 536.

[101] See below, pp. 304–305. In 1922 Hazel-Atlas, Illinois, and Owens Bottle were again seriously considering a consolidation.

TABLE 11

FACTORIES OPERATED OR CONTROLLED BY THE OWENS BOTTLE COMPANY, 1903–1920

Location of Factory	1903	1904	1905	1906	1907	1908	1909	1910	1911	1912	1913	1914	1915	1916	1917	1918	1919	1920
A. Owens Bottle Company																		
1. Libbey Street, Toledo, Ohio [1]	x	x	x	x	x	x	x	x	x	x	x	x	x	x	x	x	x	x
2. W. Toledo, Ohio [2]						x	x	x	x	x	x	?		x	x	x		
3. Fairmont, W. Virginia [3]								x	x	x	x	x	x	x	x	x	x	x
4. Westlake Street, Toledo, Ohio [4]										?	x	x						
5. Clarksburg, W. Va. [5]											x	x	x	x	x	x	x	x
6. Greenfield, Indiana [6]															x	x	x	x
7. Charleston, W. Va. [7]																x	x	x
8. Glassboro, New Jersey, #1 [8]													x	x	x	x	x	
9. Glassboro, New Jersey, #2 [7]															x	x	x	x
B. Graham (Ind.) Glass Co. [9]																		
1. Evansville, Indiana [10]														x	x	x	x	x
2. Loogootee, Indiana [10]														x	x	x	x	x
C. American Bottle Company [9]																		
1. Newark, Ohio [10]														x	x	x	x	x
2. Streator, Illinois, #1 [10]														x	x	x	x	x
3. Streator, Illinois, #2 [10]														x	x	x	x	x
D. Graham (Okla.) Glass Co. [11]																		
1. Okmulgee, Oklahoma [12]														x	x	x	x	x
2. Checotah, Oklahoma [12]															x	x	x	x
E. Charles Boldt Glass Mfg. Co. [11]																		
1. Cincinnati, Ohio [12]																		x
2. Huntington, W. Va. [12]																		x

[1] Bought from the Toledo Glass Company; destroyed by fire and rebuilt in 1912.
[2] Obtained by purchase of the Northwestern Ohio Bottle Company in 1908; consolidated in 1911.
[3] Built by the Owens West Virginia Bottle Company; consolidated in 1911.
[4] An experimental plant operated in conjunction with the Libbey Glass Company and the Westlake Machine Company.
[5] Built by the Owens Eastern Bottle Company; consolidated in 1915.
[6] Bought from Ball Brothers in 1917.
[7] Built by the Owens Bottle Company.
[8] Acquired through purchase of the Whitney Glass Works in 1915; consolidated in 1918.
[9] Wholly owned subsidiary.
[10] Established factory of subsidiary.
[11] Controlled company.
[12] Established factory of controlled company.

Glass Mould Company for $14,000 in its own preferred; [102] and on May 5, 1913, they purchased the assets of a bankrupt sand company in Toledo for $13,500. They then incorporated the Toledo-Owens Glass Sand Company.[103] In the same year they bought E. D. Libbey's and M. J. Owens' holdings of 200 shares (out of 942 outstanding) in the Buckeye Clay Pot Company (at Toledo) for $26,600. In 1918 Owens Bottle had two-thirds interest in the Mid-West Paper Box Company, which was just completing a new plant adjacent to the bottle factory at Fairmont. It made its most important move to control the things it needed for manufacturing when in 1917 it acquired leases on several thousand acres of land in a gas-producing district near Charleston, West Virginia, at a cost of over half a million dollars.[104]

The Kent Machine Company at Toledo [105] supplied all the Owens machines installed in glass factories. E. D. Libbey, M. J. Owens, W. S. Walbridge, and F. L. Geddes — all of whom were important officials and stockholders in Owens Bottle — were the majority owners of Kent and soon after 1903 became its sole owners. The Owens company had not had sufficient working capital in September, 1903,[106] to buy Kent outright; and in 1910, when both Kent and itself were beginning to prosper, it turned down a chance to do so. Instead, its directors decided that it would be wiser to employ all surplus funds in building bottle fac-

[102] Because their company did not have the necessary funds to establish its own mold department in 1903, E. D. Libbey, W. S. Walbridge, and M. J. Owens, together with one Frank Woodruff, had founded Woodruff Glass Mould.

[103] Owens Bottle originally subscribed to $114,600 of its $200,000 authorized capital. Because of unprofitable operations the works at Silica, Ohio, were closed in July, 1915. They were reopened the following year, when Owens Bottle bought complete ownership, and finally were sold to outside interests in 1920.

[104] This investment proved very profitable. Originally Owens Bottle and the Libbey-Owens Sheet Glass Company had intended to own the property jointly, but the latter company decided on February 19, 1918, to withdraw and simply buy from Owens Bottle what gas it burned.

[105] This company, moved from Grand Rapids, Michigan, in 1902, had first made match machinery at Toledo. — Owens-Illinois Glass Company, Kent Machine Company Letters, I, 19. By the end of 1904, however, the demand for bottle machines had become so great that Kent refused to accept new orders from match manufacturers and wound up its former business.

[106] See above, p. 102. For the same reason it had been unable to make its own molds.

tories and installing machines constructed by Kent.[107] Partially as a result of their putting this decision into effect and of licensing other companies between 1909 and 1914, the number of Owens automatic bottle machines used in the United States increased rapidly. Table 12 summarizes the growth. By the end of 1920 probably between 200 and 210 were at work.[108]

The financial success which crowned the efforts of the promoters and directors in the Owens Bottle Company is summarized in Tables 26–29 in Appendix C. From September, 1904, to the end of 1920, the company's manufacturing plant and equipment

TABLE 12

NUMBER OF OWENS BOTTLE MACHINES IN OPERATION IN THE UNITED STATES, 1903–1918 *

Date	*Number*	*Date*	*Number*
Sept. 3, 1903	1	Sept. 30, 1910	71
Sept. 10, 1904	1	Sept. 30, 1911	103
1905	6	Sept. 30, 1912	133
July 1, 1906	8	Sept. 30, 1913	161
Oct. 1, 1907	22	Sept. 30, 1914	187
Sept. 30, 1908	46	Jan. 1, 1918	194 **
Sept. 30, 1909	51		

* Source: Owens-Illinois Glass Company, Annual Reports of Owens Bottle Company and Owens' Private Files. The figures given in this table do not coincide with those presented by Barnett, *Machinery and Labor*, p. 88. Barnett based his information on the reports of the Glass Bottle Blowers' Association, which used different dates than the ones presented here.

** This number does not include fifteen Graham-type machines which Owens Bottle had supplied.

increased in value more than a hundred fold, but its "patents and inventions" account changed very little. In 1905 it owned no securities of other bottle companies; in 1920 such holdings were listed at $4,558,499. In all, its assets expanded ten fold, and its

[107] They contracted, however, to place orders with Kent for machinery totaling not less than $500,000 before the end of 1912. In return, Kent agreed to expand its own plant at once by at least $40,000.

[108] I have been unable to uncover exact figures for this year. During 1918–1920, however, twenty-five new machines were shipped to factories, some of which were doubtless replacements.

surplus was so large in December, 1920, as to prove embarrassing. Several important customers and employees began to criticize the company's price and wage policies in the light of its evident prosperity. The directors, after considering various remedies for the situation, finally decided in April, 1921, to capitalize part of the surplus by declaring a 50 per cent stock dividend. Stock amounting to 200 per cent of the original common shares had been distributed between 1912 and 1915, and 21,037 more shares had been issued on July 1, 1920, as a 5 per cent dividend on the amount then outstanding.

All stockholders collectively received $14,131,267 in cash [109] from 1903 through 1920. During 1907 the company paid in full the accumulated dividends on its preferred, and thereafter each share regularly secured its 7 per cent. The common stockholders first participated in profits on October 1, 1907, and the rates gradually increased from 4 per cent in 1908 to 12 per cent in 1912. For the fiscal year ending September 30, 1918, cash dividends actually amounted to 24 per cent of par value, but they fell by half in 1919–1920. Each of the 25,000 common shares issued in 1903 consequently earned $396.35 cash during the entire period, and the closing market price for one of these shares on the New York Stock Exchange, on December 31, 1920, would have been 5.45 times greater than its original par value of $100.[110] If an investor had bought his shares in 1904, when they were selling at $40 or $50,[111] and had disposed of them (and the

[109] The 18 per cent "cash" dividends declared from May 1, 1918, through October 1, 1918, however, were paid in war bonds.

[110] According to the financial page of the *New York Times* for December 31, 1920, the closing price of Owens Bottle common was $43.25 per share. Because of stock dividends and the "four-for-one" stock split effected in 1916, one of the original 25,000 shares was equivalent to 12.6 of the 1920 shares. My calculation of the total cash dividends paid from 1907 through 1920 is based upon what a holder of one share in 1903 would have received if he had kept the original share and all the others he would have acquired through stock dividends and the stock split.

[111] Owens-Illinois Glass Company, Libbey Letters, II, *passim*. The *National Glass Budget* on March 18, 1911, (Vol. XXVI) quoted from the leading Toledo newspapers that "only a few years ago, holders of the preferred Owens issue were handed presents of common. At this time one man who held three shares of the latter is said to have sold them for $5. Another gave five or ten shares to one of the family servants as a Christmas gift." It was rumored that many shares of Owens common were given away by the original recipients as tips to waiters in

additional ones he meanwhile would have received as dividends) at the closing price for 1920, he would have enjoyed a capital gain of something like 1,090 to 1,362 per cent.[112]

The annual earnings given in Table 29, in Appendix C, reveal why it was possible for the company to have paid large cash dividends while its original shares were increasing so markedly in book and market value. Net profits totaled $27,249,872 from 1903 through 1920 and approximately one-half of this was ploughed back into the business. The data also indicate the relative decline in importance of its licensing as compared with its manufacturing activities as sources of income. Annual royalty receipts increased a little better than four fold from the fiscal year ending on September 9, 1909, to that preceding January 1, 1921, whereas the yearly profit from bottle sales expanded 260 fold. The growth of the Owens Bottle Company as a manufacturer of glass containers may be measured in still another way. During 1904 it could not possibly have made more than 0.4 per cent of all bottles produced and sold in the United States (11,918,000 gross); yet in 1917–1918, it and its subsidiaries (American Bottle and Graham) contributed about 20 per cent of the total national output (42,811,927 gross).[113]

---

restaurants and to other persons. On January 1, 1905, E. D. Libbey wrote that "the common stock of this company has had quite a little boom since the first of the year," and that some sold for $71.50 soon after it was listed on the Toledo stock exchange. In March, 1911, some sold for $298 per share (*National Glass Budget*, XXVI, March 18, 1911). It was not listed on the New York Stock Exchange until July, 1916, when the big consolidations with American, Graham, and Whitney were afoot.

[112] It is perhaps needless to point out that from a purely economic point of view nearly all the $2,500,000 common outstanding in 1904 was highly "watered." Almost 90 per cent of the assets behind the stock were intangible (i.e., were patent and invention rights) and had an unknown market value which probably did not approach their book value. The company obtained little or no working capital or other tangible assets from the initial disposal of its common stock. See above, p. 102. The following sixteen years, however, proved that the "water" had been of extremely good quality.

[113] Based upon government statistics and production figures found in M. J. Owens' private files.

CHAPTER V

THE ORGANIZATION AND HISTORY OF THE GLASS FIRMS FOUNDED BY THE TOLEDO ENTREPRENEURS, 1888–1920 (*Continued*)

The Toledo Glass Company in 1903 had assigned to the Owens Bottle Company the bottle rights on its machine only for the territory and dependencies of the United States. It retained possession of all foreign rights and made no attempt to dispose of them until three American licensees had installed the Owens machine. Presumably, this delayed action was partly due to its discouraging experience in trying to introduce the semiautomatic paste-mold machine on the European market.[1] It will be recalled that some Englishmen in 1900–1901 had taken over the tumbler and lamp-chimney rights but had failed to raise enough capital to assure the success of their promotional endeavors. E. D. Libbey and his associates in Toledo Glass consequently resolved at an early date to participate directly in the creation and control of any organization which would license European manufacturers to use the bottle machine.

1. The Owens European Bottle Machine Company (1905–1920)[2]

On January 11, 1905, Libbey and one J. Clark Curtin of Hartsdale (Westchester, England) contracted with each other to form an English joint-stock company for the purpose of exploiting the patents covering the automatic production of glass containers.[3] For some reason or other the contract was rescinded in

[1] See above, pp. 100–101.

[2] The minutes of the stockholders' and directors' meetings kept by the Owens European Bottle Machine Company (3 vols.) and the Toledo Glass Company (4 vols.) have provided most of the information presented in this section.

[3] All the common shares (25,000 in number with a par value of £20 each) and $5,000 cash were to be given Toledo Glass in exchange for an exclusive European license, and Curtin was to find subscribers for the £100,000 preferred shares. On January 17, Toledo Glass even agreed to buy a twentieth of the preferred issue.

August of the same year,[4] and the directors of Toledo Glass decided to organize the Owens European Bottle Machine Company under Ohio laws. They accordingly filed a petition to incorporate, with a capitalization of $3,000,000, on October 19, 1905.

The two companies reached a mutually satisfactory agreement eighteen days later. According to its terms as modified on February 1 and October 2, 1907, Owens European received exclusive machine and revolving-basin rights for South and Central America (except Mexico), Asia, Africa, Europe, Australasia, and Cuba. In addition, Toledo Glass transferred to the new corporation all the machinery and properties which it had already acquired, built, or shipped to England and Le Havre. Owens European then paid Toledo Glass $2,487,000 in its fully paid capital stock.[5] From its remaining 5,130 shares the new enterprise gave M. J. Owens [6] and William H. Boshart 500 and 300 shares, respectively, for services rendered in connection with its demonstration plant in England and sold the rest (4,330 shares) at par for $433,000 operating capital.[7] Toledo Glass then distributed $1,487,000 of its $2,487,000 holdings as bonuses to those who had bought the 4,330 shares directly from Owens European.[8] The disposal of all 30,000 shares is summarized in Table 13.

The directors of the new corporation decided to carry through

[4] Curtin was given a twelve-months' option on the sale of the Russian rights, but apparently he allowed this to expire.

[5] The contracts actually stipulated that $150,300 of this was to be in cash. Toledo Glass, however, subscribed to $150,000 additional stock, and M. J. Owens, William S. Walbridge, and F. L. Geddes (all owners of Toledo Glass) had each bought one share. This amounted to the same thing, therefore, as paying Toledo Glass entirely with stock. Owens European later paid $709.66 in cash to cover the sums already expended in obtaining patents in South America.

[6] Apparently Owens also received about $100,000 cash in ten annual installments, for the balance sheet dated September 30, 1908, listed the item "Commission — M. J. Owens, $87,220.32" among its liabilities, and later ones included successively smaller amounts.

[7] E. D. Libbey and his associates at Toledo personally subscribed to 1,410 shares; Julius C. Prince, managing director of the Apollinaris Company, Ltd., with home offices in London, and his associates, to 1,410 shares; Adolphus Busch, a brewer and also an associate in the American Bottle Company, to 1,410 shares; and William H. Boshart to 100 shares. Apparently, payments amounting to only 87.5 per cent of these subscriptions were ever called for, and the balance due was deducted from the first dividend (return of capital). Only then did the company issue the appropriate certificates for fully paid shares.

[8] The effect of this arrangement, of course, was to remove all legal liability from

TABLE 13

DISTRIBUTION OF OWENS EUROPEAN CAPITAL STOCK PRIOR TO 1908

Recipients	Initial Distribution by Owens European	Received by Owens European from Initial Distribution	Secondary Distribution by Toledo Glass	Reason for Secondary Distribution	Net Holdings of Recipients
Toledo Glass	$2,487,000	Foreign license and properties	$1,000,000	Retained from its original holdings	$1,000,000
Adolphus Busch	141,000	$141,000 cash*	484,000	Bonus for cash supplied Owens European	625,000
E. D. Libbey and Associates	141,000	$141,000 cash*	496,500	*Idem*	637,500
Julius C. Prince and Associates	141,000	$141,000 cash*	499,000	*Idem*	640,000
M. J. Owens	50,000	Services rendered			50,000
William H. Boshart	40,000	Services rendered and $10,000 cash*			40,000
Others			7,500	Bonus for services rendered	7,500
Total Stock Distributed	$3,000,000		$2,487,000		$3,000,000

*Apparently the company called upon its stockholders for only 87.5 per cent of these amounts in cash.

the plans which the Toledo Glass Company had already fomented for the construction of a demonstration factory in England. They first formed a subsidiary, known as the Owens European Bottle Machine Company, Limited, of Manchester; then leased some land at Trafford Park; and there began to build a bottle plant in May, 1906.[9] M. J. Owens was asked to serve as general supervisor and consultant. By October 1, 1907, the investment at Trafford Park, "the best and most up-to-date bottle plant in the world," [10] totaled $257,308.[11] Although the Toledo Group built the factory primarily to convince foreign producers that the Owens machine was entirely practicable, the men there actually made and sold bottles in commercial quantities. As a matter of fact, so many bottle orders poured in that it became necessary to seek permission to run the plant for five months after it was sold. The factory cleared about $58,000 over all operating expenses during the eleven months it was worked.[12]

Anecdotes current among those who knew E. D. Libbey intimately suggest that he had encountered a cold reception upon his visit to England in 1906. Established bottle manufacturers there definitely were not interested in his new machine, and some even taunted him with his earlier failure in trying to promote the semi-automatic tumbler and lamp-chimney machine. These scoffers, however, became anxious inquirers as soon as demonstrations at Manchester proved the bottle machine to be an unqualified success. Most of them realized what probable effects the new tech-

---

the final recipients for the difference in the par value of the stock and their own cash payments in the event that the company should fail.

[9] On July 11, 1906, they entered into a contract with a Wellsburg, West Virginia, firm to supply all the skilled labor and materials needed to complete the factory.

[10] Libbey-Owens-Ford Glass Company, Owens European Bottle Company Minutes, I, 71–72.

[11] According to the balance sheet of Owens European drawn up for this date. — *Ibid.*, I, 191.

[12] From an interview with William H. Boshart, who worked with M. J. Owens as plant manager at Trafford Park. There is little reason to question the accuracy of Boshart's memory. Extant records indicate that Owens European awarded him $30,000 capital stock because of his record at Manchester and for the same reason the Owens Bottle Company made him one of its officials upon his return to the United States in 1908.

nique would have upon their own handicraft methods. Their concern greatly increased when Owens European issued the Rhein-Ahr Glasfabrik Gesellschaft mit beschränkter Haftung its first license early in 1907.[13] The Apollinaris company,[14] controlling several firms engaged in bottling and selling mineral waters from various springs in central Europe, had just incorporated this new enterprise, which was to supply it with bottles made on Owens machines. The Rhein-Ahr license remained in force throughout the life of the patents covered, and the company continued to send royalties to Toledo until after 1920.

Meanwhile three important bottle manufacturers in Germany had become particularly alarmed over the results shown at the Manchester plant and over the license granted Rhein-Ahr. They, therefore, opened negotiations in 1907 with E. D. Libbey, representing Owens European. As a result of conferences held at Frankfort and Berlin, they agreed to organize a cartel which would be known as the Europäischer Verband des Flaschenfabriken Gesellschaft mit beschränkter Haftung. This new combination was to receive exclusive right[15] to use and sublicense the automatic bottle machine throughout all countries except the United States and her dependencies,[16] Canada, Mexico, China, and Japan. The negotiating parties signed the provisional exclusive contract on August 5, 1907, in Berlin.[17] The Verband received

[13] The terms of the license, issued on February 1, were as follows: (1) The licensee was to build a bottle factory on or near the Rhine in Germany, and the licensor was to install six Owens machines and guarantee their initial operating efficiency. (2) The licensee was to pay $7,000 for each machine (plus all freight and duties) whether it was made in Toledo or Germany, but this amount was to be returned gradually through small rebates (1/40) from quarterly royalties. (3) The royalty rate was to be 36 cents per gross of merchantable output, and the licensee could not make any bottles which Apollinaris and its subsidiaries did not fill with their own mineral waters. (4) The license was not exclusive, but Owens European agreed not to grant others a license upon more favorable terms. (5) No minimum or maximum yearly outputs were stipulated. During the latter half of 1907 these terms were modified such that Rhein-Ahr did not have to install and operate more than three machines, and the royalties on output were commuted to $3,000 each year per machine. — *Ibid.*, I, 101–113, 129–130, 136.

[14] Julius C. Prince, one of those who contributed the operating capital for Owens European, was managing director of this company.

[15] Except for the license held by Rhein-Ahr.

[16] Cuban rights, however, were expressly assigned to the Verband.

[17] All the directors of Owens European approved it at Toledo on October 2,

the Manchester plant and the machines for a proposed factory in France which had been previously shipped to Le Havre. In return, it agreed to pay Owens European 12,000,000 gold marks (3,000,000 in cash on March 1, 1908, and the remainder in nine equal yearly installments).[18] Each party promised to acquaint the other, free of charge, with all improvements made on the bottle machine in the future.[19]

Several years after the Verband acquired the European rights, Owens European sold the Japanese and Chinese rights. The exclu-

1907. The final contract was signed and entered into by both parties on November 28, 1907.—*Ibid.*, I, 226.

[18] This amounted to approximately $2,850,000 on the pre-war exchange. The outbreak of the first World War naturally created complications. The British government confiscated the Verband's British patents and freed the British Association of Glass Bottle Manufacturers, Ltd., (a member of the cartel) from all its obligations to the Verband. By 1926 Owens European had not yet received the last 30,000 marks which its licensee refused to pay, and some of the other installments had been made belatedly and with depreciated currency. See the files on the *Europäischer Verband* v. *Owens Bottle Company* suit in the legal offices of the Owens-Illinois Glass Company.

[19] The sections of the contract pertaining to this provision were later the center of a prolonged and bitter lawsuit, which was finally settled outside of court by compromise in 1931. The proposal signed by representatives of the Verband and of Owens European in Berlin on August 5, 1907 (Libbey-Owens-Ford Glass Company, Toledo Glass Company Minutes, II, 201, 207) had included certain broad clauses: "The [Owens] 'European' further undertakes to effect that Messrs. Michael Joseph Owens and Edward Drummond Libbey of Toledo, likewise undertake towards the 'acquirers,' to communicate to them and let them have free of charge all cognate inventions and improvements. . . . In the event of the offer being accepted, Messrs. Libbey and Owens by their signature of the present contract, undertake personally those obligations . . . as to the communication and cession of new inventions a: 1 improvement." Elsewhere in the agreement which both parties signed it was stipulated that "for the German contract territory [information would be given free of charge] with reference to the inventions already made or to be made in the future which, through their objects, are connected with that which is the object of the said [existing] patents and thus relate to the manufacture of glasswares and of machines, furnaces, apparatus and molds serving that purpose."

Mainly on the basis of these clauses the Verband claimed, after the conclusion of the war, exclusive and gratuitous right to use the Graham bottle machine, the Libbey-Owens sheet-glass machine, and the Westlake paste-mold machine. The defendants (including Toledo Glass) based their case chiefly on three considerations: (1) The Graham and Westlake machines were not the inventions of either E. D. Libbey or M. J. Owens; (2) the sheet-glass machine did not pertain to the subject (bottles) of the patents involved in the original license granted the cartel; and (3) neither Toledo Glass nor Owens European—the two American companies which were parties to the 1907 contract—had ever owned the patents covering the Graham and Westlake machines. The Owens Bottle Company acquired the patents on the Graham machine, and the Libbey Glass and Westlake companies owned the patents on the Westlake machine.

sive license issued a Japanese firm in 1916 entailed the payment of $50,000 cash, 2,000 7 per cent debentures of 50 yen each (payable within six years), and 2,000 fully paid capital shares (each with a 50-yen par value).[20] The Chinese rights brought $25,000 around 1920.[21]

Owens European, like Owens Bottle, at first had encountered a marked reluctance on the part of established producers to accept the new machine. It had had to build a demonstration plant and license a Continental firm before the major bottle interests in Germany had become very much concerned over the future of their own business. Unlike Owens Bottle, however, the European company merely *threatened* to make bottles in competition with handicraft factories. Except for eleven months at Manchester, it did not engage in manufacturing. When Owens European issued the Verband license, therefore, its major work was completed. By February 19, 1917, it had returned to its stockholders $99.15 per share in liquidating dividends[22] and had gradually lowered the par value of its stock from $100 to $1.[23] Table 30 in Appendix C shows the extent of liquidation achieved before 1921.

2. The Libbey-Owens Sheet Glass Company (1916–1920)[24]

The Toledo Glass Company, after winding up its affairs pertaining to the Owens bottle machine, eventually became interested

[20] The yen was then worth about 50 cents. According to one provision in the contract, the licensor in effect guaranteed that the efficiency of each machine installed would permit a certain average daily production. — Libbey-Owens-Ford Glass Company, Owens European Bottle Company Minutes, III, 138–159.

[21] Owens-Illinois Glass Company, Executive Committee Minutes, II, 307. In 1911 Owens European had decided to reduce the book value of its unsold rights for Japan, China, and Manchuria from $200,000 to $2. Labor was so cheap in those countries, it was stated, that the savings in production costs which could be realized from using the machine would be very small. — Libbey-Owens-Ford Glass Company, Owens European Bottle Company Minutes, II, 82–83.

[22] By March 25, 1908, $20; by April 29, 1911, $51.50; by April 15, 1913, $72.05; by April 25, 1916, $92.65.

[23] Cash dividends amounting to $5.35 per share (or 535 per cent of the reduced par value) were also paid from 1916 through 1920.

[24] The chief sources of information presented in this section are the Colburn Records (about a dozen volumes) and minutes of the directors' and stockholders' meetings of both the Toledo Glass Company (Vols. III and IV) and the Libbey-Owens Sheet Glass Company (Vols. I and II). All of these are in the possession of the Libbey-Owens-Ford Glass Company at Toledo.

in a new process for drawing window or sheet glass which one Irving W. Colburn had been developing for several years prior to 1910. Colburn first became interested in the possibilities of manufacturing glass around 1898, when he visited a few plants employing skilled workmen to blow, slit, and flatten cylinders from which window lights could be cut. Almost immediately he discarded as impracticable his plans to roll sheets of glass and in 1899 received patents on certain methods for making tumblers and lamp chimneys. He then attempted, but unsuccessfully, to modify his machines so that they could blow window-glass cylinders. Next, he tried to develop a way of pressing and whirling disks by mechanical means in much the same way as "crown" glass had been made by hand for centuries.[25] In 1905 he was again experimenting with glass cylinders and actually received some patents which, however, proved to be of questionable value.[26] When someone asked him why he did not make window glass in sheet form, he started to re-examine the concepts which he had discarded in 1900.[27]

He and a few associates subsequently formed the Colburn Machine Glass Company in August, 1906, and built a small experimental glasshouse at Franklin, Pennsylvania.[28] They made about 4,000 boxes of inferior-grade glass on his machine before the Star Glass Company at Reynoldsville, Pennsylvania, agreed to install and test its operating efficiency under normal manufacturing conditions.[29] This trial ended in failure, and the Colburn

[25] See above, pp. 14–15.

[26] The American Window Glass Company had already obtained patents from a certain inventor named John H. Lubbers for a very similar process.

[27] In that year he had built three separate sheet-drawing machines in a little factory on Frankford Road, Philadelphia. He constructed several new models in 1904 and 1905.

[28] The company had an authorized capital of $500,000, of which only $1,000 was paid in at the time of incorporation. Seven-tenths of the stock was awarded as bonus to Colburn and the other associates. The remainder was to be sold for operating capital. Very little was. The experimental shop, costing $17,000, was completed by July 31, 1907.

[29] Colburn constructed a machine at Reynoldsville in 1908–1909. From the very first he had to make alterations in it and finally built a new machine in 1909–1910. He first drew glass on September 17, 1909. Production was suspended a month later; recommenced on December 22, 1909; suspended again the following February; begun once more in June; and finally stopped altogether in July because of inadequate funds. During these three brief periods of operation the machine made about 15,000 boxes of glass.

Company simply did not possess sufficient funds to continue experimentation.[30] On September 2, 1910, I. W. Colburn urged the stockholders to advance whatever money would be necessary to develop his machine. They refused, not only because they were unable to supply the funds without embarrassment, but also because they had lost faith in his invention. The directors of his company, however, raised about $825 among themselves. This allowed Colburn to put the machine once again in operating condition and to stage a demonstration while M. J. Owens, representing the Owens Bottle Company, was on hand. In the last quarter of the same year the owners of the Colburn Machine Glass Company learned that a group of powerful and wealthy capitalists [31] wanted to form a syndicate which would take over all the patents and supply the necessary sums to perfect the process. Since the stockholders apparently could not agree about what terms would be acceptable, their company went into receivership.[32]

M. J. Owens had seen great possibilities in the Colburn machine during the demonstration staged for his benefit the previous summer. He therefore suggested to the directors of the Toledo Glass Company about the middle of 1911 [33] that they buy the patents. They agreed, although reluctantly, and paid $15,000 for Colburn's patents at a sheriff's auction held on February 8, 1912. They then voted to spend up to $150,000 on a new iron, brick, and concrete factory at Toledo,[34] where they could test and improve the window-glass machine and also Owens' carboy machine.[35] By

[30] As is suggested in a subsequent chapter, ignorance of the physical properties of molten glass and a lack of practical experience in the industry were also important factors in Colburn's failure.

[31] I have not been able to find out anything definite about this group. Probably M. J. Owens was a member. Certainly, he had not intended at the outset to persuade the Toledo Glass Company to buy the Colburn patents. See below, p. 171, n. 86.

[32] On September 22, 1910, its total indebtedness amounted to something like $23,000. Some information on the final days of the Colburn Company is available in Clarence Brown's personal files kept by John H. McNerney of Owens-Illinois.

[33] On May 8 the directors heard that Owens had intended "to bring up, at this meeting, matters concerning the Colburn patents for window glass" but that he had been forced to visit the bottle factory at Fairmont and had not yet returned. —Libbey-Owens-Ford Glass Company, Toledo Glass Company Minutes, III, 66.

[34] On November 21, 1912, they decided to sell 691 shares of Toledo Glass holdings in Owens Bottle preferred at $116 each in order to finance the construction.

[35] Experiments with the Colburn machine were first conducted at the plant of the Buckeye Clay Pot Company (Toledo) in May, 1912. The new factory, com-

October 31, 1913, all expenditures made in connection with the Colburn process had climbed to $252,819.

The experiments showed such promise of success [36] that Toledo Glass incorporated the Libbey-Owens Sheet Glass Company on May 17, 1916, with an authorized capital of $6,000,000 (160,000 common shares with a par value of $25 each and 20,000 cumulative 7 per cent preferred with a par value of $100 each). In exchange for exclusive foreign [37] and domestic rights on the Colburn patents, Libbey-Owens paid Toledo Glass $3,999,500 in its fully paid common [38] and $325,000 in its preferred. It turned over an additional $175,000 in preferred for the experimental plant at Toledo.[39] Toledo Glass then subscribed to the remaining $1,500,000 preferred, which it immediately offered to sell at par (with four shares of common as bonus for each share of preferred) to its own stockholders.[40] Hence Libbey-Owens obtained from its $6,000,000 capital stock approximately $1,500,000 cash, a factory which it sold in the same year for $165,000, and certain valuable patent rights.

The new corporation then started the construction of a modern six-machine plant, which it intimated would cost around a million dollars, at Kanawha City, a suburb of Charleston, West Virginia.[41] Workers there pulled their first sheet of glass on Oc-

---

pleted in 1913 on Castle Boulevard, cost about $125,000. — Owens-Illinois Glass Company, Owens' Private Files.

[36] In December, 1914, 4,150 boxes of "C" grade glass made at Toledo were marketed at an average price of $1.10 each. In May of the following year a Chicago sash and door company ordered three carloads.

[37] Contrary to the procedure followed with the Owens bottle machine, Toledo Glass formed no separate European-licensing corporation.

[38] The final agreement, reached on May 31, 1916, stipulated that $600,000 of this was to be in cash. But Toledo Glass had subscribed to $600,000 of its common shares when the petition to incorporate had been filed. Hence, the final effect was the same as paying the entire $3,999,500 in common stock. The remaining $500 common was held in equal amounts by the five original individual incorporators.

[39] Libbey-Owens sold this plant for $165,000 to the Willis-Overland automobile people on August 30, 1916.

[40] Libbey-Owen-Ford Glass Company, Toledo Glass Company Minutes, IV, 154–155.

[41] Annual Report to Stockholders, December 31, 1917 (*ibid.*, Libbey-Owens Sheet Glass Company Minutes, I, 196). The president stated that "raw materials and finished glass are untouched by hand from the time material is received from the railroad up to the point of finished glass delivered at the end of the lehr ready for the cutters."

tober 17, 1917. Because of various imperfections in the furnaces, lehrs, and drawing machinery and because of the necessity of providing housing accommodations for the workers, the initial expenses actually were larger than had been anticipated. Consequently, the directors had to obtain additional funds from Toledo Glass and Owens Bottle — companies whose destinies they also guided. They told the stockholders by letter that

> the process under which we are operating is, in our opinion, a good one. We see no reason why it cannot be made a commercial success. The production obtained in our experimental plant has already been sold at market prices. Owing to certain conditions, which only experience and experimentation can overcome, we were not able to make as clear a quality of glass as, in our judgment, the market demands. Not until we have operated commercially on a large scale, at our new plant, shall we feel justified in reporting results as a success.[42]

On September 30, 1919, the president issued the statement that the "Colburn-process production has now reached the stage where it can be said to be a commercial success; this could not have been said two years ago." [43] He also announced that plans were already under way to enlarge the Charleston plant. Before the end of 1920 his company had invested $3,000,000 additional capital there.[44]

From the outset the incorporators of the Libbey-Owens Sheet Glass Company had intended that it not only manufacture window glass itself but also issue licenses to others. Almost immediately following the receipt of its corporate charter two American firms and one located in Canada made inquiries about the terms upon which they could obtain rights to use the Colburn process. The first license did not go to them but to a group of Japanese promoters. The American-Japan Sheet Glass Company, organized on February 12, 1919, obtained exclusive rights for

[42] Owens-Illinois Glass Company, Brown's Private Files.

[43] Libbey-Owens-Ford Glass Company, Libbey-Owens Sheet Glass Company Minutes, II, 26.

[44] This doubled the number of machines and furnaces. Further enlargements were made in 1923. Ten years later it was said to be the largest window-glass plant in the world. — Owens-Illinois Glass Company, "Lives of E. D. Libbey and M. J. Owens," p. 10.

Japan and agreed not to ship its output to any country in which Libbey-Owens may have taken out or assigned patents.[45] Libbey-Owens granted exclusive rights for Canada to another enterprise in 1920, and the licensee installed two machines. The Libbey-Owens Sheet Glass Development Company, Limited, created in 1919, purposed to handle the disposition of the European patents. The Toledo Group apparently bought out the British interests in this company a year later [46] and then sold all manufacturing rights for Europe and China to the Mutuelle Mobilière et Immobilière, a large Belgian company embracing the Solvay interests. At the close of 1920 the Libbey-Owens directors had just completed negotiations concerning a nonexclusive license to be issued a proposed Pacific Libbey-Owens Sheet Glass Company (at Los Angeles) [47] and were considering similar contracts with the Pittsburgh Plate Glass Company and the United States Window Glass Company.[48]

As is shown in condensed form in Table 31, Appendix C, by the end of 1920 Libbey-Owens was well along the highway leading to industrial success. Its president could scarcely conceal his jubilation in the report submitted to the stockholders on December 8, 1920. "We close the year," he wrote, "without any liabilities other than those for current bills and small accounts for construction work, and with a cash balance of $946,671.01." [49] The company had just met in full the accumulated dividends on its preferred and had declared a 25 per cent stock dividend on its

[45] Libbey-Owens received $100,000 cash and one-third the foreign company's capital stock. The Japanese factory began to pull glass on October 1, 1920.

[46] The directors of Libbey-Owens voted on June 23, 1920, to do so provided the cost did not exceed $72,000. — Libbey-Owens-Ford Glass Company, Libbey-Owens Sheet Glass Company Minutes, II, 46.

[47] Libbey-Owens at Toledo was to subscribe to 51 per cent of its preferred and common, and certain Pacific Coast groups were to take the remainder. They were to install two machines, and the licensee was to pay as royalty the cost of the machines and 10 per cent of the net selling price of glass made. The contract stipulated a minimum royalty of $30,000 a year.

[48] The plant of the latter was located at Shreveport, Louisiana. All licenses, issued or contemplated, contained certain marketing limitations. Either the licensee could not make more than a given output each year, or he had to sell his product only within specified market areas.

[49] *Ibid.*, II, 76.

$4,000,000 outstanding common shares. It did not vote its first cash dividend on common, however, until May 10, 1922.[50]

3. The Westlake Machine Companies (1907–1920)[51]

The reader may recall that the Libbey Glass Company, which E. D. Libbey had established at Toledo after the New England Glass Company had ceased operations in Massachusetts, had become an important producer of bulbs and tubing for incandescent lamps.[52] It had received exclusive rights to make all glass parts for electric lamps on Toledo Glass's semiautomatic paste-mold machine and on Owens' fully automatic bottle machine. It was almost inevitable, therefore, that Libbey Glass should seek to develop a process for blowing bulbs automatically by combining certain basic features of each machine. And it could do this most conveniently through a subsidiary agency. It accordingly incorporated the Westlake Machine Company on November 6, 1907, under Ohio laws.

The new company, with an authorized capital of $10,000, received a charter permitting it "to design, develop, demonstrate, manufacture, buy, sell, lease or otherwise acquire and handle automatic machinery of all kinds." Within a year it opened a small experimental shop on Westlake Street in Toledo and obtained authority to raise its capitalization to $100,000. The Libbey Glass Company itself, or its stockholders personally, supplied all the necessary funds. The usual procedure was to lend Westlake the sums it needed from time to time and then to convert periodically the notes held into shares of stock. By 1912 Westlake's authorized capital had increased to $1,500,000, of which $1,150,000 was covered by outstanding common shares.[53]

[50] Before the end of that fiscal year it declared cash dividends amounting to 6 per cent of the $5,000,000 outstanding common.

[51] The chief sources for information presented in this section are: (1) the Minutes of the Libbey Glass Company (Vols. II, III, and IV), (2) the Minutes of the Westlake Machine Company (1 vol.), (3) the Minutes of the Westlake European Machine Company (1 vol.), and (4) a miscellany of records and licenses preserved in three large safes in a basement of the Libbey Glass Company.

[52] See above, p. 94.

[53] The $200,000 shares issued before 1912 had apparently brought in an equivalent value in cash or other assets. On May 7, 1912, the company declared a stock dividend amounting to $400,000; placed $150,000 shares in a reserve fund to be

Work on the automatic bulb machine meanwhile had progressed so well that the directors authorized their secretary to formulate license contracts for the United States.[54] They concluded a preliminary agreement with the Libbey Glass Company on December 9, 1914. Libbey Glass had already purchased some buildings and properties at Sandusky, Ohio, and since 1910 had been making bulbs there semiautomatically. It had also promised to assume most of the cost which would inevitably be entailed in putting the Westlake machine into commercial use on a large scale. In return, Libbey Glass was to operate the machines for two years without having to pay any royalties.[55] The two companies reached new agreements on November 1, 1916, and February 21, 1917. According to the first, Libbey Glass received exclusive right to use and sublicense within the United States all of Westlake's inventions for making bulbs. It was to pay a royalty of $1 per 1,000 merchantable bulbs [56] and an initial rental on each machine covering its cost of construction plus 10 per cent.[57] According to the second agreement, Libbey Glass bought outright all rights (save those pertaining to bulbs, bottles, and jars) [58] on the Westlake machine; took over all the experimental plant and equipment at Toledo; and reaffirmed its royalty obligations to its subsidiary in connection with bulb manufacture. As a result of these transactions, the Westlake Machine Company actually received $209,475 [59] before February 21, 1917; and in addition it

---

used in rewarding well-deserving employees; offered $200,000 for sale or exchange for notes held against itself; and set aside another $200,000 to be issued as "share-for-share" bonuses to those who accepted the $200,000 offered. It actually distributed among its employees $90,000 of the $150,000 in the reserve fund and later canceled the remaining $60,000 shares.

[54] If the licensee agreed to make substantially all its bulbs on the Westlake machine, the basic royalty was to be $1 for each 1,000 merchantable bulbs. Otherwise, the royalty rate was to be twice as high.

[55] Westlake received $34,346 from Libbey Glass as royalties in 1916 and $45,538 in 1917.

[56] Of the standard size most commonly used.

[57] Westlake alone was to make the machines and to retain ownership in them.

[58] The Owens Bottle Company had acquired all bottle and jar rights by contract dated June 22, 1908.

[59] This amount included $128,036 as full payment for 8 machines installed at Sandusky, $15,000 as full payment for 1 machine at Toledo, $10,000 as full payment for an old machine at the Westlake Street shop, $55,000 as partial payment

was promised $750,000 in Libbey Glass preferred (7 per cent cumulative), about $73,000 cash for eight machines in process of completion, and the stipulated annual royalties on all bulbs made. Libbey Glass then assigned, by sublicense, all the bulb rights to General Electric on December 12, 1918.[60] Four years later it bought itself free of all royalty obligations by paying Westlake $450,000 in cash.[61]

It is apparent from the preceding account that the Westlake Machine Company received more than enough from the licensing, and later the sale, of its inventions to Libbey Glass to cover all of its own liabilities. Its expenditures prior to September 11, 1918, on patents and technical developments roughly totaled $529,000. The stockholders, who had contributed about $400,000 in cash or other valuable properties, consequently received handsome returns. The $750,000 block of Libbey Glass preferred stock was distributed among them on a pro rata basis in 1917; cash dividends amounting to $75,000 were declared between 1919 and 1921; and liquidating dividends of $511,136 were paid from 1922 through 1927.[62]

Unlike the Westlake Machine Company, which functioned entirely as a research and innovational adjunct of Libbey Glass, the Westlake European Machine Company purposed primarily to license foreign producers. As organized in 1911, it had an authorized capital of only $10,000; within ten years it disposed of its Canadian, English, and European rights. Its stockholders received no dividends until January, 1925, when $100,000 "out of monies on hand realized from the sale of property of the company owned in 1913" was distributed.[63]

---

for 8 additional machines in process of construction, and $1,439 for all accrued royalties to January 1, 1917.

[60] See below, pp. 136 ff.

[61] In effect this ended Westlake's business activities.

[62] Westlake finally liquidated on January 21, 1927.

[63] Westlake European Machine Company Minutes (held by Justin Wilcox, executor of E. D. Libbey's estate). The Westlake European Machine Company has kept only fragmentary records and at the present time is being dissolved. Once they issued the licenses, its directors, officers, and stockholders had only to see that the royalties were paid on time and that dividends were properly declared and distributed.

4. The Libbey Glass Company Once More (1895–1920) [64]

After tracing the development of the Libbey Glass Company during its first years at Toledo, it will be remembered, the account of the Toledo Group's activities after 1895 shifted to the history of the Toledo Glass Company and after 1903, to the expansion of the Owens Bottle Company and other subsequent enterprises. It now becomes necessary to pick up the thread of Libbey Glass's story and follow it until 1921. Throughout the period the company retained its primary interest in manufacturing.[65] Flint tableware of the best quality and glass parts for incandescent lamps were its major specialties. But it also played a significant role in the evolution of technology. As the preceding section has already indicated, its directors, through the subsidiary Westlake enterprise, took an active part in developing automatic bulb machinery; and the Libbey Glass Company itself later modified the Westlake machine so that tumblers, stemware, and other paste-mold items could be made on it and also perfected an entirely new process for drawing glass tubing. A description of these techniques and a discussion of how they evolved are reserved for the following chapter and Appendix B. They enter into the account presented in this section only in so far as their development affected the growth of the company.

After M. J. Owens had patented his mechanical device for blowing bulbs, lamp chimneys, and tumblers semiautomatically,[66] and while he was still in the employ of Toledo Glass, he perfected a method of pressing and fire-polishing glass "blanks" for tableware which could then be finished by skilled cutters.[67] The manufacture of "blanks" was an important business at that time, for intricately cut tableware was very much in demand. Companies

[64] The Minutes of the Libbey Glass Company and various records such as licenses, patents, auditors' reports, stock books, and court hearings (preserved in three old safes in the factory's basement) have provided most of the information presented in this section.

[65] In 1920 the Libbey Glass Manufacturing Company took over the operation of the plant, and the Libbey Glass Company thereafter received rentals and royalties on its equipment. The Owens-Illinois Glass Company obtained full control of the factory in 1937 and effected complete consolidation in 1944.

[66] See above, pp. 96–97.

[67] The basic patent, issued on July 4, 1899, was numbered 628,027.

would make variously shaped pieces of glass and sell them to small glass-cutting firms or individual workmen who cut patterns on them with the aid of copper and steel wheels. Since each finished piece embodied much labor, cut glass was expensive. Many American factories after the 1820's had pressed large quantities of tableware in engraved copper or brass molds, and the resulting products had been both beautiful and cheap.[68] The pieces did not have, however, the luster or brilliance of objects blown in paste-molds and then cut by hand, because the metal mold and plunger left the glass's surface obscured. Molten glass is always scarred through contact with metal. One could polish the surface of pressed ware on the outside by exposing it to an intense flame or by touching it up with a polishing wheel, but there was no way to restore the brilliancy of the inside surface, marred by contact with the metal plunger. M. J. Owens' device in 1899 solved the problem. The glass "blank" was first pressed in one mold, transferred to another, and there subjected to a jet of fire which played around the "blank's" inside surface.[69] Since the Toledo Glass Company was not interested in the "blank" business, it gave M. J. Owens permission to patent the process himself and to dispose of his rights as he saw fit.[70] He immediately sold them to the H. C. Fry Glass Company at Rochester, Pennsylvania.

Since the "blanks" were pressed and fire-polished in deeply engraved molds, and since semiskilled cutters had only to touch up and sharpen the patterned lines on each piece,[71] Fry was able to market the glass very cheaply. The Libbey Glass Company's sales fell sharply in the face of such competition. As more and more people began to buy the cheaper grades of cut glass,[72] the

[68] See above, pp. 18–19.

[69] One C. J. Nolan at Libbey Glass in 1904 developed a machine for cooling the outside surface of the "blank" with streams of air while the inside was being fire-polished. This made the second mold for each pattern unnecessary and consequently further reduced production costs.

[70] Libbey-Owens-Ford Glass Company, Toledo Glass Company Minutes, I, 250–251.

[71] Formerly they had had to trace the designs on each glass object with their cutting wheels — a very time-consuming operation.

[72] According to testimony given in court at a later date, the new process actually saved 40 per cent in the cost of production. — Libbey Glass Company, Documents relating to licences, etc., Safe No. 2.

preference of wealthier customers shifted to lighter artistic pieces blown by hand.[73] In February, 1903, E. D. Libbey asked H. C. Fry to allow the Libbey Glass Company to employ Owens' invention also. They reached an agreement the following year whereby the two companies were to share the patented monopoly in pressing patterns on cut-glass "blanks." [74] Libbey Glass agreed to make only 40 per cent of their combined output; and both promised not to cut selling prices [75] and to pool all existent and future patents relating to the process. Later, on June 9, 1910, they agreed to allow the McKee Glass Company at Jeannette, Pennsylvania, annually to make a limited number of pressed "blanks." The royalties received from McKee were to be divided on a 60–40 basis, with Fry getting the larger share.[76]

The Libbey Glass Company never profited very much from cutting patterned pieces itself.[77] Its chief income arising from the arrangement with Fry was the profit made from selling patterned "blanks" to independent cutting houses. Its lamp and tube department, in contrast with its glass-cutting department, however, constantly increased in importance. The company had begun to make glass parts for incandescent lamps on a commercial scale in 1891,[78] and M. J. Owens in 1894 had developed a semiautomatic machine for blowing bulbs. From 1894 through 1918 Libbey Glass and the Corning Glass Works divided the American bulb market between them. On September 27, 1894, they decided to pool their patents, to regulate selling prices, and to exchange

[73] See Henderson, "Glass-Making," *Popular Science Monthly*, XXXVII (1890), 164. Libbey's cut glass had surpassed, "in mechanical and artistic qualities, the best wares of a similar nature produced elsewhere, and it [had] become the standard of excellence in this difficult branch of the art." — Barber, *American Glassware*, p. 95.

[74] They drew up a new contract on May 19, 1908, and extended it in amended form on July 3, 1916.

[75] The agreement provided severe penalties for the other's benefit in case either company made more than its share of the output or undercut the customary price.

[76] Because McKee constantly made more than the stipulated output, the contract was canceled on December 31, 1913. Fry and Libbey later brought suit against McKee because it continued production and were awarded $113,000 damages.

[77] Even though it "sold" the "blanks" to its cutting department at a lower price than that quoted outsiders, the cutting department's profits were small.

[78] See above, p. 93.

sales and production reports.[79] A year later they both became part to a ten-year agreement with the General Electric Company and the Phoenix Glass Company at Rochester, Pennsylvania. General Electric promised not to buy bulbs from any glass company other than the three involved in the contract, but it reserved the right to undertake production on its own account. Corning, Libbey Glass, and Phoenix covenanted to supply all of General Electric's needs on a 40–40–20 basis and not to sell bulbs to any other electric company at prices below those quoted General Electric.[80] A new fifteen-year agreement, signed on April 6, 1910, modified these provisions. General Electric and the National Electric Lamp Company agreed to buy from Corning and Libbey Glass all the bulbs they would not make themselves,[81] and the prices charged were to be 21 per cent above production costs.

After the Westlake Machine Company built its fully automatic bulb machine, it licensed Libbey Glass in 1914 and 1916 to use it. Several such machines were then installed at Toledo and also in a branch factory at Sandusky, and production costs per one thousand bulbs fell by about one-fourth.[82] Largely because of this new Westlake machine General Electric in 1916 told the directors of Libbey Glass that it wanted to terminate the 1910 contract and that it wished to buy outright all of Libbey Glass's bulb business, rights, and equipment. The two parties finally reached an agree-

[79] Probably because it held some rather broad process patents, Corning threatened to sue Libbey Glass if it put the Owens machine in operation. This first agreement, therefore, represented a compromise solution to oligopolistic difficulties.

[80] The contract also stipulated the exact prices which General Electric would pay.

[81] Until the end of 1911 Corning was to supply 42.5 per cent, Libbey Glass 42.5 per cent, and the two electric-lamp companies together 15 per cent; thereafter, the proportions were to be 37.5 per cent, 37.5 per cent, and 25 per cent, respectively. The electric-lamp companies and the glass companies were to share (on a 2/3–1/3 basis) savings resulting from future reductions in production costs.

[82] According to the comparative cost statement prepared by Libbey Glass, the reductions were as follows:

Type of Bulb	Semiautomatic Machine Cost per 1000	Automatic Machine Cost per 1000	Saving per 1000
S–19–C–4	$15.96	$10.75	$5.21
S–21–E–3	16.46	11.83	4.63

One dollar per thousand bulbs went as royalty to Westlake, and the remaining $4.21 and $3.63 savings were split between Libbey (1/3) and General Electric (2/3).

ment on December 12, 1918. The Sandusky plant passed into the hands of General Electric, and Libbey Glass surrendered all its rights and good will in the bulb business. General Electric paid $67,000 cash and $90,000 in 5 per cent debentures for the machines and inventories at Sandusky. Because the directors of Libbey Glass preferred to lease their properties and rights at $113,050 a year, General Electric consented not to buy them outright at that time but merely to hold an option on their purchase for $2,261,000.[83] Libbey Glass hence ceased to make bulbs but retained its privilege of using the Westlake and other machines in manufacturing stemware, tumblers, and the like.[84]

In December, 1914, and February, 1917, Westlake had granted the Libbey Glass Company complete freedom to modify any of its automatic machine parts or principles so that they could be used in making all kinds of paste-mold ware other than bulbs.[85] The directors of Libbey Glass, therefore, voted on December 31, 1914, to initiate experiments with this end in view. The result was the Kadow machine — so called because the inventor, August Kadow, helped to perfect it.[86] In the same year they encouraged Edward Danner, one of their employees, to work on a new way of drawing glass tubing. Since his endeavors met with great success, he was able in 1917 and 1919 to patent the machine which bears his name. Only three companies in the United States secured the right to use it.[87] Libbey Glass operated four Danner

[83] Therefore, $2,418,000 was the total valuation placed on the properties and rights according to the terms proposed by Libbey Glass on December 6, 1918. I have not seen the final contract (signed by both parties on December 12), but there is no reason to believe that its terms differed from the provisional ones.

[84] Its transaction with General Electric necessitated new contractual arrangements with Westlake. See above, pp. 131–132.

[85] See above, p. 131.

[86] The Libbey Glass Company, therefore, bore the entire expense of developing the machine and owned all rights on it. It was obligated by previous contract, however, to assign all container rights to the Owens Bottle Company. The machine was not used on a commercial scale until 1921 or 1922, after which it produced most of the physical output at the Libbey factory.

[87] In 1921 the Owens Bottle Company unsuccessfully sued Libbey Glass on the grounds that tubing was often made into test tubes, which were themselves "containers." Since both Libbey Glass and Owens Bottle were under obligation to license the other without charge if their respective (present or future) machines could be used in the other company's particular line of work, Owens Bottle had claimed exclusive container rights on the Danner machine.

machines in its Toledo factory in 1919–1920 and was obligated by its 1918 contract with General Electric to allow that company to install similar equipment.[88] And the Kimble Glass Company at Vineland, New Jersey, received a license from Libbey Glass on June 15, 1918.[89] Several other licenses were later issued to foreign manufacturers.

From the preceding account it is apparent that the interests of the Libbey Glass Company shifted considerably during the years 1895–1919. Its production of fine cut glass, constantly decreasing in relative importance, was maintained only for purposes of prestige; its sale of pressed-patterned "blanks" grew in importance from 1904 to 1912 but experienced a relative decline thereafter; its bulb and tube departments expanded greatly until the bulb rights were sold to General Electric in 1918; and the manufacture of cheap tumblers and other ware on automatic machines gradually developed after 1919 and ultimately became its chief activity. In 1895 nearly all the income received by Libbey Glass came from the sale of its own output; in 1919–1920 a substantial portion was made up of domestic and foreign royalties on the Danner machine and of annual payments from General Electric for the lease of its bulb properties. Since the year it moved to Toledo the company expanded its plant beyond the limits of its original factory site only three times, and even then more or less temporarily. For eighteen months following January 1, 1891, it blew bulbs at Findlay, Ohio; it operated a demonstration shop at the Columbian Exposition in 1893; and from 1910 to 1918 it made glass parts for incandescent lamps at Sandusky, Ohio. Its directors seriously considered in 1916 the advisability of constructing a branch factory at Charleston, West Virginia, but finally decided to enlarge the Sandusky and Toledo plants in-

[88] The 1910 agreement with General Electric had stipulated that Libbey Glass would receive one-third of all future reductions in the cost of making bulbs and tubing. Since the Danner machine lowered costs on each pound of tubing by 4 cents, Libbey Glass's share would have been 1.33 cents per pound. The payments involved in the 1918 "lease-or-purchase" contract, however, covered nonexclusive rights on the Danner machine as well as exclusive bulb rights on the Westlake.

[89] Kimble, at that time a subsidiary of the Illinois Glass Company, paid the licensor annual royalties amounting, for example, to $17,241 in 1919 and $83,337 in 1920.

stead. As a result of this and other enlargements, the number of its melting furnaces was fifteen times greater in 1918 than it had been in 1889.[90]

Although the Libbey Glass Company never expanded on the vertical or horizontal plane as greatly as did Owens Bottle, its success (as summarized in Tables 32–34 in Appendix C) was nonetheless spectacular. Due mainly to the reinvestment of earnings its assets increased in value about twenty-six fold from 1889 to 1920, and the owners' equity expanded about thirty-four fold.[91] Stock dividends amounting to 890 per cent of the original shares were paid on February 3, 1892, January 17, 1905, and July 2, 1913;[92] and from 1889 through 1917 the company distributed $4,007,500 in cash to holders of its common stock. Hence the annual cash return on each original share averaged 138 per cent of its par value.[93] Such handsome dividends were possible because the company's sales and net earnings increased eleven fold and ten fold, respectively, from 1894 through 1919.[94] The net earnings totaled $6,042,567 for the years 1894–1917 and 1919, and of

[90] MacManus, *Glass Manufacture*, pp. 59–60. In 1918 there were 10 big furnaces at Toledo and 5 at Sandusky.

[91] In 1888 the stockholders had supplied $100,000 in valuable assets for an equal value of common shares. In 1917 Libbey Glass issued $1,000,000 in preferred, $750,000 of which was paid to Westlake for its machine rights, $100,000 was sold to outsiders for $105,250, and $150,000 was offered to the stockholders at par. In 1919–1920 the directors repurchased and cancelled $120,000 of this preferred. Since Libbey Glass and its stockholders personally had largely financed Westlake, an appreciable part of the $750,000 preferred paid Westlake also represented a reinvestment of Libbey's earnings.

[92] A fourth stock dividend ($250,000) was declared on February 26, 1917, but withdrawn the following November after the sale of the bulbs rights to General Electric became fairly certain.

[93] This, of course, includes the return on shares received through stock dividends. Earnings of about $5,374,000 were distributed on both preferred and common from 1889 through 1919.

[94] The year 1920 is not included because the Libbey Glass Company ceased its manufacturing activities. The group of men who had directed its affairs since the factory was first built at Toledo decided in 1917 to retire and leave the development of the new tumbler and stemware lines to their sons. A new corporation, chartered in 1920 under the name of the Libbey Glass Manufacturing Company, leased the physical properties and good will of the original company. It paid approximately $500,000 for the inventories and receivables and $40,000 rental for each year following 1921. The owners of the old company distributed the $500,000 among themselves and formed the Libbey Glass Trust to manage their other interests.

this amount about seven-tenths was distributed among the owners and the remainder ploughed back into the business.

5. The Toledo Glass Company Once More (1903–1920) [95]

A section of the preceding chapter has already pointed out that the Toledo Glass Company, chartered in 1895, was primarily interested in developing and introducing various kinds of glassmaking machinery.[96] To do this it organized the American Lamp Chimney Company and later took an active part in amalgamating it with two other enterprises to form the Macbeth-Evans Glass Company. It also sponsored the incorporation of the Owens Bottle Company in 1903, the Owens European Bottle Machine Company in 1905, and the Libbey-Owens Sheet Glass Company in 1916. In addition, its directors issued licenses to the Rochester Tumbler Company and the National Glass Company for making tumblers and certain other kinds of paste-mold ware on the Owens' semiautomatic machine; they failed in their attempt to introduce the same machine in Europe but managed to sell the rights for Canada; they disposed of their Mexican and Canadian rights on the automatic bottle machine; and they granted licenses to the Hazel-Atlas Glass Company for pressing tumblers [97] and to the Illinois Glass Company for blowing glass fonts used on oil stoves.[98] On the technological side, Toledo Glass perfected the semiautomatic tumbler and lamp-chimney machine, the completely automatic Owens' bottle machine, and the Colburn sheet-drawing machine. Its main innovational interests centered around the first in the years 1895–1900, around the second between 1900 and 1912, and around the third between 1912 and 1920.

Since the company never made glass in commercial quantities, it derived its income from the disposal of patent rights. As Table 14 indicates, it often surrendered cash or physical properties or

[95] The chief sources of information were the Minutes kept by Toledo Glass and the various companies which it sponsored. [96] See above, pp. 96 ff.

[97] The Toledo Glass Company had developed, among other things, an automatic pressing machine, for which only one license apparently was ever issued.

[98] Toledo Glass had originally sold Owens Bottle only the "container" rights on the Owens automatic machine. Since these oil-stove fonts were technically not containers but manufacturing parts, Toledo Glass could directly license other companies to make them.

TABLE 14

RECEIPTS OF THE TOLEDO GLASS COMPANY FROM THE DISPOSAL OF ITS PROPERTIES, 1895–1920

Received from	Date	Gross Receipts (in dollars) Cash	Preferred Stock (At par)	Common Stock (At par)	Total
1. Rochester Tumbler and National Glass Companies	1897–1904	$125,000	......	......	$125,000
2. Diamond Glass Company, Ltd. (Canada)	1897	15,000	......	......	15,000
3. American Lamp Chimney and Macbeth-Evans Glass Companies	1899	......	$200,000	$330,000	530,000**
4. Macbeth-Evans Glass Company	1900	......	......	11,500	11,500
5. Owens Bottle Company	1903	276,000	124,500	2,399,500	2,800,000
6. Owens European Bottle Machine Company	1905	150,300	......	2,487,000	2,637,300
7. Vidriera Monterrey	1905	100,000	......	......	100,000
8. Canadian Glass Mfg. Company	1906–1908	104,900	......	......	104,900
9. Hazel-Atlas Glass Company	1915	50,000	......	......	50,000
10. Owens Bottle Company	1917	40,000	......	......	40,000
11. Libbey-Owens Sheet Glass Company	1916	600,000	500,000	3,999,500	5,099,500
TOTAL		$1,461,200	$824,500	$9,227,500	$11,513,200

	Considerations for Gross Receipts: Patent Rights and Licenses	Cash	Plant and Equipment (Estimated Value)	Stock Disbursed on behalf of Licensees and Buyers *	Net Receipts for Patent Rights and Licenses
1.	Exclusive tumbler and stemware rights on the semiautomatic paste-mold machine	......	$10,000	......	$115,000
2.	Exclusive Canadian rights on the semiautomatic paste-mold machine	......	2,000	......	13,000
3.	Exclusive U.S. chimney rights on the semiautomatic paste-mold machine	$112,500	150,000	$43,000	224,500
4.	Stock dividend	......	......	......	11,500
5.	Exclusive U. S. container rights on Owens' bottle machine	123,750***	90,000	704,600	1,881,650
6.	Exclusive foreign container rights on Owens' bottle machine	150,300	......	1,487,000	1,000,000
7.	Exclusive Mexican container rights on Owens' bottle machine	......	......	......	100,000
8.	Exclusive Canadian container rights on Owens' bottle machine	......	......	......	104,900
9.	Exclusive press-tumbler rights	......	......	......	50,000
10.	Exclusive U.S. rights to make all kinds of things on the Owens machine	......	......	......	40,000
11.	Exclusive foreign and domestic rights on the Colburn machine	600,000	175,000	2,499,500	1,825,000
	TOTAL	$986,550	$427,000	$4,734,100	$5,365,550

* Usually as bonus to purchasers of licensees' and buyers' other securities.
** Some of the shares were sold by Toledo Glass at less than par.
*** Paid to inventors for the assignment of certain of their patents.

both whenever it relinquished some of its manufacturing rights; and in return it received payment in the form of securities or cash or both. The difficulties involved in estimating its net earnings from each transaction are great. In some cases the difference between its gross and net receipts was quite marked because Toledo Glass oftentimes operated as a secondary distributor for the purchaser's securities. It turned over at least $43,000 of the shares received from Macbeth-Evans to deserving employees of the old American Lamp Chimney Company and to the intermediary who had helped bring about the consolidation;[99] it distributed over 7,000 common shares in Owens Bottle to purchasers of the preferred and to persons holding patents which Toledo Glass bought in order to protect its own patents;[100] it gave away as bonuses about three-fifths of the securities received from Owens European; [101] and it kept for itself only three-eighths of the Libbey-Owens common it originally received.[102] These complicated arrangements are reflected in Table 14. Another source of trouble in estimating the net earnings of Toledo Glass was the difference between the market value and the par value of the securities received in payment. Of the American Lamp Chimney and Macbeth-Evans shares, for example, $125,000 preferred were sold for $112,500 cash, $100,300 common were exchanged for $78,000 in notes held by E. D. Libbey, and a few others brought only 78 per cent or 90 per cent of their nominal value.[103] Since allowance for such differences has not been made in the table, the "net receipt" figures presented should be considered only as rough approximations.

[99] See above, p. 100, n. 52.

[100] In addition, about 45 per cent of the $276,000 cash received from Owens Bottle went to inventors who assigned Toledo Glass certain of their patents.—See above, p. 102, n. 60.

[101] See Table 13, above.

[102] $1,500,000 was distributed as bonus to those who purchased Libbey-Owens' $1,500,000 preferred; nearly $750,000 went to M. J. Owens, Irving W. Colburn, and William Jacobs for services rendered in developing the sheet-glass machine and in starting production at Kanawha City; and about $250,000 was earmarked for distribution by Libbey-Owens to members of its managerial staff.

[103] The common shares of all the companies did not attain a market value equal to or greater than their par value until several months or years after Toledo Glass had accepted their securities.

From the date of its inception through 1920 the Toledo Glass Company, by disposing of its machine rights, earned around $5,000,000 (in cash or securities valued at par). It did not, however, distribute all of this in dividends to its owners. As Tables 35 and 36 in Appendix C suggest, it reinvested part of its gross earnings in later developments. It apportioned nearly all its American Lamp Chimney, Macbeth-Evans, and Owens Bottle shares among its own stockholders but retained in its treasury the shares received from Owens European and Libbey-Owens. Most of the cash acquired from Owens European in the form of liquidating dividends through 1912 was distributed, but similar receipts thereafter were used to finance experimentation with the Colburn process. Dividends on its Libbey-Owens securities (with a par value of $2,000,000) constituted the company's chief source of income after 1919. The cash received from Rochester Tumbler, National Glass, and Diamond Glass between 1897 and 1904 helped defray the cost of developing the Owens bottle machine.

After the Toledo Group had incorporated the Toledo Glass Company in 1895, it sold half of the authorized $150,000 capital stock for cash and divided the other half equally between E. D. Libbey and M. J. Owens as payment for their patents pertaining to semi-automatic paste-mold machines. At no time before 1921 did the company issue any new shares. Whenever it needed additional funds to complete a series of experiments, it took them from its own earnings or borrowed them from E. D. Libbey.[104] The

[104] At the stockholders' meeting held on September 16, 1903, "the president announced that the Owens bottle machine had been so far perfected as to manufacture bottles, and that after years of effort and experiments, many failures and discouragements, and the making of enormous expenditures (all without calling upon the stockholders generally for contributions or assistance), the machine promises to be a commercial success." — Libbey-Owens-Ford Glass Company, Toledo Glass Company Minutes, I, 263. Again, on February 2, 1904, it was recorded that "as the stockholders are aware, at this time [i.e., 1901–1902] we were working on the bottle machine. . . . Full confidence in Messrs. Owens and Bock as to the ultimate success was felt by all stockholders, and not one of our stockholders withdrew from the company. Our president, Mr. Libbey, as before, stood in the breach and whenever funds fell short generously advanced from his personal funds sufficient to complete our experiments, with ultimate success." — *Ibid.*, I, 285. On October 16, 1903, the directors agreed to pay Libbey $70,000 in Owens preferred "for advances made by him to the company to carry on its corporate business and the expenses of experimenting with and developing its glass blowing machinery . . ."—*Ibid.*, I, 278 ff.

value of each of its shares consequently soared every time it perfected a new machine and successfully concluded new licensing agreements. There are no available open-market quotations on its stock which can be used to measure the full extent of the price rise because all its shares were closely held and privately sold. In 1900 the directors agreed to pay $150 ("the same being, in the opinion of the Board, a fair and reasonable price") for each of fifty shares which they had promised to set aside as a gift to one of the men working on the Owens machine.[105] Ten years later one owner sold part of his holdings at $800 a share, and another at $1,100.[106] The company purchased several of its own shares in 1913 for $400 each and then sold them two years later at twice that figure.[107] One small stockholder was offered $5,000 for each of her shares in November, 1923,[108] and those belonging to E. D. Libbey upon his death in 1925 were each appraised at $9,000.[109] According to various rumors at Toledo, the price finally climbed to over $14,000 in the late 1920's — or 140 times higher than their original par value! [110]

6. Summary

At the outset the transfer of workmen and glassmaking equipment from Cambridge, Massachusetts, to Toledo in 1888 had not been especially auspicious. During the following thirty-two years, however, the comparatively small number of Toledo entrepreneurs who cast their lot with E. D. Libbey and M. J. Owens revolutionized technological processes, formed numerous companies, and amassed both private and corporate fortunes. Originally, they had constituted one group; later, they divided into two

[105] *Ibid.*, I, 233–234.

[106] According to William H. Boshart. Such differences in selling price, of course, were possible in an unorganized market.

[107] It had intended to set aside the shares as a gift for Irving W. Colburn in partial compensation for his services in improving the sheet-glass machine. Instead, however, it finally gave him $480,000 common stock in Libbey-Owens.

[108] Owens-Illinois Glass Company, Owens' Private Files.

[109] Walbridge, "Romance of an Invention", p. 9.

[110] The yearly cash dividends distributed in 1925–1927 averaged 306.67 per cent. They were paid almost entirely out of the earnings from the Libbey-Owens stock which Toledo Glass still held. — Libbey-Owens-Ford Glass Company, Toledo Glass Company Minutes, IV–V, *passim.*

rather distinct factions. The increasingly complicated network of companies, which has been described at some length in this and the immediately preceding chapter, stemmed from both factions, as the following outline attempts to show in summary form.

I. The first sub-group may be called the "Libbey Glass Faction."

A. It organized in 1888 the *Libbey Glass Company*, which made fine tableware, cut-glass "blanks," bulbs, and tubes, and which perfected and licensed the Kadow paste-mold and Danner tube machines. For many years following 1919 the company went under the name of the Libbey Glass Manufacturing Company.

B. In 1893 it organized a subsidiary, known as the *Libbey Glass Company of Illinois*, to demonstrate glassmaking at Chicago's Columbian Exposition.

C. It also formed in 1907 the *Westlake Machine Company*, which, existing until after 1920, developed the fully automatic Westlake bulb machine and licensed domestic producers in its use.

D. It incorporated the *Westlake European Machine Company* in 1911, which licensed foreign bulb producers until after 1920.

II. The second sub-group may be called the "Toledo Glass Faction."

A. Its members in 1895 organized the *Toledo Glass Company*, which functioned for several years after 1920. This company developed and introduced the semiautomatic tumbler and lamp-chimney machine, the fully automatic bottle machine, and the Colburn sheet-glass machine.

B. It organized the *American Lamp Chimney Company* (1898–1899), which made chimneys and low-grade tumblers before it amalgamated with two other companies to form the *Macbeth-Evans Glass Company*, which continued activities well beyond 1920.

C. It sponsored the unsuccessful *Automatic Glass Blowing Patents Syndicate, Ltd.* (1900–1902), which was supposed to introduce the semiautomatic paste-mold machine on the European market.

D. In 1903 it incorporated the *Owens Bottle Company*, which was still thriving after 1920. This enterprise at first only issued licenses permitting domestic producers to use the Owens bottle machine but later began to make glass containers itself as well.

1. The Owens Bottle Company expanded on the horizontal plane.

a. Its owners formed the *Northwestern Ohio Bottle Com-*

pany (1904–1908), which became a wholly owned subsidiary in 1908 and was consolidated in 1911.

b. E. D. Libbey participated in the organization of the *Ohio Bottle Company* (1904–1905), which amalgamated with other independent enterprises in 1905 to form the *American Bottle Company*, which in turn became a wholly owned subsidiary of Owens Bottle in 1916.

c. Some of the "Toledo Glass" group in 1905 formed a manufacturing concern known as the *Newark Machine Bottle Company*, which eventually consolidated with American Bottle.

d. Owens Bottle in 1909 incorporated the *Owens West Virginia Bottle Company*, which it completely absorbed in 1911.

e. It also had controlling interest in the *Owens Eastern Bottle Company* (1912–1915). After buying up all its stock in 1915, Owens Bottle consolidated its assets.

f. It obtained control of the *Whitney Glass Works* in 1915 and absorbed it three years later.

g. In 1916 it acquired all shares of ownership in the *Graham Glass Company of Indiana*, which in turn controlled the *Graham Glass Company of Oklahoma.*

h. It received a majority of the stock issued by the *Charles Boldt Glass Manufacturing Company* in 1919.

i. It had substantial holdings but not controlling interest in the *Hazel-Atlas Glass Company* after 1909 and until 1926.

2. Owens Bottle also expanded on the vertical plane.

a. Its most important shareholders owned and controlled the *Kent Machine Company*, founded in 1902 and continuing after 1920.

b. It purchased the *Woodruff Glass Mould Company* in 1908 from some of its own stockholders.

c. It incorporated the *Toledo-Owens Glass Sand Company* in 1913 but sold it to outside interests in 1920.

d. Together with the Libbey Glass Company it owned nearly half the stock of the *Buckeye Clay Pot Company* after 1913.

e. In 1918 it purchased two-thirds interest in the *Mid-West Paper Box Company.*

f. And it organized the *Peytona Gas Company* at Charleston, West Virginia, in 1918.

E. The Toledo Glass Company incorporated the *Owens European Bottle Machine Company* in 1905 to license foreign producers to use the Owens bottle machine. It retained one-third the capital stock of the enterprise, and its own members privately held most of the remaining shares. Owens European, not dissolved until after 1920, formed the *Owens European Bottle Machine Company, Ltd., of Manchester* (1906–1907).

F. In 1916 Toledo Glass organized the *Libbey-Owens Sheet Glass Company* to license foreign and domestic producers in the use of the Colburn machine and to make window glass itself. The parent concern retained one-third of Libbey-Owens' capital shares. The remainder was held chiefly by individual members of the "Toledo Glass" group.

1. Libbey-Owens acquired one-third interest in the *American-Japan Sheet Glass Company* in 1919.
2. And in the same year it organized, and almost immediately bought complete interest in, the *Libbey-Owens Sheet Glass Development Company, Ltd.*, and then disposed of the European window-glass rights to the *Mutuelle Mobilière et Immobilière.*

It is extremely difficult to summarize succinctly the full extent of this expansion. Perhaps the best single index would be the increase in the value of those assets owned by the operating companies. They were valued at only $151,155 in 1890 and between $50,000,000 and $60,000,000 thirty years later.[111] E. D. Libbey had arrived at Toledo in 1888 with about $100,000, which he immediately invested in the Libbey Glass Company. In 1920 the combined net worth of all the glass companies in which Libbey and his associates had a controlling interest was slightly over $40,000,000. And their annual net earnings had increased from $1,000 or $3,000 in 1890 to around $6,500,000 in 1920. The Libbey Glass Company, the only one in existence at that time, paid no cash dividends before 1892; in 1920 the operating companies distributed about $2,500,000 among their preferred and common shareholders.

[111] I have attempted to avoid such double-counting as would have resulted if all the assets of the Toledo Glass Company and the Westlake Machine Company had been included in the total figures. Most of the properties of these two companies in 1920 represented claims of ownership on the assets of the operating companies.

It is impossible to ascertain, even approximately, to what extent the Toledo entrepreneurs shared in the profits resulting from this growth. Their various companies received relatively little cash from "outside" interests, but they did acquire valuable plants, equipment, and patent rights from others in exchange for their own stock. Then, too, many of the original recipients sold all or part of their holdings (privately or on the open market) to "outsiders" and used the proceeds to finance some of the later ventures at Toledo. About all that can be said, therefore, is that the same entrepreneurs who by investing $175,000 launched the Libbey Glass Company and the Toledo Glass Company in 1888–1895 still retained controlling interest when the combined assets of their enterprises amounted to between $50,000,000 and $60,000,000.

CHAPTER VI

REVOLUTION IN TECHNOLOGY AT TOLEDO

The corporate maze erected by the entrepreneurial group at Toledo rested upon technological improvements of a truly revolutionary character. Within less than a quarter century handicraft processes which had been faithfully followed for centuries were either discarded or were adapted to machine methods. Why the mechanization of glassmaking was so long delayed and why it made such marked progress at Toledo are questions which cannot be answered satisfactorily until the new techniques themselves are fully understood.

The present chapter surveys the innovations originating at Toledo. It directs attention primarily to the nature and significance of the principles involved in the new technology rather than to a description of mechanical details.[1] The most important innovations to be considered are (1) the semiautomatic (and later the fully automatic) paste-mold machines for blowing light bulbs, tumblers, and lamp chimneys, (2) the automatic Owens bottle machine, (3) the Danner tube-drawing machine, and (4) the Colburn window-glass machine. But certain minor developments were not without importance.

1. Various Minor Inventions

The first in a long series of labor-saving devices developed or sponsored by Michael Joseph Owens was his so-called "dummy," which permitted workmen making bottles, lamp chimneys, or similar articles in paste-molds [2] to open and shut the molds without the assistance of a boy helper. The principles involved were not altogether new since other men had patented a similar contrivance in 1885. The attorneys whom E. D. Libbey later asked to

[1] Those who desire to understand all the processes are referred to Appendix B, which explains the technical details of some machines more fully than does the text.
[2] For the nature and significance of "paste-molds," see above, pp. 16–17, n. 47.

investigate the originality of Owens' invention stated in 1898 that the legality of his patent claims was, in their opinion, "extremely doubtful." In any case their validity would have to rest upon the specific construction of the machine rather than upon the principles involved.[3]

In 1899 M. J. Owens received a patent covering a method for fire-polishing the inside surface of hollow tableware in order to efface the scars and marks which contact with the metal plunger always left when glass was pressed.[4] Simple though the process was, it revolutionized the cut-glass industry[5] because it greatly reduced the amount of skilled labor necessary to form and cut the pieces. The cost of producing heavy cut glassware, accordingly, fell by at least 40 per cent.[6]

Several years later, after he had attained an international reputation as an inventor, Owens made another contribution to the art of pressing glass. On November 21, 1912, he reported to the directors of the Toledo Glass Company that he was then engaged in developing an automatic, multiple-mold machine for pressing tumblers.[7] Ever since the 1820's American factories had been making large quantities of tableware on simple side-lever

[3] Libbey-Owens-Ford Glass Company, Licenses and Correspondence of Toledo Glass Company, Vol. I, Patent Opinions Rendered E. D. Libbey by Fish, Richardson, and Storrow, March 9, 1898. The doubtful worth of Owens' patent was probably the reason why Libbey never paid Owens anything for the right to employ the "dummy" in making light bulbs. If one might judge from the remarks of those who knew both men intimately, Owens never entirely overcame the feeling that his employer had cheated him. The United States Glass Company purportedly paid Owens $5,500 for permission to use the same foot-lever mold in the manufacture of tumblers.

[4] See above p. 19. The patent, numbered 628,027, was dated July 4, 1899. According to the Owens' method, "the blanks were formed by applying a plunger against a gob of glass in a mold after which the glass article thus formed, and while still in its mold, was subjected to a blast of flame which remelted the inner surface of the article, and thus removed any scars and marks left by the plunger." — Libbey Glass Company, Documents relating to licenses, etc., Safe No. 2.

[5] See above pp. 134 ff. Apparently other glassmen did not patent successful competing methods until around 1915. — *National Glass Budget*, XXXI, April 29, 1916, p. 7.

[6] See above, p. 134, n. 72.

[7] Libbey-Owens-Ford Glass Company, Toledo Glass Company Minutes, III, 115. On May 12, 1913, he reported that the drawings for the machine had been completed and that he was ready to have one built. He estimated the cost of construction at about $15,000. — *Ibid.*, IV, 2.

presses; [8] but each of these presses had been equipped with only one set of molds, and most had been operated by hand. Owens took some of the principles of his own bottle machine which had been built in the meantime and had the company's engineers construct a large circular power-driven press equipped with many duplicate molds. Since other glassmen had already successfully used power to operate multiple-mold presses,[9] this machine may rightfully be considered as one of Owens' minor inventions. It was not nearly so extraordinary as some other products of his labors; nevertheless, it embodied enough distinctive features to be of commercial value. In February, 1915, the Toledo Glass Company, in return for $50,000, issued an exclusive license for pressing tumblers on it in the United States to the Hazel-Atlas Glass Company.[10]

2. The First Major Invention: The Semiautomatic Paste-Mold Machine

During the closing weeks of the Columbian Exposition at Chicago, Owens frequently had chatted with glassmen under his supervision at the Libbey Glass plant there and had told them that he proposed to blow glass on machines. The skilled workmen only scoffed when he derided the archaic nature of contemporary processes and expressed his conviction that epochal improvements were possible.[11] Within two years, however, his persistent efforts in the inventive field radically changed the manufacture of incandescent lamp bulbs and other paste-mold items.

[8] See above, pp. 18–19.

[9] Power presses, it seems, were introduced to the trade around 1898.—*National Glass Budget*, XIV, July 2 and September 24, 1898. One Frank O'Neill, in 1901, had built a machine in the form of a circular revolving table for pressing glass jar lids. It could make 27 lids a minute and automatically discharged the lids by tipping the molds.—*Ibid.*, XVI, April 20, 1901.

[10] Libbey-Owens-Ford Glass Company, Toledo Glass Company Minutes, IV, 48. The terms of the contract forbade Hazel-Atlas to employ the suction device of the Owens bottle machine for feeding molten glass to the molds. This company, however, had a license from one Homer Brooke to use his special feeder. See below, p. 182, n. 30.

[11] James O. Hill, one of these men, congratulated Owens on July 3, 1922, for his remarkable success and recalled how skeptical he and the others had been at Chicago in 1893. This interesting letter is preserved in M. J. Owens' private files at the Owens-Illinois Glass Company.

The semiautomatic paste-mold machine was responsible for this change. It also provided the immediate reason why the Toledo Glass Company was incorporated and why its promoters were spurred to finance new inventive undertakings.

The semiautomatic bulb machine, patented by Owens in 1894, simply incorporated the basic processes involved in handicraft production. A worker gathered glass on the end of a blowpipe and inflated it in a preliminary fashion. He then fixed the pipe, with its attached gob of glass, in an upright position on an arm of a circular rotating machine which then employed compressed air to blow the glass in carbon-coated iron molds.[12] Since each machine had more than one arm, several bulbs were always in different stages of production at any moment of time. The new technique required the services of three gatherers to collect the glass; one man to place the pipes in their appropriate place on the machine; and six small boys to lift the bulbs out of the molds, to crack them off the pipes, to keep the molds and pipes clean, and to carry the hot ware to annealing ovens. Skilled blowers were no longer necessary.

Between 1895 and 1898 the engineers of the Toledo Glass Company made some minor changes so that the machine could blow tumblers and lamp chimneys in much the same manner as bulbs. The American Lamp Chimney Company (1898–1899) made tumblers and chimneys at the same time on its Owens semiautomatics.[13] The molds were so shaped that each piece of blown glass could be cut into two parts. Since the lower section had a bottom it became a tumbler, and the upper part was sold as a chimney.

It is difficult to determine to what extent these paste-mold ma-

[12] The use of compressed air in blowing glass dates from the 1820's when a Frenchman named Ismael Robinet, employed at the Baccarat works, invented an air pump which could replace human lung power to a certain extent.— Bontemps, *Guide du Verrier*, p. 569; Peligot, *L'Art de la Verrerie*, p. 476; *Journal of the Franklin Institute*, XXIV (1833), 380. A half century later the New England Glass Company was blowing glass balls with hand pumps, and the Appert Brothers at Paris around 1884 were claiming to have perfected a way whereby compressed air could be used effectively in glassmaking.— Gaffield Collection, Glass Journal, III, 31, and Scraps, III, 130, 131.

[13] Owens-Illinois Glass Company, American Lamp Chimney Company Letters, p. 22.

chines reduced production costs. According to the *National Glass Budget*, the chief trade journal, one machine made 2,000 chimneys every five hours at a *direct labor cost* of $6.50, whereas five or six workers usually produced 200 chimneys by hand at a *direct labor cost* of $4.60. Six weeks later the same publication stated that chimneys made on the Owens machine cost only 10 cents a dozen as against 23.13 cents for those fashioned by hand "under last year's union scale."[14] These two examples suggest that the chimney machine lowered direct labor costs by approximately 86 per cent or *general production costs* by about 57 per cent.[15] After organized labor regained control of hourly wages and the amount of ware which workers in mechanized plants could produce, the cost advantage of chimney manufacturers employing the Owens machine over those who still followed handicraft methods greatly diminished.[16]

Whether the output was lamp chimneys, tumblers, or light bulbs, the machine not only used less skilled (and hence cheaper) labor than was necessary in hand shops, but it also increased markedly man-hour productivity.[17] The Toledo Glass Company in 1897 offered to demonstrate to any European manufacturer that its machine — with the aid of one feeder, three gatherers, and six small boys — would make 2,000 nine-ounce water tumblers every

[14] *National Glass Budget*, XV, June 17 and July 29, 1899. About two months earlier (May 27, 1899) the editor had stated that a machine-made chimney cost approximately one-fourth of a cent while one made by hand cost four times as much.

[15] Although the reliability of these figures is subject to question, comparisons found elsewhere indicate that they do not grossly exaggerate the reductions. The American Lamp Chimney Company is reported as having realized savings in its direct labor costs of 72 and 71 per cent for two kinds of chimneys. — U. S. Commissioner of Labor, *Regulation and Restriction of Output*, p. 643. And Victor S. Clark remarks that the cost of making a dozen lamp chimneys fell from 15 to 5 cents primarily as a result of mechanization. — *Manufactures in U. S.*, III, 262.

[16] U. S. Commissioner of Labor, *Regulation and Restriction of Output*, pp. 647–648; *National Glass Budget*, XXXIV, October 19, 1918, p. 8. In the 1920's direct labor costs of chimneys made on the semiautomatic machine were about 39 per cent less than those blown and shaped by hand and only 20 per cent less than those blown by hand in paste molds. — *Productivity of Labor in the Glass Industry*, p. 114. Little or no information has been found on the savings effected in producing bulbs and tumblers on semiautomatics. However, since bulbs, tumblers, and chimneys were all made in practically the same manner by hand and by machine, one may presume that the savings in making all three were of similar magnitude.

[17] *Productivity of Labor in the Glass Industry*, pp. 112 ff., 125 ff.

five hours.[18] And one plant in 1918 reported that six persons working with a semiautomatic made from 1,500 to 2,600 chimneys every four and one-half hours.[19] These outputs were considerably larger than three blowers, three gatherers, and six boys could produce in the old way. The wage economies and productive efficiency of the Owens machine were certainly great enough to cause many manufacturers and skilled blowers to fear that its introduction would destroy their position in the trade. Some, of course, scoffed at its possibilities and claimed that the semiautomatic would never be a commercial success.[20] But George E. Macbeth and Thomas Evans, two of the most important producers of lamp chimneys, were so alarmed over its potentialities that they agreed to consolidate their separate enterprises with the newly created American Lamp Chimney Company and to buy from the Toledo Glass Company exclusive right to make lamp chimneys on the machine.[21] The Rochester Tumbler Company, and later the National Glass Company, paid altogether about $125,000 for the tumbler rights.[22] And possession of the bulb rights proved highly lucrative to the Libbey Glass Company.[23]

3. The Owens Bottle Machine

Most unique and most important of all the technological innovations promoted by the Toledo Group was the machine which automatically gathered molten glass and formed it into containers. In 1900 all bottles and many jars made in the United States were gathered, blown, and finished by skilled workers (using techniques practiced for centuries). Two-piece chilled iron-molds, it is true, were used more than formerly in shaping the body of each container, and the snap-case had displaced the punty for holding the white-hot bottle while the worker tooled its mouth and lip.[24] Work around glass furnaces also had been gradually specialized, and the introduction of the "shop" system had entailed a marked division of labor.[25] But these and a few other

[18] Libbey-Owens-Ford Glass Company, Toledo Glass Company Minutes, I, 85.
[19] Tariff Commission, *Glass Industry as Affected by War*, p. 104.
[20] Gillinder and Evans, *Papers*, p. 17.
[21] See above, p. 99.
[22] See above, p. 98.
[23] See above, pp. 135 ff.
[24] See above, p. 17.
[25] See above, p. 22.

(Courtesy of Owens-Illinois)

FIG. 5. THE OWENS FIFTEEN-ARM "AQ" BOTTLE MACHINE, 1914
Its size may be gauged by the height of the handrail at the left.

minor improvements modified only slightly the process of producing bottles.

As the nineteenth century ended, the manufacture of wide-mouth jars (as contrasted with bottles) was undergoing really radical changes. Rather simple machines, most often designed to make only one jar at a time, were gradually coming into use.[26] They were frequently operated by hand and relied upon skilled gatherers to feed them with glass taken from the furnace.

The Owens bottle machine represented a marked improvement over these or any other devices used before 1903 in the manufacture of glass containers. Unlike all others, it was capable of producing narrow-mouth as well as wide-mouth ware; and what was more important, it displaced the gatherers as well as the skilled blowers. The real significance of this latter accomplishment may best be revealed by quoting from a brief article entitled "The Necessity of a Glass Gathering Machine" appearing in a trade journal on July 21, 1900.[27]

> That glass can be pressed and blown by machinery is a settled fact, and that it can be gathered and dropped into the molds by mechanical means is certainly a possibility, though in the present state of the art no pathfinder has yet appeared to blazon the way. . . . Eight and ten gatherings per minute is about all that mortal man can do for eight or ten hours without breaking down his constitution and by-laws, and machines going at such a speed cannot be properly called labor-savers — they are gatherer killers. A gathering machine is not only in order, it is an absolute requirement if further progress in glassworking machinery is to be made. Who will lead the way?

Supported financially by the Toledo Group and assisted by a few capable draftsmen and engineers, Michael J. Owens led the way. Each of his machines, rotating continuously around a central pillar, was equipped with from six to twenty arms. Each arm in reality constituted a separate bottle-making unit with its own set of molds. The most unusual feature of the machine was the

[26] Dos Taylor at Huntington, West Virginia, and C. L. Flaccus at Beaver Falls, Pennsylvania, around 1892–1894 were the first American producers to make jars on a machine using a press and a blow mold. — *National Glass Budget*, XXVI, May 28 and November 26, 1910. The inventor Philip Arbogast had previously assigned his basic patents on the process to the United States Glass Company. See below, pp. 178, 323–325.

[27] *National Glass Budget*, XVI.

way it gathered molten glass by sucking it up from a revolving pot connected with the central melting furnace.[28] After the glass was gathered in a small mold, it was transferred to one of regular size and there fully blown by means of compressed air. The finished bottle finally fell from the mold onto a conveyor belt leading to the annealing oven.

As one would expect in the evolution of any new device, the first machines were inferior in many ways to the models marketed around, say, 1912 or 1915. The one with which Owens first experimented between 1898 and 1900 rather closely resembled a bicycle pump or a hand spray gun. According to one of the men closely associated with the work, the hollow cylinder of this "gun" was about three and one-half feet long and enclosed at one end a piston rod equipped with a handle.

> On the other end was a hollow rod, similar to a glass worker's gathering-iron, with a two-part blank mold — the upper part being a neck of the bottle with a plunger inserted to form the finished neck — and the lower part of the mold designed to enclose a solid gather of glass.
>
> After inserting the blow rod into a furnace and drawing up the glass into the molds we have referred to, by the use of the piston rod — in other words, sucking up the glass into the molds — the glass was cut off by a sliding knife, making the bottom plate of the gathering mold. Then the entire gather was carried to a table where a finishing mold was opened, the parison being suspended by its neck in the neck mold: the finishing mold was closed around the parison — the piston rod was reversed: air was blown into the parison through the neck mold — and the bottle was blown! [29]

This simple "gun" was operated by hand and cost around fifty dollars.[30] The superficial observer doubtless could see no resem-

[28] On December 1, 1905, William S. Walbridge, a director and officer of the Owens Bottle Company, wrote Adolphus Busch, "You will kindly recall that the crux of our invention is in the revolving tank, to which we do not call special attention in mentioning the Owens bottle machine but as a matter of fact, if it were not for this revolving tank — *on which we hold the basic patents throughout the world* — the machine would not be operated without labor." — Owens-Illinois Glass Company, Walbridge Letters, I, 272.

[29] *Ibid.*, "Lives of E. D. Libbey and M. J. Owens," p. 8. "Parison" is the technical term for the gob of viscous glass. A draftsman's conception of this simple hand tool is reproduced in Fig. 6.

[30] Saunders, in *Toledo News-Bee*, October 31, 1934.

blance between it and the complicated mechanism which eventually invaded the market. But the hand "gun" embodied some of the most important principles — such as gathering by suction, utilizing the full-sized separate mold for forming the bottle's neck, and alternating the use of the gathering and blowing molds — found in later machines. Owens did not originate any one of these principles. His claim to fame rests mainly on the fact that

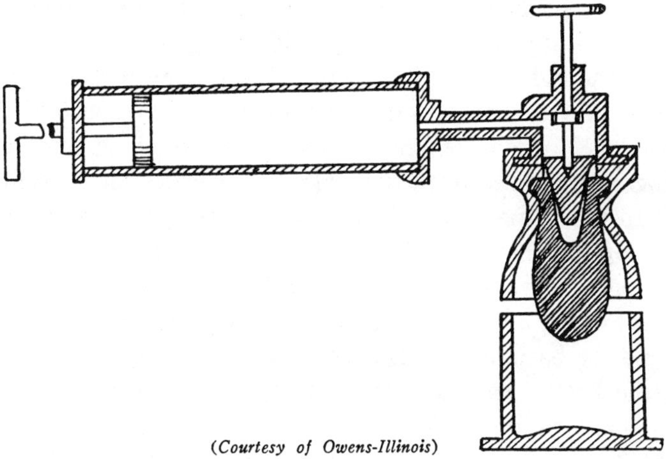

(*Courtesy of Owens-Illinois*)

FIG. 6. DRAFTSMAN'S SKETCH OF OWENS' "HAND GUN" FOR MAKING BOTTLES

he and the men working under him eventually synthesized the separate features to form an automatic mechanism.[31] Between 1900 and 1903 they built and destroyed several power-driven machines before they had one which they thought would prove a commercial success. They also made additional significant improvements before 1909, and thereafter brought out various models designed either to make special kinds of containers or to increase the rate of production.

In addition to mechanizing the gathering and forming operations, M. J. Owens and his staff of helpers improved the manufacture of glass containers in numerous ways. Prior to 1910 each machine had kept several boys busy picking up the finished bottles as they fell from the machine and carrying them to the lehr. Early in that year the Owens Bottle Company installed an

[31] See below, p. 327, n. 16.

automatic conveyor in its Libbey Street plant and a short time later in its Fairmont factory as well.[32] This conveyor displaced the "carry-in" boys because the blowing machine automatically dropped the hot bottles upside down into the cavities of iron pans. The pans then moved on a continuous belt from the side of the machine through the annealing lehr.[33] Hence it was no longer necessary to handle the bottles before they were annealed and ready to be packed. M. J. Owens in 1911–1912 designed a new type of melting furnace which could hold three times as much glass as the old furnace installed at the Libbey Street plant and which could feed two operating machines simultaneously.[34] Later on he developed automatic machines for weighing and mixing the ingredients used in making glass, conveyors for carrying the batch to the furnace, and feeders for keeping the melting tanks full.[35] Before the end of 1920 the Owens Bottle Company was operating at least one factory which was so completely mechanized that direct application of human effort was unnecessary from the time the basic materials arrived at the factory in freight cars until the bottles emerged from the annealing lehr ready for inspection and packing.

Despite the significance of subsequent improvements on the first bottle machines and of the ancillary devices mentioned in the preceding paragraph, the Owens process for blowing bottles as practiced between 1903 and 1905 constituted by far the most important labor-saving technique introduced in the bottle industry since the invention of the blowpipe. Under the handicraft

[32] Owens-Illinois Glass Company, Annual Reports of Owens Bottle Company, September 30, 1910; Owens Bottle Company Minutes, II, 46.

[33] Sometime after 1920 the company discarded the pans and allowed the bottles to fall directly on a meshed conveyor belt which carried them slowly through the lehr. This improvement, it is claimed, lowered annealing fuel costs by 75 per cent because the pans had absorbed and then lost most of the lehr's heat and also had had to be preheated at the beginning of each trip from the machine to the lehr in order to keep the hot bottles from cracking due to sudden cooling.

[34] Owens-Illinois Glass Company, Annual Reports of Owens Bottle Company, September 30, 1911; *National Glass Budget*, XXVIII, September 7, 1912, p. 3.

[35] Automatic mixers, batch conveyors, and furnace feeders were tried out at the Fairmont factory in West Virginia as early as 1909–1910. See *National Glass Budget*, XXV, *passim*. But the men with whom I have talked at Toledo and who were familiar with the Fairmont plant do not think that all of these mechanical devices proved entirely satisfactory in the beginning.

process a "shop" composed of three skilled men and three "boys" or helpers[36] could make between 15 and 20 gross of pint beer bottles every eight and one-half hours of working time.[37] One man and five boys spread over two labor shifts were all the help necessary to operate one of the first Owens machines used commercially;[38] and they made at least 100 gross[39] of pint beers every twenty-four hours, or 35.4 gross every eight and one-half hours. In other words, at least fifteen men and as many helpers, working eight and one-half hours a day, could have made by hand no more bottles than two of the early machines, under the supervision of one man assisted by five boys, produced every twelve hours. Not only did the machine save man-hours, but it also did not use expensive skilled glassmen.[40] This meant still further reductions in manufacturing costs. Justification exists for estimating that hand factories between about 1903 and 1907 had to pay out about $1.50 in wages for every gross of pint beers blown whereas manufacturers licensed to use the Owens machine found their direct labor costs reduced to approximately 10 cents.[41]

[36] Two of the men gathered molten glass and blew bottles while the third finished the mouths and lips. The boys received the bottles in a snap case, carried them to the lehr, and also cleaned the blowing pipes.

[37] The working day in glass factories around 1903 was ten hours, but the men took an hour off for lunch and two fifteen-minute periods for rest. See Barnett, *Machinery and Labor*, p. 95.

[38] One man, working twelve hours a day, could tend two machines simultaneously. The five boys, also on twelve-hour shifts, took the bottles out of the mold, put them through a fire-finishing process, and then carried them to the lehr. — *National Glass Budget*, XX, April 29, 1905.

[39] As a matter of fact, this is an understatement. The daily output on this machine soon climbed to around 125 gross.

[40] Barnett (*Machinery and Labor*, p. 104) suggests that "it would not be an extravagant estimate to put the earnings of blowers in the trade as a whole at an average of seven dollars per day." The American Bottle Company, licensed by Owens Bottle, paid its machine tenders about $1.80 a day. — *National Glass Budget*, XXII, November 24, 1906. For a more detailed comparison of wages paid machine tenders and skilled blowers after 1903, see below, pp. 204 ff.

[41] The prospectus published by the Owens Bottle Company in 1903 slightly exaggerated the differential when it estimated the machine labor costs to be approximately 6 cents per gross. — Owens-Illinois Glass Company, *Owens Automatic Bottle Machine*. The *National Glass Budget* (XXII, November 24, 1906, and February 23, 1907) demonstrated that the American Bottle Company made beer pints and quarts on the machine for about 8 to 15 cents of labor costs per gross, whereas hand-operating manufacturers paid $1.52 "shop" costs per gross. In 1910, after blowers had accepted a 20 per cent cut in wages, a gross of bottles at Streator, Illinois, cost approximately $1.15 if they were produced by hand and 11 cents if

The models that were introduced subsequently to the first one further raised worker productivity and lowered labor costs. Although the data presented in Table 15 are more often based upon general statements found in such documents as the company's annual reports than upon detailed production records of factories using the Owens machine, they aid in estimating the extent to which successive modifications of the first commercial machine increased daily output. The model introduced in 1917 apparently made roughly five times as many bottles per day as the first-type machine had done in 1905,[42] and yet fewer men operated it than had been required for earlier models.[43] Manufacturers around

TABLE 15

APPROXIMATE AVERAGE OUTPUT OF VARIOUS OWENS BOTTLE MACHINES*

Kind of Machine		Year it Was Introduced (Approximately)	Approximate Daily Output (in gross)	
Type**	Number of Arms		Pint Beers	Other Bottles, Depending on Their Size***
"A"	6	1903–1904	125	100–200
"AC"	6	1908	150	?
"AE"	6	1909	200	120–240
"AR"	10	1912	400	120–450
"AN"	10	1912	400	200–600
"AQ"	15	1914	500	180–600
"AV"	15	1917	625	300–900

* Sources: Owens-Illinois Glass Company, Annual Reports of Owens Bottle Company, Owens Bottle Company Minutes, and Owens' Private Files.

** The "AR" and "AQ" machines were designed for large ware, whereas the "AN" and "AV" machines were for smaller ware.

*** The figures in this column show the relative outputs of the different machines as of 1919, and they were taken from a letter written by H. H. Baker to the European cartel on August 27, 1919. — Owens-Illinois Glass Company, Europäischer Verband Lawsuit Files.

they were made on the six-arm Owens machine. — *National Glass Budget*, XXV, April 23, 1910. Similar comparisons may be found in V. S. Clark, *Manufactures in U. S.*, III, 259; Barnett, *Machinery and Labor*, p. 87; and Walbridge, "Romance of an Invention," p. 5.

[42] Barnett suggests that the daily average output only trebled. — *Machinery and Labor*, p. 88.

[43] The Owens Bottle directors voted on February 1, 1916, to exhort all licensees to replace their old-style machines with the latest models in order to increase both royalties and operating profits. — Owens-Illinois Glass Company, Owens Bottle Company Minutes, V, 30. The rise in machine productivity was due mainly to three things: (1) the increase in the number of arms, (2) the adoption of plural

1920 had to hire only one attendant for each machine and one foreman for every two. This meant, as M. J. Owens estimated at one time, that each fifteen-arm machine displaced from thirteen to fifty-four skilled blowers, depending upon the kind of bottles made.[44] The unskilled boys who had found employment earlier in factories equipped with Owens machines were no longer necessary after the automatic takeout, conveyor, and lehr were perfected and after a fire-finishing process was discarded.[45] According to the production figures compiled by the U. S. Bureau of Labor Statistics for 1923–1925, the fifteen-arm machines occasioned direct labor costs which averaged about 7 or 4 per cent (depending upon whether single or double molds were used) of the direct labor costs involved in making small bottles in "ideal" hand shops.[46]

These fragmentary data relating to worker productivity and direct labor costs, of course, are merely suggestive of the reductions wrought by the introduction and later modification of the Owens machine in the total costs of producing bottles by hand processes. In order to determine exactly the total savings realized, capital and other costs in addition to wages would have to be considered; but, unfortunately, suitable statistics are not available. On the one hand, it may be pointed out that the Owens process offered appreciable economies in fuel and certain overhead costs because it allowed production to be continued throughout the year and because it eliminated the small furnace or hearth used in hand shops for reheating each bottle before the mouth

---

molds on each arm, and (3) the greater speed with which the machine rotated. The development of the dipping head on each arm was primarily responsible for the increased speed. See below, Appendix B.

[44] He also stated that his latest model could produce 43,200 fruit jars every twenty-four hours, whereas one blower ordinarily made 200 in eight hours. The vice-president of the Glass Bottle Blowers' Union corroborated Owens' statement by testifying before the Tariff Commission that an Owens machine could make 350 gross of pint beers in 24 hours and displaced about 54 skilled workmen.—Tariff Commission, *Glass Industry as Affected by War*, p. 30.

[45] Owens-Illinois Glass Company, Owens' Private Files.

[46] *Productivity of Labor in the Glass Industry*, pp. 51–53. Direct labor costs on the ten-arm machines averaged about 10 and 7 per cent (for single and double molds, respectively) of the direct labor costs occasioned by hand production. The advent of the bottle machine had already caused bottlemakers' wages to tumble before these estimates were compiled. See below, p. 206.

and lip were tooled. The promoters pointed out in their 1903 prospectus that the customary summer shut-down of two months would no longer be necessary because the workmen around the machine were never subjected to intolerable heat; that bottles could be made twenty-four hours a day instead of during the usual two shifts of ten hours each (eight and one-half working hours); and that the manufacturer would no longer be subject to work stoppages caused by strikes of their skilled blowers. On the other hand, the Owens process necessitated very large initial outlays of capital for plant, furnace, revolving pot, machines, and lehrs.[47] The royalties and original license fees stipulated in each contract covering the use of the machine, furthermore, were costs to each licensee; and, as far as the Owens Bottle Company itself was concerned, interest on the capital invested in developing the machine was in the nature of a fixed charge.

4. The Graham Bottle Machine

Late in 1903, when E. D. Libbey first invited several manufacturers to witness the Owens machine in operation at Toledo, American glasshouses were still producing their entire output of bottles by hand. Subsequently various semiautomatic machines appeared on the market, and by 1915, they were offering the Owens company somewhat stiffer competition than had skilled blowers in 1903. The construction and installation of these new devices entailed a cost which was only a fraction of that involved in acquiring an Owens machine. It is true that their operation required the services of several persons, but bottle workers had already accepted a great reduction from their pre-1909 wages.[48] Manufacturers using these semiautomatics and hiring laborers at reduced rates were able to narrow the gap in production costs which had formerly separated them from producers operating under licenses obtained from the Owens Bottle Company. But a sizable gap in costs still remained. As M. J. Owens advised his fellow directors,[49] there really was no need for them to become

[47] See above, p. 103.

[48] That is, a 20 per cent reduction on many lines in 1909 and another 20 per cent cut in 1912. See below, p. 206.

[49] On June 20, 1911. Owens-Illinois Glass Company, Owens Bottle Company Minutes, II, 72.

alarmed over any semiautomatic. The only thing which could successfully challenge the favored position enjoyed by their own machine in the bottle trade would be another automatic device for gathering and forming glass. The Graham Glass Company of Indiana appeared to be on the threshold of providing just such a device in 1915.

The Graham machine, as developed somewhat earlier, was only a semiautomatic forming machine which rotated intermittently on its central axis and needed skilled gatherers to feed it. Basically, its operation was relatively simple.[50] It was neither much better nor much worse than other semiautomatics; but when its owners appeared on the verge of developing an automatic glass feeder to go along with it, it suddenly loomed as a distinct threat to the Owens machine. The Owens Bottle Company, therefore, decided on December 2, 1915, to try to purchase all patents in bottle machinery owned by the Graham Glass Company. E. D. Libbey was unable to buy the patents in question,[51] but he did arrange to purchase the Graham company outright. On June 28, 1916, a final agreement was reached.

At that time the Graham feeder only showed promise of becoming a commercial success.[52] Technically, it would be classified by glassmen as one of the "gravity-flow stream feeders." It utilized the pull of gravity for drawing a small continuous stream of molten glass from the furnace into a metal cup placed directly over the inverted mold of the forming machine. When the cup had ac-

[50] A gatherer sheared a gob of glass into an inverted blank mold, which was then righted and replaced with a blow mold. After compressed air inflated the bottle, it remained in the mold for a few seconds to harden before a worker lifted it from the machine and placed it in the lehr.

[51] Owens-Illinois Glass Company, Owens Bottle Company Minutes, IV, 282–283, 289.

[52] M. J. Owens, around 1920, wrote that the machine was really a semiautomatic in 1916 and that it was his company which converted it into an automatic.—*Ibid.*, Owens' Private Files. When the Owens Bottle group was later placed on the defensive during the lawsuit brought by the Europäischer Verband des Flaschenfabriken Gesellschaft mit beschränter Haftung (see above, p. 123, n. 19). it attempted to minimize the significance of the part it—and especially M. J. Owens—had taken in developing the machine. It pointed out that the Graham Glass Company was using a gravity or flow-fed machine for forming bottles in its Evansville (Ind.) shop before June, 1916, and that the machine was thus in commercial use despite its crudities before the Owens company took it over. But the Owens Bottle directors had not considered it a commercial success when they had authorized the Graham consolidation, and no other manufacturer had installed the machine in his plant.

cumulated a sufficient charge of glass, it divided, dropped the glass into the mold, and then closed again.[53] Since flow through space and contact with the metal cup chilled the glass, it was not a homogeneous mass when it fell into the mold but rather a lump of coils encased in thin "skins." [54] The bottles blown of glass gathered in this manner, therefore, had crinkles around their necks and streaks throughout. The Owens Bottle Company was never able to overcome completely this and other drawbacks inherent in the Graham machine. It spent approximately $409,000 trying to improve the forming mechanism as well as the feeder.[55]

After its efforts had met with some success, several bottle manufacturers in the United States asked for licenses.[56] At least three actually received them. E. D. Libbey reported to the Owens Bottle stockholders in September, 1918, that fifteen Graham machines were then in operation,[57] and a year later the foreign patent rights covering the process were appraised at $500,000 and entered upon the books of the company at that figure.[58] By 1920, however, the Owens directors had begun to doubt whether the machine would ever be able to compete as effectively as they had anticipated with feeder models developed by others. Actually it never did.[59] Only eight were still being operated in the United States in 1924, and seven of these were located at plants of the old Graham company. The other machine was at Owens' Libbey Street factory. One cannot deny that the industry wanted some such machine as the Graham. Very costly, the Owens' bottle machine was most suitable for making containers of standard sizes

[53] The two patents held by one Louis Proeger broadly covered the divided-cup principle. Graham's was really a modification of the Proeger cup. — Owens-Illinois Glass Company, Europäischer Verband Lawsuit Files, *Hearings*, pp. 369, 478.

[54] The physical properties of molten glass are such that a thin skin forms on its outer surface whenever it is chilled.

[55] *Ibid.*, pp. 466–467.

[56] M. J. Owens reported to the directors on October 17, 1916, that the Graham pouring-device machine was "very successful," that he was "designing an improvement," and that bottle manufacturers were "anxious to obtain such machines." — Owens-Illinois Glass Company, Owens Bottle Company Minutes, V, 268.

[57] *Ibid.*, Annual Reports of Owens Bottle Company, September 30, 1918.

[58] *Ibid.*, Owens Executive Committee Minutes, II, 329.

[59] Since it could not compete with other machines, American manufacturers who had obtained licenses from the Owens company quickly surrendered them. — *Ibid.*, Europäischer Verband Lawsuit Files, *Hearings*, p. 1,211.

and shapes in extremely large quantities. The Graham eight-arm machines, on the other hand, cost about one-fourth as much as the Owens fifteen-arm [60] and could be installed in relatively small factories without great expense.[61] Furthermore, it was more economical than the Owens machine in filling small orders for special kinds of bottles. When similar but more efficient machines were developed and installed by other manufacturers,[62] they crowded the Graham off the market and offered the Owens Bottle Company serious competition for the first time in its history.

5. The Westlake and Kadow Machines

To one aided by hindsight it appears to have been inevitable that the automatic gathering of glass by suction (the basic feature of the Owens bottle machine) and the principles of the semiautomatic paste-mold machine (as developed in 1894–1895) would be combined to form a completely automatic device for making bulbs and other paste-mold items.[63] The *National Glass Budget* reported on June 10, 1905,[64] that Cornelius J. Nolan of the Libbey Glass Company had done this. Although the *Budget* greatly exaggerated his accomplishments, Nolan's inventive efforts appeared so promising that E. D. Libbey and others incorporated the Westlake Machine Company in November, 1907, to carry on the experiments necessary to complete the work. By 1916 the Westlake machine was judged a commercial success.[65]

It was capable of turning out large numbers of bulbs every twenty-four hours. One model could make between 14,000 and 24,000 a day around 1918,[66] and approximately 60 per cent of

[60] The Owens fifteen-arm machine cost between $35,000 and $40,000; the Graham, between $8,000 and $9,000. — *Ibid.*, Owens' Private Files.

[61] See the voluminous data and testimony presented in the Verband Lawsuit, on file in the legal offices of Owens-Illinois.

[62] For a description of these machines, see below, pp. 183 ff.

[63] A system of cross-licensing allowed the Libbey Glass Company and the Owens Bottle Company to use each other's inventions for their respective lines of production.

[64] Vol. XXI.

[65] At first it was used entirely to make bulbs; several years later August Kadow, working for the Libbey Glass Company, successfully adapted it for blowing tumblers and other ware. See above, p. 137.

[66] *National Glass Budget*, XXXII, April 28, 1917, p. 1; XXXIV, March 8, 1919, pp. 14–15. The first successful machines built averaged close to 10,000 bulbs each day.

these would pass inspection. The Bureau of Labor Statistics calculated that the Westlake machine equipped with twenty-four molds in 1925 maintained a man-hour output which was about twenty-five times larger than that of an "average" hand shop. It occasioned direct labor costs only 4 per cent as high as those involved in handicraft production.[67] Reliable data on the labor-saving merits of the machine prior to 1920 are lacking; but the Libbey Glass Company in 1916 demonstrated that the total production costs (not just the labor costs) of one thousand bulbs were about one-fourth less on the automatic machine than on the semiautomatic developed much earlier by M. J. Owens.[68]

6. The Danner Tube Machine

The semiautomatic and completely automatic machines described thus far were based upon processes formerly performed by skilled labor. They represented merely the transformation of handicraft methods into mechanical techniques.[69] On the other hand, the Danner tube and the Colburn window-glass machines involved entirely new processes. Until 1917 all glass tubing produced in America had been made by hand. One worker took a sufficient quantity of molten glass and blew it into a short hollow cylinder with thick sides. While the cylinder still dangled from the blowpipe, a helper attached a punty to the opposite end, and the two men then pulled the glass out by walking rapidly away from each other. At any point where the diameter of the tube showed a tendency to become too small, a third worker fanned or blew upon the glass at the weak place in order to cool and harden it.[70] Obviously, this method of production rendered it extremely difficult to maintain a uniform bore in tubes and to overcome excessive breakage due to unequal cooling of the glass.

[67] *Productivity of Labor in the Glass Industry*, p. 127.

[68] See above, p. 153.

[69] The chief difference in the processes followed in shaping bottles on the Owens machine and in making them by hand was that the machine formed the mouth and lip before it blew the bottle. When skilled workmen made bottles, they finished the mouth and lip last.

[70] See, especially, the description given Thomas Gaffield by an outstanding glassman at Wheeling, West Virginia, on August 19, 1876. — Gaffield Collection, Correspondence on Glass, I, 51.

While employed by the Libbey Glass Company, Edward Danner conceived a way of drawing tubing which would avoid both of these disadvantages and at the same time allow manufacturers to dispense with the services of skilled workers. By 1920 (three years after his machine had been judged a commercial success) the Libbey Glass Company, the Kimble Glass Company, and General Electric were using it.

Glass of glutinous consistency was either ladled or allowed to flow continuously from an orifice in the melting furnace onto a downward-inclined mandrel made of refractory clay. The mandrel, slowly revolving on its longitudinal axis, thus became completely enveloped, and the glass soon began to flow slowly off its lower end. Mechanical stretchers then engaged the glass and pulled it out into the workroom on a series of rollers. There a machine cut it into appropriate lengths and sent the pieces to the lehr. Either solid glass rods or tubes could be produced in this manner. If tubes were desired, compressed air was constantly supplied through a small pipe in the center of the mandrel as the glass was drawn off.

As this brief description suggests, the tube machine was relatively small and simple. The merits of Danner's invention rest primarily upon the nature of the process evolved rather than upon some complicated integration of mechanical principles. The resultant economies in production were nonetheless important. Various estimates indicate that the Danner machine reduced the total cost of making a pound of tubing by something like two-thirds and the direct *labor* costs by 80 or 85 per cent.[71] Soon after it was perfected, it averaged 2,600 pounds of commercial electric tubing each day. About ninety workers would have been required to produce this much during their customary working days.[72]

[71] Owens-Illinois Glass Company, Owens' Private Files; *Productivity of Labor in the Glass Industry*, p. 143.

[72] Testimony in the lawsuit instituted by the Owens Bottle Company against the Libbey Glass Company in 1921 has provided much of the information presented in this section. A copy of the court hearings is preserved in some old safes in a basement at the Libbey Glass factory in Toledo.

7. THE COLBURN WINDOW-GLASS MACHINE

In 1915 all American-made windowpanes were cut from glass cylinders which had first been slit open, flattened in specially built ovens, and annealed.[73] Workmen had been doing this by hand since long before the twentieth century, and since 1903 several plants of the American Window Glass Company had operated machines which blew cylinders automatically.[74] Many inventors throughout the world at various times had imagined or attempted to develop some means of drawing or rolling sheets of glass which would prove suitable for windowpanes, but none of them had succeeded. One C. Boynton, for example, inquired in the *Scientific American* as early as March 13, 1869, why viscid glass could not be drawn out from the furnace in a thin sheet just as easily as was done in papermaking.[75] George F. Neale of the Lenox Rough Plate Glass Company replied on April 3, 1869, that the Chance Brothers at Birmingham, England, where he had formerly worked, had tried to do this about twenty years earlier. The company at that time lost "probably as much as £100,000" before it discontinued its experiments with drawing molten glass between two rollers. Neale concluded his remarks by confessing, "I believe it to be an impossibility to make sheets by passing the smelted glass between two rollers, and any one practically acquainted with the manufacture of glass would, I have no doubt, agree with me." [76] Nevertheless, others sporadically became interested in the possibility. A Belgian unsuccessfully attempted, in 1868, to interest Thomas Gaffield in a contrivance for rolling window glass on a flat table;[77] different countries had already

[73] The only other method of any commercial importance had been abandoned before the Civil War. See above, p. 16.

[74] This machine was the "Lubbers," which is described in the following chapter and in Appendix B.

[75] New Series, XX, 166. "Who is there that has capital and spunk enough to try the experiment?" he asked. "A small apparatus to run out a sheet ten inches wide, made of platinum, and set in the furnace, running out a stream of window glass, through proper orifices and annealing ovens, I believe to be a possibility, and a not far distant accomplishment."

[76] *Scientific American*, New Series, XX, 214.

[77] Gaffield Collection, Glass Journal, II, 157–159.

granted various men patents for similar devices; and throughout the remainder of the century they issued many more.[78]

At first inventors tackled the problem of rolling rather than of drawing sheets. But the rollers always scratched the glass, rendered it opaque, and left small undulations on its surface. Consequently, glass flattened in this way had to be ground and polished before it could be used in windows. In other words, workers had to subject it to the same expensive finishing processes as they did plate.[79] This drawback led some men to advocate drawing molten glass directly or indirectly from the furnace. On October 2, 1883, the United States Patent Office issued a patent numbered 285,888 to one Himan Frank of Detroit for "a process of manufacturing sheet glass, consisting of dipping the edge of a bar into the fluid glass between two diaphragms, then raising the bar vertically to the desired height by suitable mechanism, blasts of air being supplied to the sheet of glass as it is being raised." [80] Two glassmakers at Wilcox, Pennsylvania, at the beginning of the twentieth century held patents (numbered 618,851 and 635,041) which covered *continuous* drawing from a furnace.[81] Yet Irving W. Colburn, M. J. Owens, and other technicians in the employ of the Toledo Glass Company were the first to apply successfully these and other principles to the manufacture of windowpanes. The Libbey-Owens Sheet Glass Company installed several of their machines in its large new factory built near Charleston, West Virginia, in 1916–1917 and finally commenced the production of drawn window glass on a large scale.

[78] See, e.g., Fowle, *Flat Glass*, p. 45.

[79] See above, p. 21.

[80] Libbey-Owens-Ford Glass Company, Licenses and Correspondence of Toledo Glass Company, Vol. I, Report of Attorney James Whittemore to E. D. Libbey, February 21, 1902; Gaffield Collection, Scraps, III, 132–133. Gaffield remarked on the edge of the clipping: "In my opinion an impracticable idea which can never be carried out except on paper." Colburn and the Toledo Glass Company later proved his opinion faulty.

[81] Various persons tried to persuade E. D. Libbey to join a syndicate which would develop and later exploit the invention of these men. After his attorney rendered an unfavorable report on the patent situation, and after M. J. Owens discovered that the men had no machine but only some patents covering their ideas for one, Libbey refused to participate in any way in the proposed undertaking. Owens-Illinois Glass Company, Libbey Letters, I, 25, 29–30; Libbey-Owens-Ford Glass Company, Licenses and Correspondence of Toledo Glass Company, Vol. I, Report of Attorney James Whittemore to E. D. Libbey, February 21, 1902.

From the time the basic materials arrived at the factory until the finished glass was delivered to cutters at the end of the lehr, human hands touched neither the materials nor the glass. The materials were unloaded from freight cars, stored in huge bins, and in time conveyed to mechanical mixers. An automatic device then fed the batch into a huge continuous tank furnace, which melted and refined it. From the far end of the furnace the glass poured into a rather shallow tub built of refractory material. It was most essential that the glass in this tub be kept at just the proper temperature; otherwise, it would not be stretchable and pliant as the machine drew it up and over a bending roller in a continuous sheet.[82] As soon as the sheet rose from the tub, two sets of knurled rollers gripped the sheet's outer edges and thereby kept its width constant and prevented the glass from narrowing into a stringy tail. Water coolers on either side of the sheet cooled the glass and hence gave it strength for the upward pull through about two or three feet of space separating the tub from the bending roller. After it was bent over the roller into a horizontal position, the sheet was gripped along its edges by endless chains and pulled across several rollers into a long lehr. Here it traveled slowly and finally emerged at near-room temperature. A worker then broke off its edges which bore the marks of the knurled rollers and drawing chains, and another worker cut up the endless sheet into large pieces for the convenience of inspectors, pane cutters, and packers.

The operation of such a machine doubtless appears quite simple to most casual observers; and, as compared with the paste-mold and bottle machine processes, it actually is. The Colburn machine does not really form the sheet; the sheet is formed by the physical properties of molten glass as it is pulled from the tub. The machine only does the pulling and keeps the sheet at even width and thickness.[83] In spite of its apparent simplicity, however, the

[82] Until recently, the machine had to be put into reverse motion in order to start production. An iron rod or "bait" was lowered over the bending roller into the bath. When the glass adhered to it sufficiently, the bait was pulled back up, and the machine grasped the edges of the sheet. Nowadays, the bait is no longer used. Workmen simply reach into the bath with long tongs, grab the glass skin on the surface, and stretch it through the knurls and over the roller.

[83] The greater the speed of the machine, the thinner the sheet.

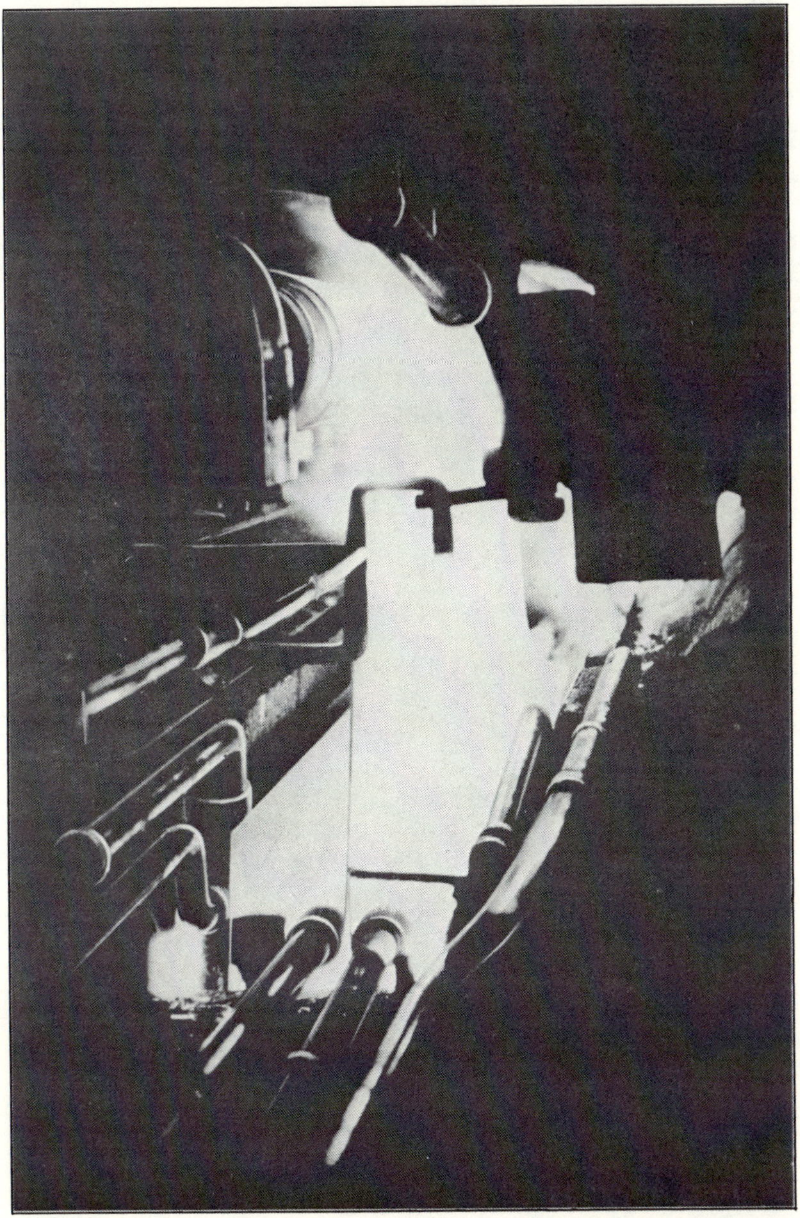

(Courtesy of Libbey-Owens-Ford)

FIG. 7. SHEET-DRAWING ON THE COLBURN MACHINE

Note the knurled rollers at the bottom and the large bending roller at the top.

process was not perfected until after several years of experimentation. How to keep the sheet of even width and how to avoid scratching or ruining the glass while it was still viscous were among the most formidable problems confronting inventors. In 1900 Irving W. Colburn had built three small machines which made sheets of such poor finish that he discarded as impracticable his idea of rolling glass.[84] Four years later he returned to his experiments,[85] interested a small group of men in forming the Colburn Machine Glass Company (1906), and subsequently constructed several new mechanical models. None of them worked satisfactorily. Marred by imperfections, the glass frequently broke in the lehr. Colburn's company became hopelessly bankrupt in 1911, and the sheriff sold its patents at public auction for $15,000 on February 8, 1912. The Toledo Glass Company was the purchaser.[86]

At that time the machine had most of its basic features. The glass was pulled up from a tub, over a roller, across a flattening table, and finally through a lehr.[87] When M. J. Owens saw it in

[84] See above, p. 125. Much of what follows is based upon data found in several volumes of records kept by Colburn and now on deposit in the Toledo offices of the Libbey-Owens-Ford Glass Company.

[85] He began to construct a new machine on April 14, 1904. This did not operate continuously, and Colburn decided to scrap it in favor of one that would. Libbey-Owens-Ford Glass Company, Colburn Records, Vol. A, pp. 114 ff.

[86] E. D. Libbey first had his attention directed to the work being done by Colburn in June, 1908, when he received a letter from H. J. Colburn, Irving's father. The writer tried to arouse Libbey's enthusiasm and financial support in Irving's invention (Owens-Illinois Glass Company, Europäischer Verband Lawsuit Files, *Hearings*, pp. 148, 644), but Libbey was only mildly interested. Eventually, it was M. J. Owens rather than Libbey who first recognized the potentialities of the Colburn process. Only when he threatened to leave the Owens Bottle Company and form his own organization to purchase, develop, and exploit the invention did Libbey and the other directors of Toledo Glass agree to buy the patents. — *Ibid.*, Owens' Private Files.

[87] The model built in 1907, together with its lehr, had been about 60 feet in length. — Libbey-Owens-Ford Glass Company, Colburn Records, Vol. VI. At first Colburn had not used knurled rollers on either side of the sheet to maintain a constant width. Instead he had had two clay disks slightly immersed in the tub. These rotated outwardly on either end of the sheet. He also had used two partly submerged rollers, extending from one side of the tub to the other, to push the glass up between them. According to Joseph Crowley, who went to work alongside Colburn and Owens in 1913, the first machine he saw in Toledo Glass's experimental shop had knurled rollers for stretching the glass but no rollers for pushing it up from the tub. This was certainly true of the 1916 model. Today, Libbey-Owens-Ford uses rollers to push up molten glass only for their plate glass machines.

operation, he recognized at once two major shortcomings. In the first place, the machine pulled the glass up about five feet before bending it horizontally over the roller. The distance was too great because the glass had time to harden too much. In the second place, the mechanical parts were not heavy and strong enough to withstand the strains and high temperatures to which they were subjected. After these faults were eliminated, others became evident. The closely meshed chain belt in the lehr, upon which the ribbon of glass rested, unduly scratched the bottom surface of the glass, and, in addition, frequently broke apart. The engineers employed by the Toledo Glass Company soon discovered that power-driven rollers, first made of clay and later of asbestos, functioned much more satisfactorily than the chain belt. The most difficult problem of all concerned the large bending roller.[88] Clay, asbestos, Monel metal, and other materials were tried, but they all scarred the glass. Finally, one made of Nichrome proved a success. All these improvements yielded encouraging results. The Libbey-Owens Sheet Glass Company, incorporated in May, 1916, soon installed drawing machines in its new plant and began production in commercial quantities on October 17, 1917.[89]

The Colburn process was much superior to all others used in making window glass at that time. Its product, devoid of ripples like those inevitably found in flattened cylinders, emerged from the lehr with a beautiful polish. The machine in operation at the experimental shop in Toledo during October, 1916, drew sheets, from 48 to 68 inches wide, at a speed of 36–60 inches per minute.[90]

[88] Technicians spent much time, effort, and money trying to pull the glass horizontally from the tub instead of up and over the roller. The clay lip of the tub, however, marked the glass and left particles of clay imbedded in it. M. J. Owens finally reported to the Toledo Glass directors on December 9, 1915, that "he deemed it useless to experiment further with drawing out the molten glass horizontally, but deemed the prospects for success in drawing it out vertically, to justify the company in building a commercial factory . . ." — Libbey-Owens-Ford Glass Company, Toledo Glass Company Minutes, IV, 64.

[89] E. D. Libbey, however, stated in his annual report to the Libbey-Owens stockholders on December 30, 1919, that the Colburn process "has now reached the stage where it can be said to be a commercial success; this could not have been said two years ago."

[90] Owens-Illinois Glass Company, Brown's Private Files; *Glass Industry*, p. 22. A machine then made daily between 200 and 250 boxes (each holding 50 square feet of window glass).

And workers at the Charleston factory late in 1920 were testing a new machine capable of pulling a sheet 84 inches wide.[91] Whereas about 5 man-hours were required to make 50 square feet of window glass by hand, the machine could produce as much while using only 0.6 of a man-hour.[92] Such economy of labor was possible because the Colburn machine was almost entirely automatic. The one developed by the American Window Glass Company in 1903 for blowing cylinders had eliminated skilled blowers, but it still needed highly paid workmen to crack the cap off and to split and flatten the cylinders.[93] The Colburn process dispensed with the services of these men as well. Each machine required no more than one or two attendants while it was forming and annealing glass, and experienced laborers had only to inspect and cut the sheets into various sizes for window lights.

8. Conclusion

From 1888 through 1920 the group of entrepreneurs and technicians at Toledo assumed the role of Nemesis as far as skilled glassmen were concerned. Workmen in one branch of the industry after another were confronted with machines which rendered their time-honored processes obsolete or reduced drastically the demand for their services. Under the duress of competition engendered by these Toledo innovations other manufacturers began to develop new devices and processes which further accelerated the trend toward mechanization in the industry. Were a person to desire at the present time to witness how glass was made, say, fifty years ago, he would be hard pressed to find a shop run in the old tradition.[94] The revolution has been extensive and is now well-nigh complete. E. D. Libbey and his associates played a leading — though by no means an exclusive — role in this tech-

[91] Libbey-Owens-Ford Glass Company, Libbey-Owens Sheet Glass Company Minutes, II, 62.

[92] Phillips, *Glass*, p. 202.

[93] See below, pp. 329–330.

[94] There are a few which make artistic tableware by hand, some which specialize in antique reproductions, and a mere handful which produce certain kinds of bottles in the old way. Side-lever presses are still employed occasionally to fill very small orders. I was unable, however, to discover a single shop which now makes crown or cylinder window glass by hand.

nological upheaval. When, in 1912, they bought the Colburn patents (which most glassmen considered absolutely worthless), the *National Glass Budget* paid them tribute. "The reputation of the Toledo people for carrying to a successful conclusion whatever they undertake is pretty generally understood, and it may be taken for granted that they would not now be backing the sheet drawing patents if they were not convinced of ultimate success." [95] This "reputation of the Toledo people" was well deserved. They were responsible for initiating and developing the semiautomatic and the fully automatic paste-mold machines, the Owens bottle machine, and the Danner tube machine. Previous experience in working glass and adapting it to machine principles enabled them to complete Colburn's work and to make the Graham bottle machine fully automatic. In regard to the former they were eminently successful; as far as the Graham machine was concerned, they had finally to admit that competitive models were superior. On the whole, their record of achievement in the field of technology was supreme.

[95] Vol. XXVIII, August 17, 1912.

CHAPTER VII

TECHNOLOGICAL INNOVATION ELSEWHERE

The members of the Toledo Group were not the only manufacturers in the United States who conducted experiments with various kinds of glassmaking machinery during the four decades following 1880. It is true that they were the most original and successful; but one should not overlook the fact that many others were also engaged in similar activities. In order to gain the proper perspective for evaluating accurately the technological contributions of M. J. Owens, E. D. Libbey, and their associates, it is necessary to understand the nature and significance of the progress achieved by others. As in all fields, technological advance in the glass industry rested upon the accumulated accomplishments of the past. "Inventive genius" consisted largely of foresight and imagination which permitted the individual to combine theretofore separate devices and principles into a workable machine.

1. The Technology of Glassmaking Around 1880

The blowpipe, invented sometime during the three centuries immediately preceding the Christian Era, was still the indispensable tool of most glassmakers in 1880. Skilled workmen blew all glass containers — such as bottles, jars, and jugs — from gobs or gathers of molten glass collected from the furnace on the end of blowpipes. They shaped the body, shoulders, and neck of each item by blowing the glass in molds of one sort or another and then fashioned the mouth and lip with hand tools. Workers cut windowpanes from squares of glass which they had first blown as cylinders and then had slit and flattened out in special ovens. Glassmakers, using simple hand tools or paste-molds, also blew and shaped lamp chimneys and other items for household use. The blowpipe was unessential in only two branches of the industry — pressed tableware and rolled plate glass.[1] Both pressing and rolling (or casting) had long been practised in glass-houses. Certain manufacturers in France had developed the art of casting plate glass in 1687–1688, and others in America had employed the side-lever press during the 1820's. These processes only indirectly paved the way for the onrush of mechanical innovation after 1890. They acquainted glassmen with some of the problems always encountered in working glass with metal machinery, and some of the early jar machines actually utilized the principle of pressing. Other developments, however, had a more direct bearing than these in timing the mechanical revolution.

The introduction and widespread use of the continuous tank furnace, patented in 1870 by Werner von Siemens' brothers in England,[2] for example, formed an important part of the setting.[3] A factory in Poughkeepsie, New York, in 1879–1880 was the first to build one in the United States. Although American producers were slower than Europeans in recognizing the relative merits of the continuous tank over the pot furnace,[4] tank furnaces became increasingly popular and numerous during the decades immediately following 1880.[5] By 1920 only those manufacturers producing plate and lead glass employed pots to any great extent.[6]

Heating furnaces with natural or producer gas in place of direct firing with anthracite or bituminous coal was another innovation which greatly favored the trend toward complete mech-

[1] For a description of these techniques, see above, pp. 19–20.

[2] See above, p. 24.

[3] For an explanation of its bearing upon the introduction of machines, see below, Appendix B.

[4] See above, p. 25.

[5] V. S. Clark (*Manufactures in U.S.*, II, 500) states that, although window-glass producers in the United States waited "more than twenty years" after Europe had introduced the tank, within six years after 1888 more than one-half the window glass made in American factories was melted in tank furnaces.

[6] Until a few years after 1920, when continuous rolling superseded the time-honored intermittent casting operation, pots were used in melting plate glass. It may be recalled that workers poured the molten glass onto the casting table directly from each pot. Direct exposure to fire reduces lead and thereby darkens all glass containing lead. Hence, to the present day manufacturers melt lead glass in covered crucibles.

anization. Not only was gas a cleaner and more efficient fuel than coal, but it also permitted manufacturers to exercise a greater control over furnace temperature than had formerly been possible. The latter advantage was extremely important. Working glass mechanically requires, among other things, that the glass have a certain viscosity when it is gathered so that each gob can be blown and then allowed to "set" or harden at regular time intervals. Window glass, especially, could be successfully blown on cylinder machines or drawn out in a continuous sheet on the Colburn machine only if it remained at precisely the temperature which caused the glass on the upper surface of the tub or pot to form a thick skin. It was relatively easy to achieve this goal with gas and, furthermore, to direct the fire to any one part of the machine or surface of glass where reheating was necessary.

An English factory in 1861 installed the first gas furnace of any commercial importance.[7] By 1880 there were twenty-one in the United States. During the following four decades Americans discovered natural gas in first one state and then another in the Middle Western area; and wherever they did, glass factories sprang up in mushroom fashion. By 1920 most plants were burning either natural or producer gas, and some had already turned to petroleum.

Still another innovation which affected the development and improvement of glassmaking machinery was the generation and harnessing of electrical power for industrial uses.[8] Electric drives are direct and hence eliminate the transmission mechanism of belts, shafts, and pulleys which are essential to steam engines. Having an individual motor for each machine, furthermore, increases the possibility of controlling the machine's speed and accuracy. Workers manipulated the first semiautomatic bottle and jar machines by means of hand levers or compressed-air treadles; but after 1900 electricity was increasingly employed to

[7] See above, pp. 24 ff.

[8] According to my former colleague Professor W. Rupert Maclaurin, electric motors were first used for industrial purposes in America around 1873, the first steam turbine was built in 1882, and the potential energy of Niagara Falls was tapped in 1895.

move the parts of all kinds of pressing and blowing machinery. Finally, it should be noted, the machine-construction and machine-tool industries in the United States had already made rapid progress. Engineers engaged in building glass machines hence did not find the synchronization of moving parts a major difficulty in their work.

2. The Development of Bottle and Jar Machines

With the exception of forming tableware by means of relatively simple side-lever presses and of casting plate glass on tables, glassmakers first applied mechanical principles to the production of jars and other wide-mouth containers. This was a logical development since the first successful jar machines employed pressing to form the "blank"[9] and finish the jar's mouth and lip and then used compressed air to blow its body to desired size. In 1881 Philip Arbogast, a practical glass manufacturer of Pittsburgh, obtained a patent covering the basic principles of this press-and-blow process,[10] but he was unable to develop a practicable machine. It was not until 1892–1893 that others succeeded where he had failed and definitely introduced into American shops the first ones for making jars.[11] Thereafter, several inventors and companies occupied themselves with developing similar machines.[12]

The tempo of their activities led the editor of the *National Glass Budget* to exclaim on June 25, 1898:

> Of the making of machinery to displace the 'gentilhomme de verre' there is now no end. Fact is that inventors, infringers and imitators

[9] The blank corresponded roughly to the partially blown bottle formed in the initial stages of hand production. The blank was an almost solid gob of viscous glass which had been partially shaped in a special mold.

[10] Davis, "American Glass Industry," p. 303; Barnett, *Machinery and Labor*, p. 68; *National Glass Budget*, XXVI, May 14, 1910; Libbey-Owens-Ford Glass Company, Colburn Records, Vol. B, 79.

[11] See below, Appendix B.

[12] In 1896 manufacturers asked the bottle blowers' union to reduce its basic wage rate by 25 per cent because the jar machine threatened their operating profits. The union refused (Barnett, *Machinery and Labor*, p. 81), but it granted a 45 per cent reduction two years later and soon thereafter removed all limits on the earnings and output of individual machine workers (U. S. Commissioner of Labor, *Regulation and Restriction of Output*, pp. 636, 637).

are tumbling over each other in their hot haste to get there first, regardless, seemingly, of how the winnowing is to end in the rock-walled palace of justice, so-called . . . Branch after branch of the glass industry is being invaded by machinery, production is being immensely increased, prices are being reduced below the living point, everything is being specialized, the industry is being concentrated, and the future is uncertain and full of dark problems upon which few can shed light.

But it is easy to invent now. So many patents have expired, so many failures have been made, and so many wrecks line the path of industrial progress, that one needs only to read up and look around, for after all, we profit most from failures — our own and others.

At the time that this exaggerated statement was published, manufacturers were making only wide-mouth jars on machines. Semiautomatics of one sort or another were responsible for most fruit and packers' jars produced in the country near the end of 1901,[13] but no one had yet been able to make bottles without employing skilled blowers.

The Owens bottle machine, which in 1903 both gathered and formed glass, was the first to produce small-mouth ware.[14] Ashley and Arnold, two Englishmen who had never worked around glass factories, in 1886 had made an earlier attempt in this direction.[15] The difficulty which they and others always encountered while trying to make bottles mechanically was literally the bottleneck. The orifice of bottles was too small for the worker to form the mouth and lip by using a plunger to press the glass up and around the mold's neck. He could do this with jars but not with bottles. Ashley and Arnold, therefore, tackled the problem from a different angle; [16] but experiments with their machine proved wholly unsatisfactory in both England and the United States.[17] Several other Europeans built

[13] *National Glass Budget*, XVII, December 28, 1901.

[14] It preceded by four or five years the first successful semiautomatic for bottles and by more than twelve years the first competitive automatic.

[15] Arnold was a mail carrier, and Ashley was manager of certain iron works. The two jointly received the so-called "Ashley" patent.

[16] See below, Appendix B.

[17] In 1893 tests were run in a plant at Woodbury, New Jersey, and possibly by the Whitney Glass Works in a branch factory at Vineland, New Jersey. — Henderson, "The Glass Industry," *Popular Science Monthly*, XLII (1893), 591–592; Atwater, "Glass-Making in America," *Engineering Mag.*, IV (1893), 885–886; Gaffield Collection, Scraps, III, 147.

narrow-mouth machines, and some were operated more or less successfully for a few years in certain Continental factories.[18] In 1900 Grote and other promoters attempted to bring one of these to the United States.[19] Despite all the earlier work of inventors and promoters on both sides of the Atlantic, however, it was not until 1907–1908 that the first satisfactory semiautomatic machine for making bottles, as distinguished from jars, was constructed in America. This was variously called the "United," "English," or "Johnny Bull" machine.[20] By 1915 the country had 114 at work in its factories. The advent of other similar types such as the "Jersey Devil" and the development of automatic feeding devices caused the United to decline in popularity. By 1919 there were only 10 such machines in operation.[21]

Whether the early semiautomatics were used for bottles or jars, they did not displace most workers previously engaged. It may be recalled that three skilled persons and four helpers composed

[18] In 1898–1902 Grote in England, Boucher in France, Hilde and Severin in Germany, and Christiansen in Sweden all tried to perfect devices for blowing bottles. — "The Johnny Bull," *Glass Industry*, XXV (1944), pp. 63–65; *National Glass Budget*, XVIII, August 30, 1902, and March 28, 1903; Rosenhain, *Glass Manufacture*, p. 100. Severin's and Boucher's, apparently, met with somewhat greater acceptance than did Grote's, Hilde's, and Christiansen's.

[19] The Green Bottle Machine Company of New Jersey was organized in 1900 by Louis Grote and some officials of the Whitall-Tatum company at Millville. — *National Glass Budget*, XVI, July 28, 1900, and March 2, 1901. The venture, however, was not successful.

[20] Pearce Davis gives 1907 as the date. — "American Glass Industry," p. 320.

[21] English patents issued in 1902 covered the United machine. On March 26, 1908, certain promoters incorporated the United Bottle Machine Company in the United States and began to license each machine for a period of seven years and for $500 annual rental. The rise and fall in popularity of the "United" is clearly reflected in the following figures showing the number used in American factories at different times:

1909–14	1912– 93	1915–114	1918–19
1910–38	1913– 94	1916– 94	1919–10
1911–71	1914–113	1917– 83	

If the United Bottle Machine Company was to compete with the Owens group, it eventually realized that it would have to develop some kind of a feeder to make its own machine completely automatic. By January 2, 1918, over $200,000 had been fruitlessly spent on engineering research. The company declared bankruptcy on May 4, 1918, and its assets were sold by receivers in 1920. — "The Johnny Bull," *Glass Industry*, XXV (1944), pp. 63–65, 78; *National Glass Budget*, XXV, July 24, 1909; XXX, September 12, 1914, p. 1; XXXIV, May 18, 1918, p. 1; *Glass Industry*, p. 211.

the "shop" of men working each pot in hand factories. Certain jar machines employed one gatherer, one presser, and two helpers;[22] the operation of the United bottle machine required two pressers, one gatherer, and two helpers.[23] The great advantage of these machines initially was that semiskilled and unskilled labor could work them, rather than that they reduced greatly the number of man-hours required to produce a given output.[24] Various improvements on the semiautomatics after 1910, however, definitely enhanced worker productivity. The number of molds in each machine increased; electrical power allowed the movement of the various parts to be synchronized; various men perfected automatic means for shearing glass from the gatherer's punty, thereby displacing a skilled worker; mechanical devices transferred the glass from one mold to another; automatic "take-outs" removed the finished product; and continuous conveyors carried the ware from the machine to the annealing lehr.[25] Each of these improvements rendered more and more automatic the process of making glass containers. The "three-man, two-boy" machine gave way to the "two-man, two-boy," then to the "one-man, one-boy," and finally to the "one-man, no-boy."[26] The best of these needed no helpers or "boys" and required only the services of one gatherer (or three gatherers

[22] These were the machines developed by the Atlas and Ball Brothers glass companies. — Barnett, *Machinery and Labor*, p. 68; U. S. Commissioner of Labor, *Regulation and Restriction of Output*, p. 636.

[23] *National Glass Budget*, XXX, September 12, 1914; "The Johnny Bull," *Glass Industry*, XXV (1944), pp. 63–65, 78; Davis, "American Glass Industry," pp. 305–306.

[24] *Productivity of Labor in the Glass Industry*, pp. 34–35.

[25] *Ibid.*, pp. 34–35; Davis, "American Glass Industry," p. 309; *National Glass Budget*, XXXIII, August 18, 1917; Owens-Illinois Glass Company, Soubier's Memorandum.

[26] The president of the Glass Bottle Blowers' Association of the United States and Canada presented data at its forty-second annual convention which showed that the one-man machine was rapidly gaining ascendency over those machines requiring two or three men.

	Number Used during the 1916–17 Season	Number Used during the 1917–18 Season
3-man machines	43	9
2-man machines	113	97
1-man machines	272	301

National Glass Budget, XXXIV, July 20, 1918, p. 1.

for every two machines). With the introduction of the "No-Boy" machine in 1917 [27] the process of shaping bottles became completely automatic. Skilled laborers, however, still had to gather glass and feed it to the machines. Until someone developed automatic feeding devices, the Owens machine remained in a class by itself.

Even before 1900 several men in both Europe and America had tried to transfer molten glass from the furnace to the molds by various means.[28] Dippers, mechanically manipulated gathering rods, spouts both with and without cocks, compressed air ejectors, and revolving cylinders with sockets for collecting glass were all patented, tested, and found wanting. The first process which held any promise of success was the one for which an American named Homer Brooke received a patent on March 31, 1903.[29] But he was not successful in putting it into practicable use. After 1907 a few companies making glass jars developed their own variation of the Brooke flow feeder.[30] Between 1911 and 1915 the Graham Glass Company, under a Brooke and a Louis Proeger license, built a feeder for its bottle-forming machine.[31] The potentialities of this device really

[27] Two machine companies, the O'Neill and the Lynch, each brought out a "No-Boy" in that year. — *Ibid.*, XXXIII, August 18 and November 17, 1917.

[28] *Glass Industry*, p. 16; *National Glass Budget*, XXXIII, February 16, 1918, p. 4; Libbey-Owens-Ford Glass Company, Licenses and Correspondence of Toledo Glass Company, Vol. I, Report on the Owens Bottle Patents by James Whittemore, October 30, 1906.

[29] *National Glass Budget*, XXXIV, June 15, 1918, p. 2; August 31, 1918, p. 1; March 1, 1919, p. 1. He had filed his patent application on March 3, 1898. According to his process, glass flowed in a small steady stream from the furnace into a mold placed directly under the spout. After a sufficient quantity had flowed into the mold, shears severed the stream of glass. Then the blowing machine revolved one notch and carried another mold directly under the furnace spout. Meanwhile, glass was collecting on the closed broad blades of the shears. Once the second mold was in place, the blades opened, the accumulated gob of glass dropped into the mold, and once again the stream from the furnace flowed into the mold unimpeded. — *Ibid.*, XXXIII, February 16, 1918, p. 4; XXXIV, August 31, 1918, p. 1; *Glass Industry*, pp. 15–16.

[30] E.g., Hazel-Atlas, Ball Brothers, and the Travis Glass Company. — Owens-Illinois Glass Company, Owens Bottle Company Minutes, IV, pp. 292 ff. See also Soubier's Memorandum. Since Brooke's patent broadly covered all stream feeders utilizing some method of interrupting and husbanding the flow of glass, they had to secure a license from him.

[31] This was the first adaptation of the Brooke feeder to narrow-mouth ma-

alarmed the directors of the Owens Bottle Company because, if successful, it would challenge the unique position long held by their own machine in the bottle — as distinct from the jar — field. They hastened to buy up all the remaining rights under the Brooke patent [32] and soon thereafter purchased the Graham company outright.[33] Meanwhile, trouble was brewing elsewhere.

The Beechnut Packing Company, the Monongah Glass Company, and a group of Hartford engineers in 1912 had incorporated the Hartford-Fairmont Company with the aim of perfecting some way to make automatically glass jars suitable for the vacuum-packing process employed by Beechnut.[34] Over a period of years Hartford-Fairmont experimented with various gob feeders,[35] all of which were superior to the different stream feeders stemming from Brooke's patent.[36] According to one of the early models, a paddle lifted the molten glass over the end of the flow-trough connecting with the furnace, and then a knife cut the gob clean of the remaining tail of glass. In a later model, a revolving "needle" or plunger, combined with the paddle, gave each gob of glass more of a pear shape. The Monongah Glass Company at Fairmont, West Virginia, first used the Hartford feeder in 1915 in the manufacture of jars and soon thereafter

---

chines. — *Ibid.*, Soubier's Memorandum. Louis Proeger had patented a dividing cup which held the glass while the forming machine shifted its molds.

[32] In December, 1915. — *Ibid.*, Owens Bottle Company Minutes, IV, 292 ff., 312 ff.

[33] See above, p. 111.

[34] From the testimony and exhibits presented in Part II of the T.N.E.C., *Hearings* and from Hartford-Empire's "Brief" in *Hartford-Empire Company et al.* v. *United States of America* (especially, pp. 16 ff.).

[35] *National Glass Budget*, XXXV, March 20, 1920, p. 1; April 3, 1920, p. 8. The T.N.E.C. hearing and exhibits, along with the various briefs submitted before the Supreme Court in October, 1943, constitute the most easily available source of information on the development of the gob feeder and the growth of Hartford-Fairmont (1912–1922). Hartford-Fairmont became Hartford-Empire in 1922.

[36] Their superiority was due to several advantages. Even the best stream feeders were messy and could not keep strings of glass from falling on the forming machine. The glass was not homogeneous when it fell into the mold but was rather a coiled mass. A thin skin always forms around glass as it flows through air, and scars are also left on it through contact with such things as metal husbanding cups or metal blades. These facts caused stream feeders to leave visible blemishes in the finished bottle or jar. Stream feeders, furthermore, did not preliminarily shape the glass in a desirable fashion. Gob feeders, on the other hand, were not messy and delivered to the molds a homogeneous mass of glass already desirably shaped like a pear.

attached it to bottle machines too.[37] It worked so well that most manufacturers came to consider the Hartford feeder better than any of the other types for which competitors had filed patent applications in increasing number after 1914.[38]

The automatic feeder showed such improvement by 1918 that its combination with any automatic bottle-blowing machine gave promise of becoming a vitally important factor in the industry's growth.[39] It has been estimated that in 1917 the Owens machine accounted for about 50 per cent of all glass containers made in the United States [40] and that skilled blowers and semiautomatics were responsible for nearly all the rest. During the production year 1922–1923 completely automatic machines, including the Owens and all those equipped with feeding devices, made around

[37] *National Glass Budget*, XXXIII, August 25, 1917, p. 1; XXXIV, March 8, 1918, p. 3; *Hartford-Empire Company et al.* v. *United States of America*, Hartford-Empire's "Brief," p. 17.

[38] Davis, "American Glass Industry," p. 310; Owens-Illinois Glass Company, "History," p. 35; *National Glass Budget*, XXXIII, June 30, 1917, p. 1; August 25, 1917, p. 1; Tariff Commission, *Glass Industry as Affected by War*, p. 137. The frantic search for good feeders after 1914 was largely due to the fact that the speed of the semiautomatics was being increased so greatly that one or two gatherers could no longer keep up with them.

[39] The success attending Hartford-Fairmont's activities spurred the Owens Bottle Company to create a strong patent position for itself in the feeder field. In 1918 it brought certain patent applications which one Clyde Lott had pending on feeding mechanisms; unsuccessfully brought suit against Hartford for infringement of the Brooke patent in 1919; and acquired exclusive license under other feeding patents (Brookfield's) in 1921. Eventually, in 1924, Owens Bottle effected a compromise with its rival. Owens agreed to issue Hartford a royalty-free exclusive license (with right of sublicense) on all its feeder patents, to refrain from licensing other manufacturers in the future, and to pay the standard royalty rates on any and all feeder devices belonging to Hartford which it might care to use. Hartford, in turn, agreed to pay Owens each year one-half of the gross royalties over $600,000 which it would receive from all licensees. Hartford did not obtain any rights on the Owens suction-fed machine. It was this compromise which formed one of the bases for the anti-trust suit recently brought against the Owens-Illinois Glass Company and the Hartford-Empire Company by the United States Department of Justice. — Owens-Illinois Glass Company, "History," pp. 30 ff.; *National Glass Budget*, XXXIV, February 22, 1919, p. 1; XXXV, December 20, 1919, pp. 1 ff.; T.N.E.C., *Hearings*, Part II; *Hartford-Empire Company et al.* v. *United States of America*, Hartford-Empire's "Brief," pp. 31 ff.

[40] Barnett, *Machinery and Labor*, p. 85. The editor of the *National Glass Budget* (XXVIII, August 24, 1912) had stated that the Owens machines installed in American plants in 1912 "could produce" 35 per cent of the nation's output of bottles and jars; in January, 1918, the Tariff Commission learned that they made about 45 per cent of the total (*Glass Industry as Affected by War*, p. 29).

80 per cent of the nation's output.[41] Much of this increase from 50 per cent to 80 per cent reflected the increased number of feeder types in use. Unfortunately, it is impossible to determine just how many were operating in American factories by the end of 1920. According to reports rendered by the president of the bottle blowers' union, ten factories had flow- and gob-feeding devices in August, 1917; eighteen had them a year later; and twenty-three had them in 1919.[42] Pearce Davis has estimated that in 1918 all Owens machines made about 10,000,000 gross of bottles and jars while other automatic types blew only about 500,000 gross.[43] Seven years later, he states, the Owens accounted for roughly 12,500,000 gross and the gob-feeders for approximately 8,500,000 gross.

As both kinds of automatic machines increased in importance, those fed by skilled gatherers declined in number. Barnett[44] has suggested that factories were working altogether approximately 150 semiautomatics for wide-mouth ware in 1903, the year that the Owens Bottle Company made its appearance, and roughly 250 in 1905. Sometime before 1915 their popularity reached its peak.[45] The reader may recall that bottles were first made on semiautomatics in the United States in either 1907 or 1908.[46] The number of such machines for narrow-mouth ware rose approximately from 20 to 300 between 1909 and 1916. Since the various feeding devices developed by Hartford and others after 1915 could be installed without much trouble or expense at factories already equipped with forming machines, manufacturers did not hesitate to do so and thereby shift from semiau-

[41] Barnett, *Machinery and Labor*, p. 85.

[42] *National Glass Budget*, XXXIII, August 25, 1917, p. 1; XXXIV, July 20, 1918, p. 1; XXXV, November 1, 1919, pp. 1 ff. The figures include factories in both the United States and Canada, but only one or two were in Canada. In 1918 Hartford-Fairmont had 18 of its feeder machines installed in 15 places, and there were at least 3 other types operating elsewhere. The editor of the *Budget* claimed that the list compiled by the union's president had omitted several factories.

[43] "American Glass Industry," p. 315.

[44] *Machinery and Labor*, p. 69.

[45] It is difficult to discover the exact year because the figures available for 1906–1915 include only those machines which were under union jurisdiction, and the union during that time was extending its control over the industry.

[46] See above, p. 180.

tomatic to completely automatic processes. Consequently, it is not surprising that the number of wide-mouth and narrow-mouth forming machines fed by hand declined from 459 in 1916 to 315 in 1920 and to 72 in 1924.[47]

Although these figures on the number of different kinds of machines used at various times are fragmentary, they reveal the basic trends. Until around 1914–1916 the semiautomatics made great progress, first, in the jar field and, somewhat later, in the bottle and narrow-mouth field. Many were making wide-mouth containers before the Owens bottle machine appeared in 1903; thereafter, they were unable to compete effectively with it since it gathered as well as blew glass. The number of semiautomatics in both bottle and jar factories, however, continued to increase after 1903 mainly because the Owens Bottle Company pursued a limited licensing policy and because it was unable to have machines built fast enough to supply the rapidly increasing demand for glass containers of all sorts.[48] Various individuals after 1915 successfully developed feeding devices which could be used to convert semiautomatic machinery into full-fledged automatics. As a result, the number of forming machines still fed by hand declined, and for the first time since its appearance in the bottle trade the Owens machine encountered formidable rivals. Production by hand meanwhile constantly dwindled in importance until only novelties and special kinds of bottles ordered in very small quantities were blown by skilled workmen.

The United States Bureau of Labor Statistics attempted in 1927 to ascertain how greatly this mechanical revolution in the bottle industry affected worker productivity, and the data it presented are useful in suggesting the relative merits of different types of machines. The Bureau first estimated the man-hour output of craft workers under "ideal" conditions such as had

[47] Barnett, *Machinery and Labor*, p. 110. These figures represent the number of machines over which the bottle union had control in both Canada and the United States. Since its control over semiautomatics was virtually complete at that time, the figures probably include most machines.

[48] The production capacity of the Kent Machine Company, which made the Owens machines, was severely taxed until around 1912.

prevailed before the hand process became practically obsolete.[49] It then expressed the average man-hour outputs for fourteen different kinds of bottles made on various machines employed in 1923–1925 as percentages of the appropriate figures for hand production. To take the data as reflecting accurately production conditions, say, in 1920 would be unwarranted because the machines were being slightly improved all the time. Nevertheless, the figures are the best ones available, and they may be used to gauge roughly the increase in worker productivity without much danger of gross exaggeration. Depending upon the efficiency of the machine they employed and the kinds of bottles they made, the man-hour output of individuals operating semiautomatics was anywhere from 18 to 334 per cent greater than the man-hour output of hand workers.[50] Automatics increased man-hour output anywhere from 229 to 4,010 per cent above that of the handicraft level.[51] Crude arithmetic averages for all makes of machines and for fourteen kinds of containers suggest that worker productivity was about twice as great for the semiautomatics and thirteen times as great for the automatics as for handicraft production. Table 16 reproduces the indexes for the most efficient machines. The savings in direct labor costs realized by manufacturers averaged about 50 per cent on semiautomatics and slightly more than

[49] "The figures given are termed 'ideal,' because they really show what a 'shop' of three experienced blowers and four helpers could produce when working on any one kind of bottles for a complete eight-hour day without changing molds and under conditions at least as favorable as those under which the machines are working today." — *Productivity of Labor in the Glass Industry*, p. 45. In every case the "ideal" figure for hand shops was greater than the "actual" figures, with the resulting tendency to understate the relative efficiency of machinery. The 1923–1925 figures for machine output tended to *overstate* the relative efficiency of glassmaking machinery in the pre-1921 period. The two tendencies to some extent, therefore, were compensatory.

[50] *Ibid.*, pp. 46–48. The lower limit was set by the Teeple-Johnson machine on pint "beers," and the upper by the Lynch on pint "whiskies."

[51] The lower limit was set by the Owens "AL" single-mold on half-gallon jugs, and the upper by the Owens "CA" double triplex-mold on four-ounce prescription ovals. The Owens "CA" double triplex mold, however, was not introduced until after 1920. The upper limit for the earlier period, therefore, must have been set by another model. It was most likely the Owens "AV" double mold, whose man-hour output for four-ounce prescription ovals in 1925 was found to be thirty times greater than that for hand shops.

Table 16

INDICES OF WORKER PRODUCTIVITY FOR SPECIFIC KINDS OF BOTTLES MADE ON THE MOST EFFICIENT MACHINES *
(Base = Man-Hour Output of Handicraft Workers)

Kind of Bottle	Hand Production	Semiautomatic Machines		Fully Automatic Machines	
	Index	Index	Name of Most Efficient Machine	Index	Name of Most Efficient Machine **
½-ounce Prescription Ovals	100	—	———	2804	Owens "AV" double with conveyor
2-ounce Prescription Ovals	100	284	One-man machine and feeder	2852	Owens "AV" double with conveyor
4-ounce Prescription Ovals	100	149	One-man machine	3025	Owens "AV" double with conveyor
8-ounce Prescription Ovals	100	160	One-man machine	2641	Owens "AQ" double with conveyor
6-dram Panel Extracts	100	—	———	2877	Owens "AV" double with conveyor
2-ounce Panel Extracts	100	300	O'Neill with feeder	2512	Owens "AV" double with conveyor
½-pint Sodas	100	344	Lynch with gatherer	1642	Owens "AR" single with conveyor
1-pint Beers	100	198	Teeple-Johnson with gatherer	1949	Owens "AQ" single with conveyor
½-pint Whiskey Dandies	100	320	O'Neill with feeder	772	Lynch with feeder
1-pint Whiskey Dandies	100	434	Lynch with gatherer	742	Lynch with feeder
1-pint Milks	100	273	Miller with gatherers	1475	Hartford-Empire with double feeder
1-quart Milks	100	299	Miller with gatherers	1449	Hartford-Empire with double feeder and conveyor
½-gallon Packers' Jugs	100	147	Jersey Devil	825	Owens "AQ" single
1-gallon Packers' Jugs	100	141	Jersey Devil	815	Owens "AQ" single

* Source: *Productivity of Labor in the Glass Industry*, pp. 46–48. The data mainly are for 1923–1925, and the indices — especially for the fully automatics — have to be reduced somewhat if they are to reflect the situation in 1920.

** The Owens "CA" double triplex-mold machine in 1925 was the most efficient for the four classes of prescription ovals, but I have not listed it because it was introduced after 1920. The Hartford-Empire with double feeder also came after 1920, but it was the only Hartford listed. Even the Owens Bottle Company admitted that the machine made by Hartford in 1920 was more efficient in the production of milks than any of its own models. For these reasons it was thought desirable to include the post-1920 Hartford machine in the table for comparative purposes.

90 per cent on automatics as compared with 1925 "hand-shop" costs.[52]

Although the data collected by the Bureau of Labor Statistics show well enough the superiority of the automatics over the semi-automatics and of both over the hand process, they do not adequately reveal the respective merits of the Owens machine and the gob-fed machine. It appears from Table 16 that the Owens was more efficient than any of its rivals in the manufacture of most kinds of containers. Nonrevolving machines using gob feeders, however, could make "whiskies" and "milks" more cheaply and faster than could the revolving Owens machine. The glass sides and bottoms of milks and whiskies were thick and took considerable time to harden before the bottles could be taken from the molds. Suction gathering from a revolving pot, the process used with the Owens machine, furthermore, required that molten glass in the furnace be kept at a higher temperature than was necessary for gob feeding.[53] For these two reasons the Owens machine had to revolve slowly whenever it made milk and whiskey bottles. For nearly every other kind of container, however, it was technologically — if not always economically — superior to rival models. Its technological advantages became evident whenever manufacturers had to fill very large orders for bottles of standard size and shape; for short orders requiring oddly shaped molds the gob-fed machines were cheaper to operate than the Owens.[54] This, the main economic advantage of such machines as the Hartford, largely accounts for their increasing popularity after 1918. Even the Owens Bottle Company found it expedient to install some Hartford equipment in its factories during the 1920's.

3. The Development of Window Glass Machines

The American Window Glass Company introduced to the trade the first successful machine for blowing window-glass cylinders in

[52] *Ibid.*, pp. 51–53. Long before 1925 handicraft wages had tumbled as a result of machine competition. If the labor unions had been able to maintain the differential advantage which they had guarded for skilled glassmen before the twentieth century, the reductions in direct labor costs due to machinery would have appeared even greater in 1925 than they actually did.

[53] Owens-Illinois Glass Company, Owens' Private Files.

[54] See above, pp. 164–165.

the same year that the Owens automatic bottle machine appeared on the market. Almost fifty years earlier a Frenchman named Loup had received a patent on a process rather similar to that utilized by the American concern, but he had planned to draw and inflate glass cylinders with devices operated by hand.[55] Martin Andreas Oppermann, a Belgian manufacturer, in 1886 secured German patents on a cylinder glass machine, which, however, he was never able to develop satisfactorily.[56] Other persons, both in the United States and abroad, also tried to devise ways of displacing skilled blowers.[57]

Finally, around 1896, John H. Lubbers, an American window-glass flattener, began to experiment with cylinder machines [58] and soon attracted financial support from men interested in the American Window Glass Company.[59] The American Window Glass Machine Company, incorporated as a subsidiary enterprise on March 7, 1903, acquired all of Lubbers' patents and then licensed the parent company. It was not until 1905, however, that the machines showed enough improvement to permit effective competition with hand blowers.[60] And only after surmounting some major difficulties which subsequently emerged and after bringing to an end a fierce competitive war waged by hand producers [61] did the American Window Glass Company become profitable. It

[55] *National Glass Budget*, XXXIV, November 23, 1918, pp. 1 and 12; *Glass Industry*, p. 21; Davis, "American Glass Industry," pp. 272–273. Robinet, another Frenchman, had first invented an air pump for blowing glass in the 1820's. See above, p. 152, n. 12.

[56] *Glass Industry*, p. 21; Libbey-Owens-Ford Glass Company, Colburn Records, III, 21. The Window Glass Workers' Union in 1874 forbade its members to work with a similar machine developed by a man named Ralph Gray. — *Pottery and Glassware Reporter*, December 3, 1891, found in Gaffield Collection, Scraps, III, 148.

[57] See, especially, *National Glass Budget*, XXXIV, November 23, 1918, pp. 1 and 12.

[58] Fowle, *Flat Glass*, p. 43.

[59] A large number of window-glass firms had combined in 1895 to form the American Glass Company. Four years later it reorganized and changed its name to the American Window Glass Company. — U. S. Commissioner of Labor, *Regulation and Restriction of Output*, pp. 601–602.

[60] *Glass Industry*, p. 21; *National Glass Budget*, XXXIV, November 23, 1918, p. 1. The machines had been developed amid the greatest secrecy (U. S. Commissioner of Labor, *Regulation and Restriction of Output*, p. 612), and the first ones tested on a commercial scale had been installed in the company's plant at Alexandria, Indiana, in 1902 (*National Glass Budget*, XXXII, April 14, 1917, p. 1).

[61] See below, pp. 218 ff.

(*Courtesy of Libbey-Owens-Ford*)

FIG. 8. MANUFACTURE OF WINDOW GLASS ON THE LUBBERS CYLINDER MACHINE

began to pay the cumulative dividends on its preferred in 1913 and annual dividends on its common in 1918.[62]

Lubbers' invention simply mechanized the hand process of blowing cylinders which at that time all window-glass factories in the United States were employing.[63] Almost as soon as his appeared on the market, other inventors began to build similar machines. Two — the Healey and the Douchamp — were out by 1908, and various factories installed four other types before 1916.[64] All of these, including the Lubbers, were eventually rendered obsolete by sheet-drawing processes developed after 1910.

At about the same time that Irving W. Colburn in the United States first began his experiments with pulling glass in the form of sheets,[65] Émile Fourcault in Belgium was working independently on vertical drawing.[66] After receiving various patents between 1902 and 1904 [67] and before the outbreak of the first World War, he made appreciable progress in perfecting his machine. Whether success had been fully attained by 1914, however, is

[62] *National Glass Budget*, XXXIII, October 20, 1917, p. 1; XXXIV, June 22, 1918, p. 1. Monro told the Tariff Commission in 1918 that the American Window Glass Company and the American Window Glass Machine Company had spent "millions of dollars" in perfecting the machine and had piled up a loss before 1911–1912 "that would have staggered a banker." — Tariff Commission, *Glass Industry as Affected by War*, pp. 75, 79. The balance sheet for the parent company on September 1, 1911, showed a deficit of over $3,000,000 (or almost one-seventh of its combined assets). — *National Glass Budget*, XXVII, October 21, 1911, p. 1.

[63] See below, Appendix B. The *National Glass Budget* (XVIII, March 21, 1903), nevertheless, exclaimed that "nothing, in our opinion, has ever been introduced in any branch of the glass industry which is more far-reaching, more radical in its departure from century-old methods, or more revolutionary in its effect than the window glass blowing machine." The Owens bottle machine appeared about six months later. Reference to Figure 8 may aid in understanding the Lubbers process.

[64] These were the Okmulgee, the Douchamp-Henshaw, the Frink, and the one built by the Pittsburgh Plate Glass Company. — *Glass Industry*, pp. 21, 191. The American Window Glass Company and its subsidiary successfully sued the manufacturers and users of at least two of these machines for patent infringement in 1914–1919. — *National Glass Budget*, XXX, May 30, 1914, p. 2; XXXIV, November 2 (p. 1) and 16 (p. 4), 1918; XXXV, October 18, 1919, pp. 1 ff.

[65] See above, pp. 125 ff., 171 ff.

[66] Around 1890 Paul Simon, another European, had been trying to pull glass horizontally from the bottom of a melting tub. — Gaffield Collection, Scraps, III, 37.

[67] *National Glass Budget*, XXI, June 10, 1905, and March 31, 1906; Davis, "American Glass Industry," p. 399; Turner, "Twenty-One Years," *Jour. of Soc. of Glass Technology*, XXII (1938), 109–111.

doubtful.[68] Shortly after the establishment of peace various Belgian, English, and Czechoslovakian factories installed his machines; and between 1921 and 1923 a few manufacturers in the United States began to use them.[69] Whereas Colburn drew his glass up a few feet and then over a bending roller, Fourcault pulled his vertically all the way.[70]

All of the machines introduced to make window glass on a commercial scale were much superior to the hand process. Skilled workmen could blow cylinders of about 5 or 6 feet in length and 12 or 14 inches in diameter; [71] the ones blown on the Lubbers machine around 1917–1920 were at least five times as long and about 21 or 24 inches in diameter.[72] Although the handling of each large cylinder required the services of more men than were needed for the smaller ones produced by handicraft methods, the Lubbers machine made it possible to replace the highly-paid blowers and gatherers with unskilled or semiskilled laborers.[73] The reduction in direct labor cost involved in making window glass (especially that of double strength)[74] hence was somewhat greater proportionately than the increase in man-hour output. The Bureau of Labor Statistics has estimated that, as compared with hand production, the Lubbers machine raised worker produc-

[68] *National Glass Budget* (XXV, April 23, 1910) reported that Fourcault's factory was idle and that his financial backers had "got cold feet with the result that his process is now in the has-been class." M. Eugène Gentil of the Saint-Gobain Plate Glass and Chemical Company, however, said in 1914 that the Fourcault machine was currently described as making great progress in Belgium. — *Ibid.*, XXX, June 27, 1914, p. 13. See also *ibid.*, XXXV, August 23, 1919, p. 1. Turner gives 1919 as the year which witnessed its first commercial use. — "Twenty-One Years," *Jour of Soc. of Glass Technology*, XXII (1938), 110.

[69] *National Glass Budget*, XXXV, February 14, 1920, p. 3; *Productivity of Labor in the Glass Industry*, p. 6; Tariff Commission, *Flat Glass*, p. 31; Jerome, *Mechanization in Industry*, p. 99.

[70] See below, Appendix B.

[71] Libbey-Owens-Ford Glass Company, Colburn Records, Vol. B, 41–42; III, 6–7; see above, p. 15, n. 42.

[72] *Productivity of Labor in the Glass Industry*, p. 154. The early models in 1903 had made cylinders which measured only about 10 or 15 feet in length.

[73] *Ibid.*, p. 154.

[74] "Double" and "single" strength refer to the thickness of the glass. Single-strength glass weighed 16–19 ounces per square foot and was about one-tenth or one-twelfth of an inch thick; double-strength weighed 24–26 ounces per square foot and was one-eighth or one-ninth of an inch thick. — Tariff Commission, *Flat Glass*, p. 27.

tivity by 133 per cent for single-strength glass and by 73 per cent for double-strength; it lowered the direct labor costs by 57 and 47 per cent respectively.[75] The Colburn and Fourcault machines not only made better glass than the Lubbers, but they also displaced additional workers of appreciable skill — the cylinder "cappers," "splitters," and flatteners.[76] The Bureau of Labor Statistics was unable to obtain cost and production figures from the Libbey-Owens Sheet Glass Company for 1925–1926, but it calculated that the Fourcault machine, as compared with handicraft processes, reduced the direct labor costs by 69 per cent for single-strength glass and by 67 per cent for double-strength. The output per man-hour was found to have been increased by 161 and 228 per cent, respectively.[77]

Because of their superiority, the Lubbers, Colburn, Fourcault, and other window-glass machines eventually crowded handicraft methods off the market. The trend is readily discernible in the data presented in Tables 17 and 18. That the hand process lingered as long as it did may be explained by the "live-and-let-live" policies pertaining to pricing and marketing their output which manufacturers using machines adopted, by the reduction in wages

[75] *Productivity of Labor in the Glass Industry*, pp. 159–160. The calculations were based upon 1925 wage rates. Immediately after the machine was introduced, the wages of skilled blowers tumbled, and various restrictions on their daily output were flagrantly disregarded. (See below, pp. 222 ff.) It has been variously estimated that, as compared with the pre-1903 period, the machine increased worker productivity from five to eight fold and lowered direct labor costs by 85 per cent. — V. S. Clark, *Manufactures in U. S.*, III, 259; Libbey-Owens-Ford Glass Company, Colburn Records, Vol. B, 41–42; *National Glass Budget*, XVIII, March 21, 1903.

[76] Although the technological superiority of sheet-drawing over cylinder-drawing must have been obvious to any glass manufacturer in 1917–1920, the extreme cautiousness of the Bureau of Labor Statistics led it to conclude in 1927 that there was no clear indication which of the two would "finally prevail in the industry." — *Productivity of Labor in the Glass Industry*, p. 162.

[77] *Ibid.*, pp. 159–160. This meant that the Fourcault compared with the cylinder machine as follows:

	Index Number of Man-Hour Output	Index Number of Direct Labor Cost
For single-strength glass		
Cylinder machine process	100	100
Fourcault process	112	74
For double-strength glass		
Cylinder machine process	100	100
Fourcault process	132	62

and increase in productivity of craft workers, and by the increase in prices, curtailment of imports, and rapid expansion of exports during the first World War.[78]

4. The Development of Paste-Mold Machines

Between 1894 and 1898 M. J. Owens and the Toledo Glass Company developed the first semiautomatic machines which successfully formed incandescent lamp bulbs, tumblers, and lamp chimneys in paste-molds; [79] about twenty years later the Libbey Glass and Westlake Machine companies put on the market the first completely automatic machine for making lamp bulbs.[80] This led other concerns and inventors to build competing models;

TABLE 17

APPROXIMATE PERCENTAGE OF THE TOTAL AMOUNT OF WINDOW GLASS PRODUCED IN THE UNITED STATES BY VARIOUS MANUFACTURING PROCESSES FOR SPECIFIED YEARS *

Year	Hand Cylinder Process (per cent)	Machine Cylinder Process (per cent)	Sheet Drawing ** Process (per cent)
1900	100	None	None
1904	95	5	None
1916	40	60	None
1919	25	65	10
1923	10	52	38
1926	2	59	39
1929	None	20	80
1932	None	5	95
1935	None	None	100

* Reproduced from Tariff Commission, *Flat Glass*, p. 39. The estimates made by Pearce Davis ("American Glass Industry," p. 283) differ only slightly from the figures presented here.

** Includes the Libbey-Owens (Colburn), Fourcault, and Pittsburgh Plate sheet machines.

[78] See below, pp. 217 ff., 246 ff.

[79] No evidence has appeared which suggests that Owens' invention was not the first successful machine of its type. According to the editor of the *National Glass Budget* (XIV, June 25, 1898), "Mike Owens struck an original lead far from the beaten, footworn path of experimenters, and demonstrated that what for ages had been deemed impossible, was entirely within the reach of the practical glass blower of inventive bent of mind. . . ."

[80] See above, pp. 131 ff., 165 ff.

but since they embodied no basically important or radically novel features, it will not be necessary to describe their technological features. The superintendent of the mold shop at the Bryce Brothers' Glass Company at Mount Pleasant, Pennsylvania, invented a nonrotating, four-mold tumbler machine in 1898; [81] and the original Owens model underwent many improvements.[82] In 1916 a typical machine with five molds could make twelve or fourteen tumblers a minute, and some with double-cavity molds produced them at a more rapid rate.[83] By 1922 there was no fully

TABLE 18

NUMBER OF MACHINES USED IN THE AMERICAN WINDOW-GLASS INDUSTRY, 1903–1920 *

Year	Number of Cylinder Machines of All Makes	Number of Colburn Sheet Machines	Number of Furnace Pots or "Rings" Worked by Hand
1903	24	—	?
1904	?	—	2,280
1905	124	—	2,342
1906	?	—	2,772
1907	130	—	2,881
1908	?	—	2,699
1909	160	—	2,391
1910	116	—	2,315
1911	128	—	2,438
1912	128	—	2,257
1913	162	—	1,564
1914	155	—	1,452
1915	284	—	1,625
1916	308	—	1,913
1917	321	6	1,826
1918	312	6	1,903
1919	344	6	2,108**
1920	329	6	2,367**

* Reproduced from Jerome, *Mechanization in Industry*, p. 100.

** As a result of a scheme to curtail production, about half of these pots were worked the first six months of each year, and the remainder the second six months. —*Ibid.*, pp. 99–101.

[81] *National Glass Budget*, XIV, June 18, 1898.

[82] The Toledo Glass Company had sold all its rights and interests in the machine in a series of transactions extending from 1897 to the latter part of 1904. It, therefore, did not participate in the later improvements. See above, pp. 98 ff.

[83] *Glass Industry*, p. 63.

automatic tumbler machine except the Kadow, which the Libbey Glass Company at Toledo had developed.

The lamp-chimney machine, developed by Toledo Glass and sold to the Macbeth-Evans Glass Company, spurred the superintendent and mechanical engineer of the Pittsburgh Glass Company (a Westinghouse subsidiary) to bring out a supposedly less complicated one in 1898–1899.[84] Neither machine was ever improved to the point that it could dispense with skilled gatherers, and several factories continued to hire workers to blow and shape chimneys off-hand or in separate molds.[85] Complete mechanization never invaded this branch of the glass industry because it definitely fell into a declining state as electric lighting increasingly supplanted oil and gas lamps in American cities. Manufacturers, therefore, did not deem it desirable to tie up any more of their funds in capital equipment than were already invested in buildings and furnaces.[86]

The Empire Machine Company, a subsidiary of the Corning Glass Works, in 1915 brought out a semiautomatic bulb machine operated by electricity.[87] This machine, apparently, was not combined with a feeder (and thereby converted into a completely automatic) until after 1920.[88] The Owens semiautomatic for making bulbs, on the other hand, had appeared in 1894, and the Libbey Glass Company had begun in either 1916 or 1917 to use the automatic machine which August Kadow and the Westlake Machine Company had developed from the semiautomatic.[89] Com-

[84] *National Glass Budget*, XV, June 17, 1899.

[85] With the index of the man-hour output for the off-hand process taken as 100, similar indices for the hand paste-mold and the semiautomatic machine were estimated in 1927 to be 139 and 142, respectively.—*Productivity of Labor in the Glass Industry*, pp. 10, 113. The direct labor costs occasioned by fashioning chimneys off-hand were said to have been reduced by about 21 per cent whenever skilled blowers used paste-molds and by about 37 per cent when machines were employed.—*Ibid.*, p. 114.

[86] Cf. *ibid.*, p. 114.

[87] *National Glass Budget*, XXX, March 20, 1915, p. 1; XXXIV, June 15, 1918, p. 2. Corning, since its agreement with the Libbey Glass Company in 1894, had installed bulb machines either identical or very similar to the Owens. See above, pp. 135 ff.

[88] Turner ("Twenty-one Years," *Jour. of Soc. of Glass Technology*, XXII [1938], 110) gives 1923 as the year for the first commercial use of the automatic Empire.

[89] See above, pp. 96–97, 131–132, 136 ff.

pared with handicraft processes, the Empire semiautomatic slightly more than doubled worker productivity and reduced direct labor costs by about 70 per cent. The Westlake automatic made roughly twenty-four times as many bulbs per man-hour as hand blowers and lowered direct labor costs by about 96 per cent.[90]

5. Summary

Whereas the American glass industry in 1880 had lagged definitely behind the European in matters of technology, by 1920 it had forged to the front in respect to developing and using efficient machines, producing on a large scale, and lowering manufacturing costs.[91] The report rendered by M. J. Owens upon his return from England in 1901 could easily have been the report of a fictitious English glassmaker who might have visited the United States twenty years earlier. Owens said:

> It is hard for one to get into the England factories. They are all surrounded by high brick walls and have rules against the admittance of outsiders. I learned, however, that much of their machinery is crude as compared with that of America. The English manufacturers are slow in adopting new inventions, preferring to stick to old methods of doing things.[92]

Manufacturers in the United States perhaps had a greater incentive than Europeans to adopt labor-saving devices because of the high wages and limited outputs maintained in America by strong craft unions.[93] But this chapter has already presented enough material to show that domestic inventors could not always claim priority in sketching the outlines and principles of glass machinery. Much of the exploratory work done in the bottle and window-glass fields had its origin in Europe. Some of the first semiautomatics for bottles were built in France, Germany, and England, and various individuals in the same countries held many

[90] *Productivity of Labor in the Glass Industry*, p. 127.

[91] Several improvements of a less revolutionary nature than those mentioned in this chapter are described briefly in Appendix B.

[92] *National Glass Budget*, XVI, March 9, 1901. Owens had accompanied E. D. Libbey to look after their interests in the paste-mold semiautomatics sold to a British syndicate. See above, pp. 100–101.

[93] See below, pp. 200 ff.

patents for rolling or drawing sheet glass. Success in incorporating the technological principles into workable and efficient machines, however, most frequently occurred in the United States. In nearly every case its manufacturers appear to have applied themselves with more zest to innovating and experimenting and to have achieved greater commercial success than did foreigners.

It has also been shown that the entrepreneurs at Toledo, brilliant as their accomplishments were, by no means had a monopoly on inventive genius in America. Arbogast blazed the trail which finally led to the mechanization of jar plants, and the American Window Glass Company perfected the cylinder machine. Their achievements spurred many others to devise additional or alternative ways of displacing skilled workmen and of reducing costs. Technological ferment and change throughout the industry characterized the period 1880–1920. That most of the innovations introduced by E. D. Libbey, M. J. Owens, and their associates were novel, nevertheless, remains undisputed. There is no reason to believe, for example, that the Owens semiautomatic paste-mold machines were not the first to make bulbs, chimneys, and tumblers successfully. His automatic bottle machine was constructed more than a dozen years before any other which gathered and formed bottle glass without the intervention of craft labor. Although Fourcault in Belgium and Colburn in Pennsylvania were working on their sheet-drawing processes at about the same time, Colburn carried on his research and experiments independently and attained success first. Finally, the Westlake automatic bulb machine and the Danner tube machine had no immediate rivals anywhere.

CHAPTER VIII

CONTROL OF LABOR AND PRODUCT MARKETS

Technological improvements at Toledo and elsewhere seem to have been the chief foundations of the sweeping industrial changes in glassmaking outlined in Chapter III. But there were other factors. Since historical change usually involves an intricate interrelationship of both cause and effect, it is not always possible to determine exactly to what extent a particular factor was a cause and to what extent it was an effect in any given development. Variations in competitive conditions, in wage structures, in the organization of workers, and in glassware prices, for example, were at once conditioning factors as well as manifestations of industrial and technological changes. High wages and prices, sustained by strong labor unions in conjunction with manufacturers' associations and stiff tariff barriers, presented a favorable opportunity for reaping unusual profits if entrepreneurs and inventors could succeed in perfecting labor-saving machinery. The reaction of the unions and less efficient manufacturers to the innovations affected the extent to which the innovations themselves influenced output, price, and wage rates. The history of the Owens bottle machine illustrates such a cause-and-effect relationship.

The high wages exacted by skilled workers provided one of the prime incentives of M. J. Owens and E. D. Libbey in their work on an automatic machine for blowing bottles. After their endeavors had met with success, the machine was installed in a few factories and operated by unskilled and semiskilled labor to make beer and milk bottles. Bottle prices, however, were maintained at approximately their former level to permit high royalties to the Owens Bottle Company. Following 1905 a series of technical modifications allowed the machine to be used in forming large and small containers of many different types. Hand manufacturers and workers consequently became increasingly uneasy lest all branches of the bottle industry be taken out of their grasp. Hence they attempted to check the introduction of the machine by reducing prices and wages. In 1909 and 1912 the union agreed to accept successive wage reductions totaling 40 per cent of the former level, and manufacturers cut selling prices and hastened to install the best available semiautomatics. The Owens licensees accordingly revised their price lists, and the Owens Bottle Company felt compelled to scale down its royalty rates.[1] Hence conditions in bottle-making around 1903 favored the introduction of the machine, which in turn had marked repercussions upon the industry's wage and output structure; and these repercussions were then partially responsible for subsequent developments.

Since the present study is concerned primarily with the factors responsible for technological and industrial change rather than with the effects of such change, attention in this and the following chapter is focused upon the influence exercised by organized labor, manufacturers' associations, foreign competition, and shifts in consumer demand upon the revolution in production after 1880.

1. The Organization of Workers and Manufacturers in the Container Industry

Labor unions in the American bottle industry did not become very important until the 1880's.[2] In earlier decades whatever formal organizations had existed had been either local and regional in character or had endured for only a few years. The first successful union on a national scale was the American Flint Glass Workers' Union, organized in 1878 as one unit of the Knights of Labor.[3] By 1880 two other independent groups — the eastern and the western — had emerged. Both became affiliated with the Knights of Labor in 1886, but they did not merge to form the United Green Glass Workers' Association until four years later.

[1] See above, p. 107.

[2] See above, pp. 37–38.

[3] *Glass Industry*, p. 310; U. S. Commissioner of Labor, *Regulation and Restriction of Output*, p. 624. The union, covering several distinct trades in the industry (prescription bottles, tableware, lighting goods, etc.), later became affiliated with the American Federation of Labor.

In 1891 upon its withdrawal from the Knights of Labor, the name was changed to the Green Glass Bottle Blowers' Association of the United States and Canada, and sometime after 1900 the word "Green" was dropped.[4] From 1890 to 1913 the American Flints and the Bottle Blowers both claimed jurisdiction over bottle workers.[5] Since the Bottle Blowers' Association had a lower piece-rate wage schedule than did the Flints, bottle production tended to shift to those factories over which the Association had control.[6] The advent of the semiautomatic machines only increased the tension which had naturally developed between the two unions. In the struggle which ensued, the Bottle Blowers' Association gradually gained the ascendancy, and the conflict was terminated in its favor by a jurisdictional agreement in 1913.[7]

By 1890 both the green and the flint bottle workers had secured rather extended control over wages, hours, apprenticeship, and working conditions through a system of annual wage conferences with manufacturers. The first such conference between representatives of the Green Bottle Blowers and their employers was held in 1886, and the Flints participated in their first one about a year later.[8] Manufacturers and workers soon came to accept the principle that neither side could decide independently matters pertaining to the trade. According to the system evolved by the Bottle Blowers and the National Bottle Manufacturers' Association, the employers presented their requests to representatives of the union

[4] *Glass Industry*, pp. 303–304. One writer states that the name was changed from the United Green Glass Workers' Association to the Glass Bottle Blowers' Association in 1895.—Barnett, *Machinery and Labor*, p. 73.

[5] At the outset there had been some reasonable distinction between the spheres of influence which each sought. Most of the flint bottles made at that time were blown of lime instead of lead glass, but the batch was melted in covered crucibles to protect the glass from discoloration. Green glass, on the other hand, was melted in open pots. After gas furnaces replaced those burning coal, it was possible to melt glass in open pots or in tanks more easily and cheaply than in covered crucibles, because less fuel was needed for fusing the materials and for refining the glass. Flint bottle glass, consequently, came increasingly to be melted in tanks. This development removed the technological basis distinguishing the fields covered by the two unions.

[6] Barnett, *Machinery and Labor*, p. 73.

[7] The reader is referred to Barnett's excellent study (pp. 76–79) for an account of the struggle between the two unions.

[8] *Glass Industry*, p. 303; Davis, "American Glass Industry," pp. 211–212, 228.

at a preliminary conference held in May; then the requests were submitted for discussion at the union's annual convention in July. The following month representatives from the two sides met again to conciliate, if possible, whatever differences existed. If no agreement was reached, both parties were pledged to abide by the existing rules during the following year.[9] It was within this framework of control that the union tried to cope with the problems created by the advent of bottle-making machines.

The Arbogast patent for pressing glass in a small (or "blank") mold and then blowing it to full size in another mold [10] had been purchased by the D. C. Ripley Glass Company in 1885. The American Flint Glass Workers' Union, with jurisdiction over this company, set such high wages and such limited outputs for the operatives that the company soon discontinued using the machine.[11] Two nonunion factories then began to make vaseline jars on it or similar devices in 1892 and 1893; and another, the Atlas Glass Company, began producing large quantities of wide-mouth ware on a machine which it had built. Hence it appears that union antipathy was partly responsible for driving the first jar machines to unorganized shops.

The Bottle Blowers' Association, like the Flint Union, was at first inclined to oppose the introduction of machinery into the trade. In 1891 it forbade its members to work on the Ashley or any other machine,[12] but apparently it did not enforce this rule.[13]

[9] Barnett, *Machinery and Labor*, pp. 93–94.

[10] See above, p. 178.

[11] "Glass Workers and the Introduction of Labor-Saving Machinery," *Pottery and Glassware Reporter*, December 3, 1891, found in Gaffield Collection, Scraps, III, 148. Ripley himself offered the following testimony in the suit brought by the United States Glass Company against the Atlas Glass Company: "The policy of the labor organization has been that no device that lessened the amount of skill would be operated except for the same amount of money as was paid for the full exercise of skill in the manufacture of the article by new processes. Also any new device which enabled the workmen to make more with less labor could not be used to advantage from the fact that the numbers [of jars made] were limited to the number determined by the labor organization." — Quoted by Barnett, *Machinery and Labor*, p. 73, n. 5.

[12] "The Glassblowers' Warning and Challenge," *The Pottery and Glassware Reporter*, December 15, 1891, found in Gaffield Collection, Scraps, III, 149. Cf. Henderson, "The Glass Industry," *Popular Science Monthly*, XLII (1893), 598–599.

[13] The vice-president of the union stretched the truth when he told the Tariff Commission in 1918 that "since we were organized . . . we have not resisted in

Beginning in 1898 it set out to gain jurisdiction over all jar machines and to persuade manufacturers to allow its skilled workers to operate them.[14] From that time on the Bottle Blowers' Association did not attempt to fight the machines by restricting their use in plants over which it had jurisdiction or by setting maximum output quotas for the operatives.[15] Nevertheless, in an effort to check the spreading use of wide-mouth machines, it adopted in the same year a 45 per cent reduction in the piece-rate wages paid hand blowers making fruit jars.[16] Even a wage cut as drastic as this did not arrest the march of the semiautomatics. The machines were simply too efficient to be pushed aside. In fact, they made better jars than could be blown by hand. The union then decided that the salvation of its members lay in accepting and controlling the machine rather than in trying to fight it through additional wage reductions.

When the Ball Brothers Glass Manufacturing Company, one of the largest fruit-jar producers, began to mechanize its plant at Muncie, Indiana, the Bottle Blowers' Association persuaded the management to allow the displaced skilled blowers to become machine tenders if they proved capable. These machines were of the Arbogast type and combined pressing with blowing. Consequently, Ball Brothers continued to employ large numbers of press operatives who belonged to the American Flint Glass Workers' Union. As a result, jurisdiction over the machine sections in the plant was shared by the two unions until 1900. In that year the prescription-ware division of the Flints voluntarily broke away from its old organization and joined the Bottle Blowers' Association. This allowed the Association gradually to obtain full jurisdiction over wide-mouth machines because it was then in a position to supply manufacturers with both pressers and blowers.[17]

---

any way the introduction of machinery." — Tariff Commission, *Glass Industry as Affected by War*, pp. 31, 140.

[14] Barnett, *Machinery and Labor*, p. 79.

[15] "The Glass Bottle Blowers and the prescription branch of the Flint Glass Workers are alike, in that there has never been in recent years, with two or three temporary exceptions, any . . . limitation of output." — U.S. Commissioner of Labor, *Regulation and Restriction of Output*, p. 625.

[16] *Ibid.*, p. 636; *National Glass Budget*, XV, July 22, 1899.

[17] See Barnett, *Machinery and Labor*, pp. 73, 75–79.

All limitation on the earnings of both kinds of machine workers was removed, and by 1904 the output per presser had increased from 3,200 to around 4,000 or 4,300 jars per day.[18]

Although the number of wide-mouth semiautomatics in operation around 1905 represented a potential displacement of about 25 per cent of the skilled bottle blowers employed in the United States in 1897, almost no displacement resulted.[19] The output of all kinds of glass containers increased by nearly 50 per cent from 1899 to 1904. This allowed hand blowers, who found themselves confronted by machines in jar factories, to seek employment in bottle shops. Comparatively few of them elected to become machine operators. In fact, as Barnett has indicated, the Bottle Blowers' Association sometimes found itself embarrassed by an inability to find enough machine pressers to supply the needs of jar manufacturers.[20] It had earnestly striven from 1896 to 1907 to extend its control over nonunion factories, and its efforts had met with gratifying results. Whereas in 1896 about two-thirds of the 6,229 bottle blowers had been unionized, a decade later 9,627 out of 10,997 skilled workers and machine operators were union members.[21] The jurisdictional conflict with the American Flint Glass Workers' Union had already led to the virtual supremacy of the Bottle Blowers' Association in the container industry by 1905. Hence the Association was in a better position than formerly to deal with the problems raised by the continued spread of the semiautomatic machines, and it doubtless would have done so had not the Owens automatic appeared on the market in the meantime.

The Owens machine, unlike the semiautomatics of that day, required neither skilled gatherers nor pressers.[22] Its operatives, working in twelve-hour shifts, were drawn from the ranks of the unskilled and were paid accordingly. They received twenty cents

[18] U. S. Commissioner of Labor, *Regulation and Restriction of Output*, p. 637.

[19] Barnett, *Machinery and Labor*, pp. 69–70.

[20] *Ibid.*, p. 81.

[21] *Ibid.*, pp. 83, 93. There were still some nonunion shops and some workers under the jurisdiction of the Flint union, but these two groups were not very important.

[22] See above, pp. 155 ff.

an hour in 1906; at the same time hand blowers were earning about $7.00 a day and workers on the semiautomatics from $4.00 to $5.00.[23] The machine, furthermore, ran with the help of very little labor. The union could not hope to secure re-employment in the factories using Owens equipment for all its members who would probably be thrown out of work. The temporary depression in the bottle market in 1908–1909 and the increasing output of Owens licensees forced the union in 1909 to formulate a definite program for action — which it consistently followed thereafter.

The program embodied four major policies.[24] (1) The union urged hand manufacturers to institute three eight-hour shifts in place of the customary two shifts of ten hours each. In this way, it argued, employment could be spread, the working day reduced, and economies realized from continuous operation. Factories using semiautomatic machinery introduced the eight-hour day in 1911, and hand plants agreed the following year to make the change as soon as sufficient blowers and helpers became available in their localities. (2) The union insisted that the number of apprentices be drastically reduced. In return for heavy concessions in wage rates, manufacturers reluctantly consented not to take on any new apprentices in 1909 and again in 1912. The following year the union agreed to return to the customary ratio of one apprentice for each fifteen journeymen, but at the same time it obtained a 50 per cent increase in the wages paid apprentices.[25] This almost entirely removed the employers' incentive to take apprentices in preference to journeymen. (3) Producers were encour-

[23] *National Glass Budget*, XXXIV, July 20, 1918, p. 1. See also Barnett, *Machinery and Labor*, p. 104. The Owens Bottle Company, in 1918, paid its machine operatives from $3.00 to $3.36 per day with opportunities to earn something more under the bonus system. — Tariff Commission, *Glass Industry as Affected by War*, pp. 140–141.

[24] For a more detailed analysis of the four policies than is presented here, see Barnett, *Machinery and Labor*, pp. 95 ff.

[25] Formerly, apprentices had received half as much as journeymen and had been required to spend five years in "learning the trade." After the first year, however, an apprentice usually became almost as productive as a journeyman. Manufacturers, therefore, had found it profitable to take on as many apprentices as they possibly could. In the 1913 agreement, the period of apprenticeship was reduced to four years and the wage rate raised to 75 per cent of the piece-rate paid journeymen. — *Glass Industry*, p. 308.

aged to install the most efficient semiautomatic machines so that they could compete more effectively with the Owens group. The union recognized that, as far as it was concerned, the semiautomatic was a lesser evil than the Owens automatic. And (4) piece-rate wages were cut in 1909, 1912, and 1914. As early as 1905 hand manufacturers had urged the union to lower its wage scale so that they could fight the machine. The president of the union, however, had stoutly maintained that no reduction however great would allow hand blowers to compete with the machine in those lines of bottles in which the machine was already successfully at work. The pressure had become so strong by 1909 that a 20 per cent reduction was instituted for beers, sodas, brandies, catchups, and other similar heavy ware, and a 10 per cent reduction for whiskey ovals and the like. In 1912 the union agreed to a further cut of 20 per cent on all ware which the Owens machine was then making. Two years later, after it had been incontestably proved that the machine could produce small bottles as well as large ones, the wages paid blowers of prescription ware were also lowered by 20 per cent.[26] This was the last time the union attempted to meet the competition of the Owens machine by reducing wages.

Most of the first flow and gob feeders appearing on the market following 1915[27] were installed in nonunion shops, and many of the displaced hand gatherers sought positions as machine attendants. The union formulated no definite policy in regard to the new development until 1918. At that time it secured agreements with a few employers and thereafter sought to bring the remainder under its influence.[28] Wages were set at fifty cents per hour in 1918, at sixty cents in 1919, and at sixty-six cents in 1920. Under these rates the men earned appreciably less than when engaged in the much more arduous task of gathering glass by hand to feed the semiautomatic machines, but they received more

[26] *National Glass Budget*, XXV, April 23, 1910; XXVI, February 4, 1911; XXVIII, June 8 and 29, August 24, and September 7 and 14, 1912; XXX, September 5, 1914, p. 1; Owens-Illinois Glass Company, Owens' Private Files; Barnett, *Machinery and Labor*, pp. 100–103.

[27] See above, pp. 182 ff.

[28] Tariff Commission, *Glass Industry as Affected by War*, p. 139; Barnett, *Machinery and Labor*, pp. 111–112.

than similar workers in nonunion shops. Hand gatherers in 1918 were paid about $7.00 or $8.00 for an eight-hour day, whereas attendants for feeding and forming machines in nonunion plants received thirty-five cents per hour. In addition to fixing the basic wage rates, the agreements concluded by the union stipulated that the manufacturers would use the operatives it supplied provided the latter were satisfactory and experienced. If they were not, then the manufacturers could hire men from any other source, but the operatives would have to join the union after working six months. Barnett has roughly estimated that about one-third the workers who had tended and fed the semiautomatics in 1917 found employment as operatives of the feed devices before 1925.[29]

The foregoing account of the attitude of bottle-makers to machine technology may now be summarized. It appears that organized labor was inclined to fight the first wide-mouth semiautomatics, that the Glass Bottle Blowers' Association after 1898 accepted the machines and attempted to gain jurisdiction over them, that it saw no hope for its members in the Owens machine and consequently tried to prolong the life of hand shops by reducing wages and encouraging manufacturers to install the latest semiautomatic devices, and that after 1917 it attempted to place its hand gatherers as operatives on automatic feeding devices. At no time following 1898 did the union restrict the output of its members employed on machines or try to prevent union plants from mechanizing their processes. Its attitude may appear rather unusual to some readers since craft unions frequently direct all their resources to fighting mechanical innovations. An explanation of its position, however, may be found in the rapidity of the technological change confronting it, the marked superiority of machine processes over handicraft methods, the jurisdictional conflict between the American Flints and the Bottle Blowers at precisely that period when the machines first gained a toehold, and the expanding market for all kinds of containers which postponed the full impact of the new technology upon workers' welfare until it was too late to do much to ameliorate the situation. The Glass Bottle Blowers' Association did manage to preserve the

[29] *Machinery and Labor*, p. 112.

system of bargaining collectively with representatives of the National Bottle Manufacturers' Association,[30] but the power it had wielded around 1890 over wages, hours, and conditions of work rapidly diminished. The preferred position which its members at one time had enjoyed in the labor market consequently disappeared.

Both union and nonunion plants in 1890 had observed the tradition of the "summer stop." Because of intense summer heat, work around bottle furnaces was suspended during July and August.[31] In 1908 the union reluctantly permitted employers to run their semiautomatics throughout the year and secured a stoppage of only two weeks in the agreement signed with them in 1912. At the same time the summer stop for hand workers was reduced from two months to one. The competitive pressure exerted by the Owens machine, which worked all twelve months, caused the union finally and definitely to accept in 1917 the principle of continuous operation for both hand and machine plants. During the hottest months of the summer, however, the workers enjoyed staggered vacations lasting four weeks for hand blowers and two weeks for machine operatives.[32] Primarily as a means for spreading employment, during the period of mechanization most union factories reduced the number of working hours per day from ten to eight.[33] But they did not increase the hourly earnings proportionately.

As a matter of fact, the daily earnings of bottle workers relative

[30] "At the present time [i.e., 1916] it is unlikely that there could be found a better example of successful collective bargaining than that which is carried on between the Glass Bottle Blowers' Association and their employers." — *Glass Industry*, p. 304. The bottle industry had not experienced a national strike since 1884, and even local interruptions of work were infrequent.

[31] Nonunion plants usually closed down for only four or six weeks. — U. S. Commissioner of Labor, *Regulation and Restriction of Output*, p. 639. The union does not seem to have used the summer stop as a means for controlling output as did the window-glass workers' union.

[32] Barnett, *Machinery and Labor*, pp. 108–109; *Glass Industry*, p. 308; *National Glass Budget*, XXVII, July 15 and 29, 1911.

[33] Around 1890 bottle blowers worked in two shifts of 10 hours each (with 8.5 hours of actual working time). Those who operated the early jar machines worked 8 hours; and by 1916 twenty hand factories had adopted the 8-hour day while the remainder retained the old system of two 10-hour shifts. The union introduced the 8-hour shift in those factories using automatic feeding devices with which

to those of other groups declined drastically. In the decade 1870–1880 bottle workers had received anywhere from two to three times as much as ordinary laborers;[34] by 1920 the great majority of them had been reduced to jobs where their skill counted for little and which were not very much more remunerative than employments open to common workmen.[35] Barnett has estimated that the daily earnings of hand blowers fell from about $7.00 in 1905 to around $4.50–$5.00 in 1914 and rose to approximately $6.00 in 1917.[36] Indices of hourly wages rates in other organized industries increased by 27 per cent during the same twelve years. Union operatives on semiautomatic machines in 1917 earned on the average $6.78 a day.[37] Those tending automatic feeding and forming machines in union factories in 1918 received $4.80 per day as against $3.00 for those employed in shops with which the union had not then concluded working agreements.[38] The Illinois Glass Company, a nonunionized licensee of the Owens Bottle Company, was paying its machine operatives twenty-five cents an hour prior to a strike initiated in September, 1916.[39] The American Bottle Company, another Owens licensee, had paid its men about twenty cents an hour after it was incorporated in 1905 and before it was unionized. One of its wage agreements with the Glass Bottle Blowers' Association stipulated in 1918 an hourly wage of 37.5, 40, and 45 cents, depending upon whether the man was employed on a machine with six, ten, or fifteen arms.[40]

The advent of the first semiautomatics had not affected very much either the hours or the earnings of bottle-makers. The union secured increases in hand blowers' wages after 1898 which brought their level up to the standard which had prevailed earlier,

---

it concluded working agreements in 1918–1920.—U.S. Commissioner of Labor, *Regulation and Restriction of Output*, p. 637; *Glass Industry*, pp. 306–307; Tariff Commission, *Glass Industry as Affected by War*, p. 136.

[34] See above, pp. 32–33.

[35] Cf. Davis, "American Glass Industry," pp. 322–323, 332.

[36] *Machinery and Labor*, pp. 104–105.

[37] Tariff Commission, *Glass Industry as Affected by War*, p. 140. Hence the daily income of workers on the semiautomatics was actually higher than that of hand blowers in 1917. They had been about equal three years earlier.

[38] *Ibid.*, p. 139.

[39] *National Glass Budget*, XXXII, September 16, 1916, p. 1.

[40] *Ibid.*, XXXIV, July 20, 1918, p. 1.

and enlarged outputs on improved jar machines — together with an increase in the piece-rate wage in 1903 — permitted operatives to earn as much per day in 1905 as hand workers.[41] The period from 1907 through 1915, however, witnessed drastic wage cuts and growing unemployment as the Owens machine became an increasingly important factor in the trade. It has been estimated that the working force of skilled workmen — including both hand blowers and semiautomatic operatives — fell from around 8,000 in 1905 to 4,000 in 1917.[42] During the war years the earnings of bottle-makers increased but not as much as in industry generally, and the number of skilled workmen employed continued to decline.[43]

In the light of the preceding discussion, it is possible at this point to appraise the effects which these trends in the labor market had upon the developmental and promotional efforts emanating from Toledo. Control over apprenticeship, and high wages maintained by the union through collective bargaining created a strong incentive for manufacturers to adopt labor-saving devices. During the latter part of the nineteenth century payments to labor accounted for approximately 40 per cent of the total cost of making bottles,[44] and skilled glassmen received a large portion of these wages. In the prospectus published in 1903 under the title, *The Owens Automatic Bottle Machine*,[45] the promoters tried to capitalize on the conditions prevailing in the labor market at that

[41] Barnett, *Machinery and Labor*, pp. 82–83.

[42] *Ibid.*, pp. 105–107. Unemployment first became acute around 1909; and almost a fourth of the 7,500 or 7,700 members in the Bottle Blowers' Association were unemployed in 1911–1912. — *National Glass Budget*, XXIV, April 17, 1909; XXVI, April 15, 1911; XXVII, September 23, 1911, and January 20, 1912; XXVIII, August 3, 1912.

[43] In 1924 there were only 749 skilled blowers at work in American factories who belonged to the bottle union; seventeen years earlier, membership in the union had numbered well over 9,000 skilled workers (including blowers, gatherers, apprentices, and semiautomatic operatives). — Barnett, *Machinery and Labor*, pp. 106, 114. The effect which mechanization had upon workers in the container industry may be pictured in another way. In 1899 over 28,000 wage earners were employed; in 1925, less than 22,000. And skilled workers of all kinds constituted a much smaller percentage of the total in 1925 than in 1899. The output of glass containers in the meantime had increased more than three fold.

[44] See above, pp. 30–31.

[45] Found in M. J. Owens' private files at the Owens-Illinois Glass Company.

time. They claimed that their machine would lower the direct labor costs of production from about $1.50 to around 6 cents per gross of bottles;[46] that no labor organization would be able to control its use and output because it did not require the services of skilled glassworkers; that the furnaces could be worked until necessary repairs forced the fires to be extinguished and would not have to lie idle during the summer months merely because of rules promulgated by the Glass Bottle Blowers' Association; and that manufacturers would be relatively free from strikes because their help would be unorganized. The general claim that the machine would greatly reduce the importance of labor as an item in total production costs was certainly justified. The survey of the glass industry conducted by the United States Bureau of Foreign and Domestic Commerce in 1916 reveals that hand factories paid out 51.49 per cent of the sales value of their annual product in wages and enjoyed an annual net profit (after allowance for depreciation and interest charges) of 2.55 per cent. The total wages and machine royalties expended by licensees of the Owens Bottle Company amounted to only 31.38 per cent of the sales value of their output, and the companies themselves averaged 12.15 per cent net profit.[47]

The Toledo Group had never intended that the machine should upset the wage — and especially the price — scale already established in the industry. The original plan was to set royalty rates at such a level that the Owens Bottle Company and its licensees would share equally in the resulting decreases in direct labor costs. Actually the machine's introduction did not disturb the bottle market very much during the first years. Several factors were responsible for this. In the first place, the Owens Bottle Company had found only three licensees and had been able to supply them with only twenty-two machines by October 1, 1907. Secondly, an increased demand for glass containers absorbed the enlarged output without any difficulty. And, thirdly, even though members of the Bottle Manufacturers' Association had urged as

[46] This seems to have exaggerated somewhat the machine's efficiency. See above, p. 159, n. 41.

[47] *Glass Industry*, pp. 104–105.

early as 1905 that wages be cut so that they could undersell the licensees, the blowers' union had been strong enough to enforce its refusal. Thus the conjuncture of circumstances allowed the Toledo Group to place the Owens Bottle Company securely on a profitable basis before the end of its fourth year.[48]

By the close of 1909, however, the bottle trade was definitely unsettled. The number of Owens licensees had suddenly jumped to eight during the year,[49] and over fifty-one machines had been installed as compared with twenty-two in 1907. The market demand for glass containers also had slackened. Even though machine bottles had fallen little if any in price, buyers had come to prefer them to the hand-made variety because of their uniform weight, content, and quality. In other words, the Owens machine had begun to cause sales of hand-produced items to decrease. As has already been pointed out, wages of skilled workers fell by 20 per cent in 1909 and by a like amount again in 1912. The Owens Bottle Company did not view the reaction of handicraft workers and producers with equanimity. M. J. Owens himself wrote the officials of the Glass Bottle Blowers' Association in 1912 and urged the union to reject the manufacturers' demands for the second wage cut. Such a reduction would only tend to "demoralize" the trade, he argued, and his machine was so efficient that the only way hand blowing could survive was through the "co-operation" of all bottle manufacturers and workers.[50] Because the union remained unconvinced and accepted the wage cuts, the Owens Bottle Company had to reduce the royalty rates stipulated in its licenses issued between 1909 and 1914 by one-half or two-thirds the rates previously planned and to lower substantially the royalties of its original licensees.[51]

The purpose behind the union wage reductions, of course, was to permit hand manufacturers to lower selling prices in order to

[48] See above, p. 116.

[49] See above, p. 105.

[50] *National Glass Budget*, XXVIII, August 24, 1912.

[51] See above, p. 107. See also *National Glass Budget*, XXV, May 23, 1910. The first licenses were redrawn and the royalty rates reduced from 50 and 54 cents per gross to 40 cents for the first million bottles made in any one year. On all output above 1,000,000 the royalty was to be 25 cents per gross.

preserve their individual markets.[52] According to the statistics presented by Shirley P. Austin in his special report on the glass industry in the *Twelfth Census*,[53] beer, whiskey, and soda bottles were valued at $3.75 a gross in 1899. Co-operation among producers had kept the price at approximately the same figure until 1909. Then hand manufacturers, in spite of the efforts made by Owens licensees to keep them in line, reduced prices and thereby called forth retaliatory measures. During the latter half of 1911 the American Bottle Company, the largest firm equipped with Owens machines at that time, offered to sell its pint beers for $2.75 a gross.[54] Before the "price war" ended with the re-establishment of co-operation, the price fell to a low of $2.60.[55] E. D. Libbey reported to the stockholders of the Owens Bottle Company in 1911 that "competition is still active and has resulted in reduced cost of production and reduced prices for bottles. As compared with former prices [i.e., for 1908] the present price of bottles made by your company and its subsidiary companies show an average reduction of twenty per cent."[56] By 1914 the price list was only 16 per cent below the 1908 level, but the following year it fell again to the 20 per cent mark.[57] During the war years

[52] The editor of the *National Glass Budget* (XXVI, February 4, 1911) summarized the "period of truce" which had existed earlier in these words:

> . . . instead of going about with a chip on their shoulder hand manufacturers worked in common with machine interests, with the result that a profitable market has been maintained [;] bottle blowers have been earning high wages, and a much more satisfactory condition of affairs exist for all concerned than would be the case had hand manufacturers in conjunction with the workers attempted to dispute the right of way of twentieth century methods [;] and even though the Bottle Blowers' Association is maintaining in idleness about one-sixth of its membership, the amount of money that is being paid out weekly in wages is double or treble what it would be had they taken it into their heads that they could beat out the mechanical blowers.

[53] *Twelfth Census*, Vol. IX, Part III, 999.

[54] *National Glass Budget*, XXVII, September 9, 1911.

[55] According to the testimony presented by the vice-president of the Glass Bottle Blowers' Association before the Tariff Commission (*Glass Industry as Affected by War*, p. 138). President Hayes of the union had reported in July, 1911, that beer pints were then selling for $1.00–$1.50 less per gross than before the Owens machine was introduced.—*National Glass Budget*, XXVII, October 21, 1911.

[56] Owens-Illinois Glass Company, Annual Reports of Owens Bottle Company, September 30, 1911.

[57] *Ibid.*, September 30, 1914, and September 30, 1915. Barnett (*Machinery and

prices rose throughout the industry and surpassed $3.75 a gross.[58]

There can exist no doubt that the attitudes of the workers' union and of the hand manufacturers' association affected the prices set for containers made on the Owens machine. The Owens Bottle Company kept "price-control" clauses in its license contracts until 1911,[59] and thereafter it maintained rather definite "understandings" with its licensees on the matter. Hence users of the Owens machine did not enter into price competition with each other and actually wished to avoid it with manufacturers using hand and semiautomatic methods. The U. S. Tariff Commission in 1918 concluded that "notwithstanding the great difference in the cost of production, there is no competition in the domestic market between the product of hand labor and that of the [Owens] automatic machine."[60] The Owens company attempted to rationalize its position by stating around 1920 that "the introduction of the . . . machine has not been so serious to the glass workers as it would have been if the company had not pursued the policy of practically limiting the machines to supply the increased bottle demand."[61] This is true, but it is equally true that the company's high rate of earnings prior to 1920 could not have been maintained if prices had been determined in a competitive market.

The policies pursued by the Bottle Blowers' Association not only influenced selling prices but also affected the Owens Bottle Company in other ways. By encouraging hand manufacturers after 1909 to install semiautomatic machinery, the union indirectly sped up the process of developing more efficient bottle-making devices. This, in turn, spurred M. J. Owens and his associates to increase the efficiency of their machine. The Owens models with ten and fifteen arms instead of six — introduced after 1911

Labor, p. 130) has concluded that "from 1905 to 1914, the decrease in the price of bottles made on the Owens bottle machine was, as nearly as can be ascertained, about 15 per cent."

[58] Tariff Commission, *Glass Industry as Affected by War*, p. 138.

[59] See above, p. 108.

[60] *Glass Industry as Affected by War*, p. 43. The Owens Bottle Company and its licensees did not export to foreign countries because of the licensing agreement concluded with the Europäischer Verband in 1907. See above, pp. 122 ff.

[61] Owens-Illinois Glass Company, Owens' Private Files.

— increased greatly the daily output per machine.[62] Improvements in the semiautomatics, furthermore, paved the way for the automatic feeders to appear after 1915. Feeders simply could not have been used with such simple, yet clumsy, forming machines as the "Johnny Bull" and "Jersey Devil." It was only after the first semiautomatics were replaced by more efficient types that automatic feeding became practicable.[63] And it was the flow and gob feeders which constituted the first real threat to the favored position of the Owens machine in the bottle trade.

The union's policy of shifting to an eight-hour day after 1909 also seems to have had its repercussions upon the Owens Bottle Company. Late in 1914 E. D. Libbey recommended to the Board of Directors that the company reduce the hours of work for its plant employees from twelve to eight. M. J. Owens, as general manager, however, opposed the change because he felt that it would "involve serious complications." [64] He was then asked to reconsider the matter and render a report on working conditions at a later date. In the 1916 Annual Report Libbey told the stockholders that their company recently had granted an increase in wages and had also adopted the three eight-hour shift system in its plants.[65] One cannot prove that this action resulted directly from the increasing popularity of the shorter working day in hand and semiautomatic factories, but it is improbable that the two were wholly unrelated. The Owens Bottle Company had never experienced any serious labor trouble, and it was anxious to preserve this record. If unionized workers enjoyed a shorter day, its own men might become dissatisfied with their twelve-hour day. And labor organizers usually find fertile soil wherever dissatisfaction with working conditions exists.

The Libbey Glass Company had been a union plant from the outset,[66] and M. J. Owens had once been a power in the flint-glass

[62] See above, Table 15, p. 160.

[63] Since the 1915 forming machines had to be still further improved after workable feeding devices were developed before the combination could make bottles very successfully, it follows that the earlier semiautomatics would have been even less adaptable.

[64] Owens-Illinois Glass Company, Owens Bottle Company Minutes, IV, 76.

[65] *Ibid.*, Annual Reports of Owens Bottle Company, September 30, 1916.

[66] And it so remained. Production by hand at Libbey Glass, never entirely

union.[67] But the bottle machine did not require the services of skilled glassmen, and its advent spelled the doom of bottle blowers. When the machines were first installed in the Newark plant of the Ohio Bottle Company, the workers there grew sullen and caused appreciable damage through sabotage.[68] The Owens Bottle Company, therefore, had considered it safer (and cheaper) to start with unskilled workmen. Although Owens personally remained sympathetic toward labor unions in principle,[69] he did not wish to share with any outside organization control over the men who operated his machines. When the directors of the company in 1915 received an invitation to send representatives to the annual meeting and preliminary wage conference of the National Bottle Manufacturers' Association, they refused. They desired to "co-operate" with manufacturers but did not wish to expose themselves in any way to collective bargaining.[70] The company's officials in December, 1918, denied the union's request for permission to organize their workmen,[71] and seven months later they promised to consider joining the Manufacturers' Association only if the company could be assured that such action would not entail relationship with any labor union.[72]

The National Bottle Manufacturers' Association, organized during the latter half of the nineteenth century, concerned itself primarily with labor questions and represented the individual employers in all negotiations with the union. Its existence, of

---

discontinued for some ware, was replaced only gradually by machine processes. Hence the transition never occasioned a violent and permanent break with the union.

[67] See below, pp. 275–276.

[68] According to William H. Boshart, former president of the Owens Bottle Company.

[69] According to Thomas W. Rowe, who for many years was president of the American Flint Glass Workers' Union and at the same time remained one of Mike's closest friends. The mold workers employed by Owens Bottle were unionized.

[70] The American Bottle Company and, apparently, the Graham Glass Company had been unionized sometime before they were purchased. In 1920 their plants were the only ones belonging to the Owens Bottle Company which dealt with the union.

[71] Owens-Illinois Glass Company, Owens Bottle Company Minutes, VI, 284; Executive Committee Minutes, I, 272.

[72] *Ibid.*, Executive Committee Minutes, II, 333.

course, promoted a "community of interests" among the member firms, but apparently it did not wield undue power over prices and output. Certainly there is no clear-cut indication that it regularly fixed production quotas for each firm. In 1920 its aims included the collection of statistics in the bottle trade, the maintenance of "fair prices" during times of depression by encouraging manufacturers to reduce output rather than sell below costs, the introduction of "good" and uniform cost-accounting systems in all firms, and the establishment of "one-price" and "open-price" policies for its members. The Owens Bottle Company applauded the program to improve trade conditions in the bottle line,[73] but it did not join the Association. Instead, in July, 1919, it sponsored the organization of the National Glass Container Association for the purpose of combating the inroads of tin-can manufacturers in the glass-container field,[74] of developing export markets,[75] and of creating an agency entirely free of labor-union connections which the Owens company could use to apply pressure in the bottle trade.

2. The Organization of Workers, Manufacturers, and Jobbers in the Window-Glass Industry

In no branch of the industry during the period 1880–1920 did organizations of laborers, employers, and wholesale dealers wield as much monopolistic power for as long a time as did those involved in the production of window glass. Except for intermittent periods when stiff competition prevailed, problems relating to such matters as wages, apprenticeship, output, and prices were resolved by group action. Consequently, the pattern of this industrial control and the way it functioned greatly influenced manufacturing and technological developments.

Many window-glass workers had belonged to local or regional unions before 1880, but all attempts to create a permanent na-

[73] *Ibid.*, II, 396–397.

[74] This was to be accomplished by ascertaining through experimental research the best methods for preserving foods in glass and by conducting extensive advertising campaigns.

[75] *Ibid.*, Owens Bottle Company Minutes, VII, 63.

tional organization had proved unsuccessful.[76] In 1877 or 1878 a gatherers' association at Pittsburgh joined the Knights of Labor as Local Assembly 300,[77] and before the end of 1880 unions of blowers, flatteners, and cutters had joined forces with it. During the following twenty years membership in the Window Glass Workers of America (as Local Assembly 300 was called) included nearly all skilled glassmen employed in window-glass factories throughout the country. Manufacturers were brought together in a national organization by representatives from twenty-two plants meeting at Cleveland on July 9–10, 1879, and electing E. L. Day of Kent, Ohio, as their president.[78] On January 14, 1880, the American Window Glass Manufacturers' Association formally adopted a constitution and set of by-laws.[79] The extent of its market coverage is best revealed in a statement made before the U. S. Tariff Commission in 1882 to the effect that the Association's membership comprised all but six or seven window-glass manufacturers in the country.[80] In co-operation with the labor union it maintained rather effective control over the industry until the business recession of 1893–1895. The American Glass Company was formed in 1895 as a selling pool for approximately 85 per cent of the existent window-glass factories. This organization was succeeded[81] four years later by the American Window Glass Company, a corporation with $17,000,000 capital and with fifty-three plants representing about 70 per cent of the nation's furnace capacity.[82]

The first years of the twentieth century were stormy. The

[76] Davis, "American Glass Industry," pp. 175–176.

[77] See above, p. 37.

[78] Gaffield Collection, Glass Journal, III, 230–232.

[79] Gaffield Collection, Minutes of the Proceedings of the American Window Glass Manufacturing Association.

[80] Tariff Commission, *Report of 1882*, p. 2006.

[81] The inherent weaknesses of the pool as a device for creating and maintaining monopoly control in an industry appear to have been responsible for the short life of the American Glass Company.

[82] In genesis and position as "price leader," the American Window Glass Company was to the window-glass industry what the United States Steel Corporation (1901) was to the steel industry. Although the American Window Glass Company is still (1947) in existence with home offices at Pittsburgh, it no longer dominates the industry as it did during the early years of the twentieth century.

amount of capital required to set up a window-glass factory was relatively small — especially if pot furnaces were built instead of the more costly tanks.[83] Hence glassworkers and outsiders readily built new plants whenever the selling price of window glass rose in response to output restrictions instituted by established manufacturers. Even in 1899 there were approximately thirty-one independent factories with a production capacity equivalent to about 462 pots. The American Window Glass Company, with its fifty-three factories, had a capacity estimated at 1,912 pots.[84] Yet the output of the independents, when added to annual imports, amounted to about one-half of total consumption on the domestic market.[85]

In order to squeeze the independents, the American Window Glass Company immediately entered into a contract with Local Assembly 300 whereby $500,000 of its par-value stock was set aside in trust for the labor union.[86] Local Assembly 300, in return, promised to supply all the gatherers, blowers, flatteners, and cutters which the American Company might need. Since restrictive apprenticeship rules enforced since 1880 had created a relative shortage of labor compared with the increased number of manufacturing enterprises, the contract threatened to cut off the independents from the available labor supply. At the same time the American Window Glass Company fostered an organization embracing most distributors in the country and entered into an arrangement whereby its own output would enjoy a favored posi-

[83] The fact that all the co-operative shops springing up around 1901 were equipped at the outset with pot furnaces rather than with tank furnaces was singled out by government investigators in 1904 as a significant indication of the practical glassmaker's distrust of technological innovations. — U. S. Commissioner of Labor, *Regulation and Restriction of Output*, p. 608. It seems more likely, however, that the fact reflected the difference in the cost of constructing tank furnaces as compared with pot furnaces. Glassworkers who formed these co-operatives were nearly always short of capital.

[84] Twenty-three of American's 53 factories operated continuous tanks whereas only 4 of the 31 independents did. The capacity of melting tanks was calculated in terms of so many pots and included in the total capacity figures. — Davis, "American Glass Industry," pp. 263–264.

[85] *Ibid.*, p. 264.

[86] The stock was to remain in trust until the accumulated dividends on it equaled its par value.

tion on the wholesale market.[87] These threats to their freedom in manufacturing and selling caused the independents to expand output and lower selling prices. For a short time intense competition prevailed.[88] Meanwhile, dissension was developing within the labor union over the $500,000 trust fund set up under the terms of the contract with the combination. After a little more than twenty years of relative peace, Local Assembly 300 split; and the dissentients, forming the Amalgamated Window Glass Workers' Association of America, undertook to supply the personnel needed by the independents.

In December, 1900, the independent manufacturers organized their own selling agency, the Independent Glass Company.[89] As most frequently happens in cases of duopoly, the American Window Glass Company decided the following month to cease fighting its rival and to inaugurate some kind of a "share-and-share-alike" arrangement. The National Wholesale Glass Jobbers' Association (the distributors' association closely affiliated with American) accordingly agreed to divide its purchases between the Independent Glass Company and the American Window Glass Company. It immediately raised the price at which it bought glass by 30 per

[87] The National Wholesale Glass Jobbers' Association, as the new organization was called, was monopolistic. Every important window-glass jobber in the country took stock in the association and agreed to abide by all its rules governing purchase and price. The association purchased glass from manufacturers and then apportioned the quantities among individual jobbers. If a distributor failed to abide by the association's rules, he was dropped from membership and left to satisfy his needs as best he could by buying in small quantities and at high prices.—U. S. Commissioner of Labor, *Regulation and Restriction of Output*, pp. 602–603.

[88] The wholesale price of American single-strength glass of top grade (sizes 6" x 8" to 10" x 15") dropped from $3.15 per box of fifty square feet in November, 1899, to $2.10 in January–March, 1900.—U.S.B.L.S., *Wholesale Prices*, Bull. 39, p. 429. The *National Glass Budget* (XVII, December, 1901) reported that this "ruinous competition" had forced the American Window Glass Company to sell at or below costs for several months.

[89] The member firms bought stock in the Independent in proportion to the number of pots they operated and sold their entire output to it each year at a price which all had previously agreed upon. The Independent Glass Company then marketed the combined output of its members as best it could without discriminating between individual producers. The Independent's board of directors, composed of one representative from each member firm, decided at the beginning of the production year how many months all the furnace fires were to be lighted. If a member failed to conform to this decision or sold any part of its output directly to wholesalers, it forfeited its shares of capital stock in the Independent.

cent and in March, 1901, by another 25 per cent.[90] Apparently, "co-operation" was definitely beneficial to both combinations of manufacturers. But the drastic increase in price proved to be the occasion for fresh trouble.

The number of co-operative enterprises established by glassworkers who chafed under output limitations imposed by the two combinations again multiplied,[91] and even outsiders hastened to build furnaces. Whereas at the end of 1900 the American Window Glass Company and the Independent Glass Company together had controlled about 90 per cent of the pots and tanks used in melting window glass, their jurisdictional coverage had shrunk to around 60 per cent by January, 1902. Furthermore, many of the co-operatives, staffed by members of Local Assembly 300, had also formed their own selling agency known as the Federated Window Glass Company. In 1902 the Independent Glass Company rescinded its agreement with the American Window Glass Company concerning control of output and prices, and the American and the Federated then joined forces. Together they agreed to raise wages by 10 per cent if Local Assembly 300 would consent to a further reduction in the length of the production year.

This action, however, evoked conflict between the two labor unions, Local Assembly 300 and the Window Glass Workers' Association of America. The latter, it should be remembered, supplied workers to members of the Independent Glass Company and also to those plants which did not belong to any one of the three combinations of manufacturers.[92] Local Assembly 300 was in

[90] U. S. Commissioner of Labor, *Regulation and Restriction of Output*, p. 603. The monthly prices given by the Bureau of Labor Statistics (U.S.B.L.S., *Wholesale Prices*, Bull. 39, pp. 428–429) suggest that the cost of window glass to retailers increased by something like 55 or 60 per cent between January and April, 1901.

[91] Trade journals in Germany took notice of this tendency on the part of American workers in 1901–1903 to leave their employers and organize co-operative factories. See the *National Glass Budget*, XIX, June 20, 1903. Co-operatives were not a new thing in the industry. See above, p. 61. Throughout the nineteenth century, groups of American glassworkers had frequently organized their own ventures, supplied what little capital had been necessary to build a small plant, worked the furnaces themselves, and shared in the profits and losses. Most ventures of this sort were short-lived. Either they failed from lack of capital or adequate entrepreneurship; or, if successful, a few of the originators bought out the interest of the others and reorganized the firms on a nonco-operative basis.

[92] A letter addressed to Irving W. Colburn and dated April 16, 1903, reveals

favor of helping the combinations to restrict output, but the other union opposed such a policy. Consequently, a wage and price war raged. Local Assembly 300 agreed to reduce wages 5 per cent below any wage cut adopted by the rival union. The resulting chaos caused the National Wholesale Glass Jobbers' Association to suspend activity and stop all its expenses. In the same year (1903) the American Window Glass Company created additional consternation by introducing to the trade its cylinder or Lubbers machine, which could make window glass without the aid of skilled gatherers or blowers.

From its inception, Local Assembly 300 had attempted to prevent the adoption of labor-saving devices,[93] and both it and the Window Glass Workers' Association in 1903 had by-laws which forbade their members to work in any factory equipped with window-glass machines.[94] But fratricidal strife weakened greatly the ability of either to enforce its rule.[95] Confronted with the possibility of having to accept drastic wage reductions amounting to

---

that at that time jurisdiction over the window-glass industry was divided as follows:

	Number of Companies Controlled	Total Pot Capacity	Percentage of Pot Coverage
American Window Glass Company	38	1660	43
Federated Window Glass Company	44	870	22
Independent Glass Company	16	507	13
"Outsiders"	46	832	22

Two of the "outsiders" had not reported their pot capacity. — Libbey-Owens-Ford Glass Company, Colburn Records, III, 10.

[93] See an article entitled, "Glass Workers and the Introduction of Labor-Saving Machinery," *Pottery and Glassware Reporter*, December 3, 1891, found in Gaffield Collection, Scraps, III, 148.

[94] Section 18 of Article I of the by-laws of Local Assembly 300 was as follows: "No manufacturer or company will be allowed to operate any invention or machine for the purpose of making window glass, at any time, where the scale of L. A. 300, K. of L., is in force. In case this is done the wage scale of L. A. 300, K. of L., shall be cancelled, and the members will immediately cease work." The Window Glass Workers' Association of America had an identical rule. — U. S. Commissioner of Labor, *Regulation and Restriction of Output*, p. 612.

[95] Nevertheless, the general manager of the American Window Glass Company told the U. S. Tariff Commission in 1918 that "at the time the machines were introduced we had a great deal of opposition from the glassworkers' organization and the men [i.e., hand manufacturers] who were in sympathy with them. They endeavored in every way to prevent the machines from becoming a success." — Tariff Commission, *Glass Industry as Affected by War*, p. 44.

something like 50 per cent (according to union estimates), the two unions united the following year as the National Window Glass Workers of America. Due to the co-operation of hand manufacturers who looked to organized labor for help in meeting the threat of mechanized production and due to the fact that the Lubbers machine was not then an unqualified success, the new union was able to obtain wage increases in 1905 and 1906. At the same time, however, it decided to discontinue the time-honored restrictions on worker output.[96] Competitive conditions were again intensified in the years 1907–1909, following certain technical improvements in the machine and a marked increase in its use.[97] Both wages and prices fell after the union then consented to tie its wage rates to a sliding scale based upon the selling price of window glass. In order to arrest the downward trend, fifty manufacturers in April, 1909, agreed to form a new selling agency known as the Imperial Window Glass Company.[98] The American Window Glass Company did not participate directly in the new organization, but it offered to co-operate in all attempts to maintain the market.[99] Between April, 1909, and October, 1910, the wholesale price of window glass shot upward; and Imperial is reported as having made over a million dollars' profit within ten months.[100] In November, 1910, however, the Federal Govern-

[96] Davis, "American Glass Industry," pp. 279–280. By 1916 the union was once again enforcing output restrictions in hand shops.—*Glass Industry*, p. 40.

[97] In 1903 there were only 24 cylinder machines in the United States; the number rapidly increased to 130 in 1907 and 160 in 1909. See above, Table 18, p. 195. Davis suggests that the machines in 1909 were capable of supplying almost half the annual domestic consumption.—"American Glass Industry," p. 279. He also states that in the same year about half the hand manufacturers were in bankruptcy.—*Ibid.*, p. 280.

[98] It did not begin business until January, 1910. Over fifteen factories were then leased at high rentals and kept idle in order to decrease output.—*Glass Industry*, p. 235.

[99] The Imperial and the American together controlled approximately 97 per cent of the country's pot capacity.

[100] *Glass Industry*, p. 235. According to this bulletin, the price of window glass increased by 70 per cent, whereas the wholesale prices quoted by the U.S.B.L.S. reveal that the rise for American single-strength glass of first grade (sizes 6″ x 8″ to 10″ x 15″) amounted to only 35.7 per cent.—*Wholesale Prices*, Bull. 87, p. 539, and *Wholesale Prices*, Bull. 93, p. 456. One should not hastily conclude that these figures are contradictory, for they may have reflected price changes in different stages of the marketing process. The higher percentage probably referred to the increase in price received by glass manufacturers; the lower one measured the

ment began an antitrust suit against the company's officials and directors; the defendants pleaded *nolo contendere* and were fined; and the Imperial Window Glass Company was dissolved.

Until the summer of 1912 competitive conditions once again prevailed. At that time J. R. Johnston, who had been an important broker in the trade and secretary of Imperial, organized the Johnston Brokerage Agency and handled the output of about half the hand plants and five machine plants in the country. The American Window Glass Company then assumed the role of "price leader" for the industry, and the Johnston Agency followed its prices. Johnston himself took an active part in promoting further co-operative relations among jobbers, workers,[101] and manufacturers. Largely through his efforts the National Glass Distributors' Association was formed in 1914, and at about the same time producers organized the National Association of Window Glass Manufacturers. Prices were coaxed upward; and when the Libbey-Owens Sheet Glass Company brought out its Colburn machine in 1916–1917, chaos was successfully avoided.[102]

From the preceding discussion it is apparent that the manufacture of window glass was subjected to monopolistic controls during most of the years falling between 1880 and 1920.[103] Part

change in price which retailers had to pay wholesalers. In any event, the increase was marked. As was to be expected in an industry where the entry of new firms was relatively easy, the high prices caused the number of independent companies to increase from five (not counting American) to fourteen before the end of 1910, and five additional ones were then under construction.

[101] In addition to the National Window Glass Workers of America (which, as a legatee of Local Assembly 300, embraced blowers, gatherers, flatteners, and cutters in hand shops), there were two other unions in the industry during and after the period of the first World War: (1) the Window Glass Cutters' and Flatteners' Association of America, and (2) the Window Glass Cutters' and Flatteners' Protective Association of America. The former maintained control over the skilled workmen still employed in independent shops using cylinder machines similar to the Lubbers; the members of the latter (probably a "company" union) worked for the American Window Glass Company.

[102] The foregoing account of the organization of the window-glass industry is largely based upon four sources: U. S. Commissioner of Labor, *Regulation and Restriction of Output; Glass Industry; National Glass Budget*, XXIV–XXXV; and Davis, "American Glass Industry."

[103] Nourse and Drury (*Industrial Price Policies*, pp. 23–25, 172–176) justly concluded that "though this technique of curtailment was for some years effective in holding up prices and also profits, in the end it ceased to do either." I am indebted to Earl J. Hamilton for calling this reference to my attention.

of the time (for example, 1880–1899) the controls were lodged in the hands of two parties — the workers and the manufacturers — thereby creating a "bilateral" monopoly. At other times (for example, 1899–1903 and 1914–1920) the power was shared by three groups — the workers, the manufacturers, and the distributors. This "bilateral" or "trilateral" monopolistic situation, sporadically upset by intense competition, exerted a decided influence upon the development of the window-glass industry. The workers in no other branch of glassmaking were as completely unionized as were those producing window glass,[104] nor were there any who were more successful in maintaining co-operative relationships with most manufacturers.[105] Of course, the split which occurred around 1900 in the ranks of Local Assembly 300 weakened the effectiveness of union control during the critical years which followed, and the advent of the cylinder machine (1903) and sheet-drawing machine (1916–1917) presaged the doom of both skilled workers and hand manufacturers. As compared with Local Assembly 300 in its palmiest days, the labor unions in 1920 wielded little power. And entrepreneurs who continued to employ handicraft methods in their shops must have realized that they were at the mercy of those operating mechanized plants. However, since the powerful chose to "co-operate" with hand producers and with organized labor until after 1920, the full impact of the new machines upon conditions of work was somewhat tempered.

Because of their strategic position in the industry's monopoly framework, window-glass blowers and gatherers enjoyed much higher hourly earnings and shorter working days than did workers of similar skill in bottle and flint-glass factories. From 1890 through 1903, window-glass blowers averaged 81.8 cents an hour and worked only 36.7 hours a week during the production season, whereas flint-glass blowers and green-glass (i.e., bottle) blowers averaged 54.8 cents and 53.5 cents, respectively, per hour and worked 50.3 hours and 51.7 hours, respectively, a week. During

[104] Cf. p. 217, above.

[105] For a few years following 1880 the union maintained a differential wage scale in favor of producers in areas where fuel was costly in order to offset their competitive disadvantage. — U. S. Commissioner of Labor, *Regulation and Restriction of Output*, pp. 600, 615.

the same period window-glass gatherers received on the average 54.5 cents an hour for a week of 36.7 hours, while flint-glass gatherers averaged only 28.7 cents an hour and worked about 49.5 hours each week.[106] In 1906–1907 blowers of window glass earned about 16 per cent more per hour than bottle blowers and 38 per cent more than flint-glass blowers; window-glass gatherers averaged approximately twice as much each hour of work as did those employed in gathering flint glass.[107]

[106] Calculated from the data on wages and hours found in *Wages and Hours of Labor*, pp. 93–97. Flatteners and cutters of window glass averaged 44.5 cents an hour (for a 68.5 hour week) and 40.6 cents an hour (for a 58.8 hour week), respectively. It appears from the data presented in the report prepared by Aldrich for the U.S. Committee on Finance (*Wholesale Prices, Wages, and Transportation*, IV, 1566, 1572) that window-glass and bottle blowers in New Jersey had approximately the same weekly earnings in 1881–1889, but the former presumably worked fewer hours each week.

[107] *Wages and Hours of Labor in Manufacturing Industries, 1890–1907*, pp. 40–41. The trend of hourly wages paid skilled window-glassworkers (blowers, gatherers, flatteners, and cutters) compared with the trend of hourly earnings of other skilled glassworkers (bottle blowers, flint blowers, flint gatherers, and flint gaffers) during the critical period 1890–1907 is revealed by the indices constructed from wage relatives (1890–1899 = 100) published in the same bulletin (pp. 85–87):

Year	Skilled Window-Glassworkers	Other Skilled Glassworkers	Year	Skilled Window-Glassworkers	Other Skilled Glassworkers
1890	109	98	1899	111	103
1891	104	98	1900	123	106
1892	107	100	1901	116	110
1893	103	99	1902	143	110
1894	97	102	1903	151	112
1895	77	99	1904	112	114
1896	89	101	1905	136	111
1897	95	101	1906	152	115
1898	109	99	1907	166	117

The sharp decline in window-glass hourly wage rates in 1894–1895, compared with the steady trend in the rates paid skilled workers in other branches of the industry, reflected the disintegration of union-employer relationships following 1893, which ended with the organization of the American Glass Company in 1895. The upward climb from 1896 through 1900 (due to the restoration of "co-operative" arrangements between union and manufacturers) received a marked setback in 1901 because of the split in Local Assembly 300 and the deterioration of unified control in the industry. Hourly earnings of window-glassworkers soared to new heights in 1902–1903, after the two unions and the three combinations of manufacturers initiated "harmonious" relationships. The drastic reduction in the wage index from 151 in 1903 to 112 in 1904 resulted mainly from the competitive wage cutting of the two unions and the uncertainty engendered in the trade by the introduction of the Lubbers cylinder machine. The rapid rise of the index to 166 in 1907 stemmed chiefly from the realization on the part of workers and hand

Unfortunately, data are insufficient to allow similar comparisons for later years. There is cause to believe, however, that the differential advantage of window-glass blowers disappeared completely as competition with machines increased and as the union's strategic position deteriorated. After the National Window Glass Workers of America in 1907 adopted a sliding scale of wages tied to selling prices, the piece-rate earnings of blowers fell until in 1912–1913 they were only one-third of what they had been in 1903.[108] The general manager of the American Window Glass Company in 1918 stated that skilled blowers had received only about $15 per week during the dark days of 1911–1912 when price and wage competition was fiercest.[109] During the last year of the first World War, cutters and flatteners in machine plants were paid the same wages as those employed in hand factories, but skilled blowers earned no more (and sometimes less) than did those men tending the blowing machines.[110]

Variations in monopoly controls were reflected not only in the relative trend of wages received by skilled glassmen but also in the attempts of workers and manufacturers to restrict output and in the resultant fluctuations of wholesale prices. The U. S. Department of Commerce and Labor reported in 1904 that "the regulation of output [of window glass] . . . by the joint action of skilled workmen, manufacturers, and jobbers . . . has been brought about in three ways, namely, by limiting the maximum output per week of each blower, by limiting the number of blowers so that there are not enough to operate all the pots and pot capacity, and by alliances of the unions with leading manufacturers to force a shut down of the independent manufacturers when stocks accumulate." [111] From the outset Local Assembly 300 had set limits on the number of cylinders each blower could make during

---

manufacturers that the machine had not yet been perfected commercially and that the American Window Glass Company did not intend to use it for starting a price and wage war.

[108] Tariff Commission, *Glass Industry as Affected by War*, p. 44.

[109] *National Glass Budget*, XXXIV, September 7, 1918, p. 1.

[110] Tariff Commission, *Glass Industry as Affected by War*, p. 79.

[111] U.S. Commissioner of Labor, *Regulation and Restriction of Output*, pp. 599–600.

the week; [112] and until 1902–1903 [113] the workers abided by the rules. The new union organized in 1904 lifted the restrictions for several years but later reinforced them after co-operative relationships were established between hand and machine plants.

In order to limit the supply of skilled labor, the unions retained complete control over the number of applicants accepted as apprentices. Only the sons and brothers of legitimate union members were eligible to apply.[114] In 1879 the membership fee was only $1.50; ten years later it was $100; and in 1903 it was reduced to $25 for those who completed the prescribed apprenticeship period. Foreign glassmen had to pay $500.[115]

Window-glass manufacturers meeting in Cleveland in 1879 had decided to work their furnaces only nine months in the year in order to limit competition, curb over-production, and decrease bankruptcies.[116] In January of the following year the members of the American Window Glass Manufacturers' Association agreed to limit their output from year to year by as much as a two-thirds majority should decide and to refrain from making glass during the months which the Association should specify.[117] Local Assembly 300 co-operated with the manufacturers on this matter, and by 1885 the blast of nine months had become standard in window-glass factories.[118] The three combinations of manufacturers (American, Federated, and Independent) openly attempted to control output during the first few years of the twentieth cen-

[112] Libbey-Owens-Ford Glass Company, Colburn Records, Vol. B, 41, 75; Davis, "American Glass Industry," pp. 177–178.

[113] When rumors about the forthcoming Lubbers cylinder machine were flying thick and fast around glass furnaces, hand blowers apparently began to disregard their unions' weekly output limits of 30 boxes for double-strength glass and 48 boxes for single-strength.

[114] See Davis, "American Glass Industry," p. 179; U.S. Commissioner of Labor, *Regulation and Restriction of Output*, p. 617.

[115] The immigration of workers at times had become alarming, and the union in 1885 promoted the formation of the Universal Federation of Window Glass Workers of the World in the hope that through such an organization it could stem the tide. — U.S. Commissioner of Labor, *Regulation and Restriction of Output*, pp. 614–616.

[116] Gaffield Collection, Glass Journal, III, 232.

[117] Gaffield Collection, Minutes of the Proceedings of the American Window Glass Manufacturers' Association.

[118] U.S. Commissioner of Labor, *Regulation and Restriction of Output*, p. 601.

tury by varying the length of the annual blast. It may be recalled that one of the reasons for the temporary breakdown of monopoly control in 1903–1904 was the unwillingness of one of the two labor unions to accept the further reduction in the working year suggested by two of the combinations.[119] Between 1914 and 1917 hand factories were working about six or seven months each year, and mechanized plants were running from eight to ten months.[120] It would appear, therefore, that whenever effective organization in the industry permitted, output was restricted.

The demand for window glass was relatively inelastic[121] and depended largely upon the volume of new construction in the country. Since the cost of windowpanes probably did not amount to much more than 1 or 2 per cent of the total expenses involved in erecting buildings, great changes in price should not have occasioned proportionate variations in the amount of window glass purchased on the market. Manufacturers, therefore, were correct in thinking that they could increase their total net profits by restricting output. The profits of the first selling pool (the American Glass Company) were reported as being $700,000 in 1896, $1,750,000 in 1897, and $2,100,000 in 1898.[122] And the fact that the Imperial Window Glass Company, another selling agency, cleared over $1,000,000 in ten months during 1909–1910 has already been mentioned.[123]

The prices of window glass quoted on the wholesale market to merchants selling at retail unmistakably reflect changes in the degree of monopoly power exercised in the industry.[124] Table 19 presents wholesale price indices for (1) two grades of American window glass, (2) five other kinds of glass (nappies, pitchers, tumblers, and two sizes of polished plate glass), and (3) all commodities included in the Bureau of Labor Statistics' general index.

[119] See above, pp. 221–222.

[120] Tariff Commission, *Glass Industry as Affected by War*, p. 21. See also *Glass Industry*, p. 194.

[121] There were few substitutes for window glass. Plate glass, the best of these, was much more expensive than ordinary window glass.

[122] Davis, "American Glass Industry," pp. 294–295.

[123] See above, p. 223.

[124] As one should have expected on a priori grounds, these monthly averages presented by the U.S. Bureau of Labor Statistics were not sensitive enough to reveal immediately and fully all variations in prices received by manufacturers.

TABLE 19

WHOLESALE PRICE INDICES FOR WINDOW GLASS, OTHER KINDS OF GLASS, AND GENERAL COMMODITIES, 1890–1920 *

Base: 1913 = 100

	General Commodities	Window Glass	Other Kinds of Glass
1890	81	81	156
1891	80	81	158
1892	75	73	142
1893	77	78	142
1894	69	73	131
1895	70	61	129
1896	67	69	131
1897	67	85	102
1898	70	101	111
1899	75	104	110
1900	81	102	125
1901	79	150	135
1902	84	117	128
1903	86	97	135
1904	86	106	115
1905	86	99	113
1906	89	105	117
1907	94	102	118
1908	90	86	100
1909	97	84	103
1910	101	107	107
1911	93	82	97
1912	99	81	95
1913	100	100	100
1914	98	99	100
1915	101	111	96
1916	127	125	121
1917	177	166	179
1918	194	267	193
1919	206	284	220
1920	226	299	272

* Source: U. S. B. L. S., *Wholesale Prices*, Bull. 320.

During the years 1892–1895, when the control exercised jointly by Local Assembly 300 and the American Window Glass Manufacturers' Association weakened, the wholesale price of window glass dropped more considerably than did that of general commodities.[125] The formation in 1895 of the selling pool known as the

[125] The marked decline in the price of "other kinds of glass" (from an index of 156 in 1890 to one of 129 in 1895) was due mainly to the decrease in polished plate-glass prices, a result of technological improvements.

American Glass Company and in 1899 of the manufacturing combination called the American Window Glass Company permitted the price of windowpanes to be pushed upward, not only absolutely but also relatively to other commodities. The period of price competition in 1900 between American and the independents was of such short duration that it did not cause the annual index number to decline very much. After the "share-and-share-alike" agreement with the newly formed Independent Glass Company was inaugurated by the American Window Glass Company and its affiliate, the National Wholesale Glass Jobbers' Association, the price index for window glass soared in 1901 and declined almost as precipitously the following year, when monopoly controls deteriorated. Uncertainty occasioned by rumors of the new cylinder machine was mainly responsible for the further drop in 1903. When this Lubbers machine failed in its first year to justify the mutual fears of hand workers and employers, confidence and "stability" returned. Accordingly, the price index rose by more than 9 per cent. The reappearance of competitive conditions in 1907–1909 and in 1911–1912 caused prices to fall again each time. The activity of the short-lived Imperial Window Glass Company was reflected in a marked price rise in 1910; and the "co-operative" relationships effected by the Johnston Brokerage Agency, the National Glass Distributors' Association, and the National Window Glass Manufacturers' Association were partially responsible for the rise in the annual index after 1912.[126]

The chief result of collusion among producers was accurately stated in 1918 by the Tariff Commission:

> Summarizing American productivity efficiency in the making . . . of window glass by machines, together with the marketing methods employed, it is evident that the competitive effectiveness of . . . [this branch of the glass] industry at home and abroad is practically nullified by the policy of restricting output to suit the requirements of hand production and of marketing automatic and machine production under conditions that encourage the continuation of antiquated methods and of

[126] War-time conditions, of course, were mainly responsible for the tremendous increase, but it should be noted that the price of window glass reached a higher peak than the prices of wholesale commodities generally or of other kinds of glass.

profits for machine production based on the cost of production of hand-made glass.[127]

The very high wage rates exacted by Local Assembly 300 prior to 1903 and the authoritative power wielded by organized labor created a strong incentive to perfect the cylinder machine as soon as possible. After introducing the machine, the American Window Glass Company preferred to maintain prices and widen its profit margin rather than pass the resulting economies on to the consumer through lower prices. William L. Monro, general manager of the company at that time, testified before the Tariff Commission in 1918 as follows:

I do not mind telling you that we have had no effect [!] on the wages of the hand blowers in this country, as a matter of policy we have had no desire to drive the hand blowers off the map. There is no other machine on the market today that can drive them out, other than our machine,[128] and we probably could; but in driving anybody out of a line of business you have to take some of the medicine you administer to them, and from a business standpoint we thought it was much more profitable for us to be satisfied with a reasonable share of the country's business rather than to drive out operatives from an industry that has existed as long as this one had existed. The policy of live and let live has been our policy throughout, not only in production, but in prices.[129]

One entrepreneur who still followed the hand process for window glass attempted to justify the policies of organized labor before the same commission:

I am not condemning union labor because in the glass business especially I do not know that they could hold the business unless they were unionized; they would be helpless. *They work with us to maintain the trade.*[130]

The execution of such co-operative policies by the three power groups — the largest mechanized company, the hand manufac-

[127] *Glass Industry as Affected by War*, p. 49.

[128] Monro was here showing a natural unwillingness to concede the possible superiority of the Colburn sheet-drawing machine over his company's cylinder machine. The Libbey-Owens Sheet Glass Company, it is true, had not been able to pronounce its Colburn method an unqualified success before Monro testified. See above, p. 128.

[129] Tariff Commission, *Glass Industry as Affected by War*, pp. 47–48.

[130] *Ibid.*, p. 89. My italics.

turers, and the unions — kept the market more or less stable while the entrepreneurs at Toledo perfected their new process. The potential profit margin, consequently, must have appeared greater to E. D. Libbey, M. J. Owens, and their associates in the years 1912–1916 than it would have otherwise. It is true that the cost differential in favor of machine producers had been reduced somewhat because the wages of skilled blowers and gatherers relative to other workers were less than they had been before 1903. But one group of skilled laborers — the flatteners — had not been forced to accept long-range drastic wage cuts. Their services remained essential whether window-glass cylinders were blown by machine or by hand. Their hourly earnings increased from 1903 to 1919 by approximately the same percentage as did the general wage index compiled by the Bureau of Labor Statistics.[131] The Colburn process eliminated flatteners as well as blowers and gatherers; hence it enjoyed an even greater labor-cost advantage than did the Lubbers process.[132] It would indeed be rash to state that Irving W. Colburn would not have begun work on the sheet-drawing method around 1906 or that the Toledo Group would not have invested time, effort, and capital between 1912 and 1916 in bringing his machine to the market if wages in the trade had been lower than they actually were, or if competitive conditions had prevailed. But one may safely assume that the profit incentive would have been less and that the Libbey-Owens Sheet Glass Company probably would not have been able to begin paying dividends to its stockholders as soon as it did, had the selling price of window glass been determined in a competitive market.

The Toledo Group was no more anxious to upset the trade in window glass than it had been desirous of disturbing the bottle market earlier.[133] The *National Glass Budget* [134] reported in 1917 that M. J. Owens had recently attended the window-glass jobbers'

[131] The general wage index rose from 80 in 1903 to 184 in 1919. — U.S.B.L.S., *History of Wages, Colonial Times to 1928*. Window-glass flatteners averaged 58.68 cents per hour in 1903 and $1.343 per hour in 1919. — *Monthly Labor Review*, X, (1920), p. 102; *Wages and Hours of Labor*, pp. 93–97.

[132] See above, p. 173.

[133] See above, pp. 107–108, 211 ff.

[134] Vol. XXXIII, September 15, 1917, p. 3.

annual meeting and had addressed the gathering briefly. "He stated that it was the purpose of his company [i.e., the Libbey-Owens Sheet Glass Company] to co-operate in every way with those at present engaged in the manufacture of window glass; that it would be the policy of his company to pursue a course such as was carried out in the bottle trade on the advent of the Owens automatic machine." The same trade journal stated that a joint committee of Libbey-Owens and the National Window Glass Manufacturers' Association met on October 21, 1918, and decided upon the output quota to be allotted Libbey-Owens for 1918–1919.[135] Even before this the directors in the new company had entered into a contract with the Johnston Brokerage Agency whereby the latter organization was to market the output of Libbey-Owens in much the same way as it marketed the products of most firms in the industry.[136] Therefore, the fact that approximately one-fourth the window glass produced in this country in 1919 was made by hand, 65 per cent by cylinder machines, and only one-tenth by the superior sheet-drawing process,[137] partially reflects the collusive practices in manufacturing and marketing.

3. Organization in Other Branches of the Industry

Although tableware, lighting goods, and plate glass were not manufactured under truly competitive conditions throughout the decades 1880–1920, it seems hardly to be worth-while to describe in great detail the variations of organizational control in these branches of the industry. Such a description would shed little additional light on the probable cause-and-effect relationship between the restrictive policies of labor unions or manufacturers' combinations, on the one hand, and the activities of the Toledo Group, on the other. Wages were maintained at relatively high levels wherever workers were strongly unionized, and output was also restricted by means of the summer stop and, especially, through limitations upon worker productivity. With the introduction of semiautomatic machines for blowing tumblers, lamp chim-

[135] *Ibid.*, XXXIV, October 26, 1918, p. 11.

[136] Libbey-Owens-Ford Glass Company, Libbey-Owens Sheet Glass Company Minutes, I, 160.

[137] See above, Table 17, p. 194.

neys, and bulbs the pattern of industrial control changed. As manufacturers making pressed glassware improved their machines by applying electric power and automatic feeding devices, they gradually forced hand pressers to specialize in fashioning novelties and other goods marketed in small quantities. The majority of employees in plate-glass factories were unskilled or, at best, semi-skilled. They remained unorganized until after 1920 and, consequently, were powerless to exert much influence in the trade. On the manufacturers' side, the most interesting developments were the unhappy experience of two groups of tableware producers who tried to strengthen their position in the market by merging into what proved to be unwieldy combinations, and the successful attempt of the Pittsburgh Plate Glass Company to attain a position of dominance in its own field.

Flint and lime glassworkers formed their first successful national organization in 1878. This American Flint Glass Workers' Union was composed of several departments, each of which embraced those workers specializing in a particular line of ware such as lamp chimneys, pressed tableware, or bottles and vials.[138] Until 1887–1888 the union was unable to institute uniform wages and working rules in each line of production for all sections of the country. It carried on collective bargaining regionally. Wages were set, and limitations upon the amount of ware each worker could make during a "move" (the technical term for half a day's work) were everywhere practised; but regional differences persisted. For example, piecework wages were lower and output limits on individual workers higher in eastern than in western shops.[139] Producers along the seaboard had argued that without this special treatment they could not compete with manufacturers in the interior because of marked differences in fuel costs. By the end of 1888, however, the union had succeeded in abolishing the regional wage and output differentials and had persuaded all manufacturers to cease work completely for six weeks during the summer.[140] It was agreed, furthermore, that wage rates and other questions

[138] The activities of the bottle department have already been treated in the first section of this chapter.

[139] Gaffield Collection, Scraps, III, 8–10; U.S. Commissioner of Labor, *Regulation and Restriction of Output*, pp. 627, 653.

[140] *Ibid.*, pp. 628–629, 653.

would be settled by means of annual conferences composed of representatives of the union and of manufacturers.[141] In this way the two sides avoided industrial strife until 1892, when probably the greatest strike in the history of American glassmaking occurred.

In the previous year fifteen firms making blown and pressed ware out of lime and flint glass had merged to form the United States Glass Company. The new corporation, capitalized at over $4,000,000, cleared almost $3,000,000 profits its first year.[142] Almost immediately it demanded that the union remove all restrictions upon worker output. Compliance would have been unprecedented; long before they were organized, skilled glassmen had practiced control of output. Quite naturally, therefore, the union refused to comply with the request. The ensuing strike lasted four years and cost the American Flint Glass Workers' Union around $2,000,000 in strike benefits.[143] The United States Glass Company succeeded in staffing its shops with greenhorns and nonunion blowers, but its officials later agreed that the company never recovered from the losses suffered during the conflict. Although in 1918 it was still operating twelve plants,[144] it rarely ever had shown any profit except in its first year.[145]

The great strike of 1892–1896 also destroyed much of the union's power. In 1896 the pressed tableware department voted to rescind all restrictions on worker productivity in those plants over which it had control.[146] The summer stop likewise was abol-

[141] *Ibid.*, p. 660.

[142] J. Gillinder, in *One Hundred Years of American Commerce*, pp. 282–283. Gillinder pronounced it "without question" the largest flint-glass company in the world.

[143] U.S. Commissioner of Labor, *Regulation and Restriction of Output*, pp. 653, 659.

[144] Tariff Commission, *Glass Industry as Affected by War*, p. 127; *National Glass Budget*, XXXIV, October 5, 1918, p. 5.

[145] The company probably should be considered as a child of the combination movement sweeping through American industry around 1891. It is still (1947) in existence, but its size has greatly decreased. It now operates only one small factory (located at Tiffin, Ohio) which produces a rather good grade of crystalware known in the trade as "Tiffin Ware." Boxes upon boxes of records relating to its earlier activities are stored away in a dark, grimy attic at the plant.

[146] The output of a unit ("shop") of workers, consequently, almost doubled.—See U.S. Commissioner of Labor, *Regulation and Restriction of Output*, p. 659.

ished.[147] Consequently, when new machines made their appearance in the trade, labor was unable to fight them effectively. Those workers and those manufacturers who, for one reason or another, preferred to continue their handicraft methods rather than swing over to machine production simply diverted their energies to new products as new techniques invaded their old lines. The demand of American consumers for specialties with distinctive quality and design provided craftsmanship with something like a safety valve. Because such items could be produced economically only in small quantities,[148] it was not feasible to use expensive glassmaking machinery. In spite of the fact that their real wages declined, craftsmen specializing in pressed and blown ware were among those least affected by technological change.[149]

Manufacturers in the field were organized into various trade associations,[150] but not one of them separately nor all of them collectively were strong enough to control the market. New firms could enter the industry with relative ease because the amount of capital required was not prohibitive. Competition, consequently, remained stronger and monopoly controls weaker in the production of tableware than in any other branch of the American glass industry. The United States Glass Company, which was mentioned earlier, proved unprofitable partially because the technology employed in blowing or pressing fine tableware was such that the economies to be derived from combination on the horizontal plane were more than offset by managerial diseconomies. This same weakness was also revealed in the experience of another combination. On November 1, 1899, nineteen important firms, representing roughly half the available productive capacity in the general line of tableware, tumblers, and novelties, merged to form the

[147] Davis, "American Glass Industry," p. 361; U.S. Commissioner of Labor, *Regulation and Restriction of Output*, pp. 629, 653; *Productivity of Labor in the Glass Industry*, p. 100.

[148] *Productivity of Labor in the Glass Industry*, p. 99; Davis, "American Glass Industry," pp. 338, 342, 358.

[149] This is also the opinion of Davis ("American Glass Industry," pp. 358, 360).

[150] The 1905 *Directory of the Glass Trade* published by the *Commoner and Glassworker*, for example, lists the National Association of Manufacturers of Pressed and Blown Glassware, the American Association of Flint and Lime Glass Manufacturers, and the Glass Association.

loosely knit National Glass Company.[151] By 1905, however, the company had turned back all its factories (with the exception of four which were then lying idle) to their original owners or to other entrepreneurs who operated them independently under leases.[152]

Lamp-chimney workers, like those making tableware, were also organized as a department of the American Flint Glass Workers' Union, and until the twentieth century they were quite successful in maintaining high wages, limitations upon output, and the six-weeks' summer stop.[153] The advent of the Owens semiautomatic chimney machine in 1898 occasioned grave concern; and, according to a statement made by a union official in 1903, his organization attempted to buy outright the machine "with all its patents and protections." "Had we succeeded," he stated, "the machine would have been laid aside. We were prepared to give $300,000 for these monopoly rights, but we could not buy it." [154] The union, nevertheless, was determined to control its operation in order to minimize the cost differential between hand and machine plants.[155] The Macbeth-Evans Glass Company, which had purchased the exclusive chimney rights from the Toledo Glass Company, maintained a union shop and constantly complained that the union's restrictions upon employees operating the machines were depriving it of all opportunity for profit. Open conflict developed in 1904–1905, after the union refused to modify its policies, and Macbeth-Evans emerged as an open shop.[156] Since the company was the largest single producer of lamp chimneys, the break was a serious blow to the union. Complete demoralization in the trade, however, was avoided through a realization on the part of both workers and manufacturers that the market for lamp chimneys was dwindling.

[151] S. P. Austin, "Glass," *Twelfth Census*, Vol. IX, Part III, 978.

[152] *Directory of the Glass Trade* (published by the *Commoner and Glassworker*, 1905), p. 21.

[153] A prolonged strike over the use of a patented crimping device in 1879–1880, however, had almost ruined the union. — U.S. Commissioner of Labor, *Regulation and Restriction of Output*, pp. 642–643. See also above, p. 38.

[154] *Ibid.*, p. 650. However, I have found no indication whatever in the company records at Toledo that the union offered to buy the machine rights from the Toledo Glass Company at any time.

[155] *Ibid.*, pp. 644 ff.

[156] Tariff Commission, *Glass Industry as Affected by War*, p. 104.

Employers were unwilling to tie up any more of their funds in labor-saving devices, and labor recognized that militant tactics on its part would be of little avail. Consequently, this unofficial truce permitted hand plants to subsist alongside those which were mechanized.[157]

Since the Toledo entrepreneurs did not concern themselves with plate glass until after 1920, the organization of this branch of the industry had little effect upon their activities. It will suffice to state that the number of firms making plate glass remained small (never more than eleven before 1920), and that one — the Pittsburgh Plate Glass Company — was so much larger than the others that it was able to assume the role of "price leader" shortly before or after 1900.[158] The New York City Plate Glass Company, organized in 1881 under the leadership of Captain John B. Ford and his son Edward, built a large factory at Creighton near Tarentum, Pennsylvania. In August, 1883, the company changed its name to the Pittsburgh Plate Glass Company and began building new plants. In 1895, through a reorganization, it acquired all but three of the factories making plate glass in the country.[159] There were ten independents in 1908–1911 and six in 1920.[160] There is no evidence, however, that the size and power of the Pittsburgh Plate Glass Company deterred the Toledo entrepreneurs, under the leadership of E. D. Libbey and M. J. Owens, from entering this branch before 1920. It should be remembered that this entrepreneurial group, in trying to produce sheet glass on the Colburn machine, faced about as many technical and organizational problems as it could handle successfully. As soon as these were solved, it could, and did, turn its attention to the profitable opportunities awaiting any mechanized producer in the plate-glass field.[161]

[157] As late as 1927 it was estimated that from two-fifths to one-half of the chimneys made in this country were blown by hand. — Davis, "American Glass Industry," p. 356; *Productivity of Labor in the Glass Industry*, p. 114.

[158] See Davis, "American Glass Industry," p. 395.

[159] Its plants were located at Creighton, Tarentum, Ford City, Cochran, and Charleroi in Pennsylvania, Elwood and Kokomo in Indiana, and Crystal City in Missouri. — Libbey-Owens-Ford Glass Company, Colburn Records, IV, 31.

[160] Tariff Commission, *Flat Glass*, p. 92.

[161] Owens himself shortly before his death experimented with certain devices (e.g., the wood-block bending belt) which he hoped could be used successfully in making plate glass on a modified Colburn machine.

CHAPTER IX

FOREIGN COMPETITION, DOMESTIC CONSUMPTION, AND PLANT RELOCATION

In addition to the influence exerted by organized labor and manufacturers' associations, other factors affected the growth of glassmaking and improvements in its technology. At least three of these were important enough to justify separate treatment. Throughout most of the nineteenth century American glasshouses enjoyed tariff protection from lower-cost producers abroad, and import duties remained high until 1913. Furthermore, the outbreak of the European war the following year more than counteracted the tariff reductions incorporated in the Underwood Act. The population of the United States meanwhile was increasing in size, experiencing a general rise in its standard of living, and shifting its consumer preferences in favor of glassware. Manufacturers in the decades 1880–1920 not only faced this expanding domestic market but also discovered in natural gas their cheapest and best fuel. Accordingly, they shifted the location of their furnaces to the natural-gas areas of the Middle West. Successive sections in the present chapter, therefore, analyze (1) the extent of foreign competition and the effect of the protective tariff, (2) changes in the price of glassware and shifts in consumer demand, and (3) the availability and cheapness of basic materials and their influence in guiding the relocation of glass factories.

1. Foreign Competition and the Protective Tariff

In 1880 the manufacture of pressed glassware apparently was the only branch of the industry in which American manufacturers were able to compete with foreign producers abroad and everywhere at home.[1] The introduction of the side-lever hand press and the development of cheap lime glass had so lowered manufacturing costs that Americans were able to sell their pressed ware

[1] See above, pp. 34, 51–52.

throughout South America and even in Europe. The high cost of labor and the relatively backward state of technological methods caused domestic producers in other branches to have higher unit costs than foreign competitors. High tariff barriers to some extent counteracted this disadvantage. In 1880 the duties collected upon all dutiable glass imports amounted to almost 55 per cent of their value.[2] And for the same year the *ad valorem* equivalents of specific duties ranged from 20 to 103 per cent for unsilvered, polished plate glass, depending upon its size; from 35 to 40 per cent for unfilled bottles and jars; from 46 to 57 per cent for unpolished window glass, depending upon its size; and from 35 to 40 per cent for pressed and blown glassware.[3] Such tariff rates offered substantial protection to American manufacturers and therefore help to account for the industry's size in 1880. But did domestic glassmaking remain in the relatively high-cost field from 1880 through 1920? Did the government continue to favor it by maintaining high import duties? And if so, was the tariff in any way responsible for the mechanical revolution following 1890? These are the questions with which the present section is concerned.

Despite the reductions in rates of pay after 1903, the wage bill in America remained at a higher level than the wage bill of European producers. A comparison of thirty-seven American factories with one European plant in 1890 showed that the labor costs involved in making window glass were three times higher in the United States; the labor cost involved in a box of handmade window glass in a Belgian shop was 42.1 cents in 1909, and for thirty-five American shops in 1916 it averaged $1.11.[4] Another comparison may be useful even though the year for which it was drawn comes after the terminal date of this study. It has been estimated that in 1926 the direct labor employed in making a pound of window glass in six American plants equipped with Fourcault or Colburn sheet-drawing machines amounted to nine-tenths of a

[2] *Glass Industry*, p. 335; Gaffield Collection, Scraps, III, 134.

[3] Davis, "American Glass Industry," pp. 189, 214, 231–232, 252. Davis apparently calculated his *ad valorem* equivalents by dividing the average specific duty for each grade of glass by the corresponding average import value. See *ibid.*, pp. 130, 145 n. 2, 187 n. 3.

[4] Tariff Commission, *Glass Industry as Affected by War*, p. 45.

cent while in ten Belgian factories using the Fourcault it amounted to only thirty-six hundredths of a cent.[5] Throughout the hearings held by the Tariff Commission at Pittsburgh in 1918 manufacturers complained of the much higher wages they had to pay compared with foreign competitors.[6] And upon the same occasion the report of representatives of the American Flint Glass Workers' Union who had visited Europe around 1905, was quoted as stating the fact that "in the cost of manufacture of glass, *aside from the skilled labor*, there is very little difference, if any, on either side of the Atlantic." [7] Existing wage differentials help to account for the fact that American factories were not producing window glass as cheaply as were those in Belgium.[8]

Not only did Belgian window-glass producers have lower unit manufacturing costs than Americans, but they also benefited from lower transport charges to markets along the seacoast of the United States. Before the first World War they could ship their glass to the Pacific coast for only 35 cents per hundredweight and to New Orleans for 14 cents; the overland freight costs on a like quantity sent from any factory in the American Middle West was around 90 cents to the Pacific coast and (from Pittsburgh) 43 cents to New Orleans.[9] Because of such differentials combinations of American producers, as early as the turn of the century, had divided the domestic market into six or more pricing districts. Prices were set at a higher level in interior regions close to manu-

[5] Tariff Commission, *Flat Glass*, p. 66. The comparable figure for sixteen machine-cylinder plants and two hand-cylinder plants in the United States was 1.61 cents; for three hand-cylinder plants in Belgium, 1 cent.

[6] Tariff Commission, *Glass Industry as Affected by War*, *passim*.

[7] *Ibid.*, p. 53. My italics.

[8] From 1923 through 1926 a square foot of window glass could be manufactured in Belgium at about half the cost in American factories. — Tariff Commission, *Flat Glass*, pp. 66, 103. It is probable that the cost differential was at least as great prior to 1920 (and certainly before Belgium became embroiled in the first World War) because window-glass machinery was then much less prevalent in American factories than in 1923–1926. See Table 17 p. 194.

[9] Testimony given before the 1918 Tariff Commission by William L. Monro of the American Window Glass Company. — *National Glass Budget*, XXXIV, September 7, 1918, p. 4. As late as 1936, according to the Tariff Commission, freight costs on Belgian window glass shipped to American ports along the Gulf, Pacific, and Atlantic coasts were much less than railroad charges from domestic centers of production to these points. — Tariff Commission, *Flat Glass*, p. 71.

facturing centers than in coastal areas. Around 1900 purchasers at Pittsburgh had to pay 14 cents more for fifty square feet of glass than did customers at Boston and 20 cents more than did those at Philadelphia.[10] Even the *National Glass Budget*, a trade journal loyal to American manufacturing interests, admitted in 1901 that in no other industry had prices been adjusted so carefully to the costs of imported ware in different sections of the country.[11] Hence it seems conclusive that, at least through 1920, the costs of making and marketing window glass in the United States remained higher than abroad — especially in Belgium — even though important technological innovations after 1903 had decreased the differential.

The manufacture of polished plate glass in 1880 could reasonably have been classed as an "infant industry" whose unit costs were considerably higher than those of foreign producers.[12] The head of the plant at New Albany, Indiana, told the Tariff Commission in 1882 that the duties on large plates (amounting to approximately 80 per cent) were probably sufficient but that those on the smaller sizes should be raised above their 20 per cent level.[13] At that time the techniques employed in American factories were inferior to those used in the most efficient European works. Before 1920, however, domestic producers increasingly had become recognized as important innovators as well as leaders in the reduction of costs through large-scale operations.[14] Since this branch of the industry employed only unskilled and semiskilled workers, American manufacturers did not have to pay wages which were as high, compared with European rates, as did producers who found skilled glassmen indispensable. The differential cost barrier lessened during the decades 1880–1920 even though it did not entirely disappear.[15] It is significant that, starting in 1908, the statistics compiled by the Bureau of Foreign and Domestic Com-

[10] Cf. Davis, "American Glass Industry," pp. 292, 294.

[11] *National Glass Budget*, XVII, May 25, 1901, p. 4 (also referred to by Davis, "American Glass Industry," p. 291).

[12] See above p. 52.

[13] Gaffield Collection, Scraps, III, 123.

[14] See below, pp. 334–336.

[15] This is also Davis' opinion ("American Glass Industry," p. 393).

merce list separately the exports of American plate glass even though the quantities remain small until 1915–1920. The fact that exports decreased markedly while imports greatly increased with the return of world peace [16] suggests that the domestic industry still suffered from high costs.[17]

As has already been indicated, the United States enjoyed a competitive advantage in the production and sale of pressed glassware. Firms in this field were the only ones making glass which did not seek increased tariff protection in 1882 but merely asked for a retention of the old 35 per cent duty, which was itself largely superfluous.[18] The development of rotary, power-driven presses and automatic take-outs and conveyors following 1900, furthermore, helped American producers maintain their superior position. In the blown-ware branch, however, even a 60 per cent duty could not equalize foreign and domestic costs, and in this connection the most responsible factor seems to have been the high wage rates paid American glassmen. As semiautomatic and automatic machines for making tumblers, light bulbs, lamp chimneys, and glass tubing appeared on the market, domestic plants attained costs equal to, if not below, those of foreign competitors. Wherever handblowing and the consequent employment of skilled labor continued necessary, unit costs remained much higher than the selling price for similar ware drawn from abroad.[19]

Until the twentieth century this was largely true of glass con-

[16] Exports of plate glass rose from around $40,000 in 1914 to about $4,220,000 in 1919 and fell to slightly more than $1,000,000 in 1922. Imports increased from around $4,000 in 1918 to approximately $2,500,000 in 1920 (and $16,300,000 in 1923). — *Ibid.*, Appendix R.

[17] At least three other factors besides the end of the war were involved in the sudden increase of imports: (1) the great expansion in the demand for plate glass in the automobile and building industries, (2) the dislocation of exchange rates, and (3) the existence of a European plate-glass cartel. Since 1904 foreign producers had been organized as the International Convention of Plate-Glass Manufacturers, "one of the strongest and most successful trusts known to the commercial world, controlling absolutely and arbitrarily the output of its members." — Tariff Commission, *Glass Industry as Affected by War*, p. 61. With a firm control over the European market, it appears to have set low discriminatory prices on glass sold in America. Cf. Davis, "American Glass Industry," p. 392.

[18] Davis, "American Glass Industry," p. 233.

[19] Before the first World War about half the thin-blown fine tableware and scientific glassware employed in university and research laboratories in the United States had been imported. See *ibid.*, p. 363.

tainers as well. But several articles in the *National Glass Budget* contended that by 1900 the United States led the world in the manufacture of fruit jars and that by 1920 her bottle industry also had attained world supremacy. Semiautomatic and, especially, automatic labor-saving machinery as well as increases in the scale of operations were almost entirely responsible for the change. The fact that exports of bottles and jars jumped from negligible values in the 1880's and 1890's to something like $9,500,000 in 1920 (and averaged about $3,000,000 annually during the years 1921–1930) supports the conclusion that the industry's competitive status in the world market improved markedly as a result of mechanization.[20]

It would appear from this brief survey of the industry's several branches that in the case of window glass, plate glass, and blown glassware American factories continued through 1920 to have higher costs of production than did Europeans, that those making containers came to enjoy lower costs, and that those pressing glassware were able to maintain the favorable position achieved earlier. Since the Federal Government had built up a system of protective tariffs for most kinds of glass long before 1880, it is not surprising that the heavy import duties remained in force thereafter. Vested interests, once created, resist destruction vigorously. Until 1913 the tariff levied on flint and lime glassware varied from around 40 to 60 per cent; on green glassware (chiefly bottles and jars), from 30 to 40 per cent until 1890 and from then until 1913 from 53 to 67 per cent; on unpolished, unprocessed window glass, from 60 to 105 per cent (depending upon the size of the panes imported) until 1890 and averaged from over 50 to 90 per cent for all sizes from 1890 to 1913; and on unsilvered polished plate glass, from 15 to 103 per cent (depending upon size) until 1890 and roughly about 70 per cent on all sizes until 1913.[21]

[20] Cf. *ibid.*, pp. 333, 334–335.

[21] These figures are based upon the calculations and conclusions of Davis ("American Glass Industry," pp. 189, 214, 231–232, 252, 287–288, 327, 367–368, 388–389). The duties on rough or unpolished plate glass averaged about 25 per cent on sizes measuring 24″ × 30″ or less and around 42 per cent on larger pieces. Manufacturers needed less protection from foreign competitors on their rough than on their polished plate because much less labor was required to make rough plate than to make and also polish it.

The Underwood Tariff of 1913 was the first general statute since the Civil War to lower most duties drastically. The rate on flint and lime glass averaged 45 per cent (with light bulbs taxed at 30 per cent); on bottles and jars, 30 to 45 per cent; on unprocessed window glass, about 18–26 per cent; and on unsilvered polished plate glass, around 20 per cent.[22] The most drastic reductions, hence, were for window and plate glass; the least, for lime and flint glassware. The trend of protection may be roughly gauged by changes in the ratio of collections at the custom houses to the value of glass imported. These percentages are presented in Table 20. They cannot be taken as indicative of the height of the tariff wall because the higher the duties, the smaller was the amount of glass imported. If any of the rates were prohibitive, they did not affect the percentages. Nevertheless, the percentages increased following the passage of new legislative acts raising the duties and decreased whenever reductions were initiated. The outbreak of the first World War largely nullified the effect which the lowered rates incorporated in the Underwood Tariff normally would have had upon imports.

TABLE 20

RATIOS OF DUTIES COLLECTED ON GLASS TO THE VALUE OF GLASS IMPORTS, 1867–1916 *

(In per cent)

1867	52.2	1877	48.9	1887	61.8	1897	47.3	1907	53.2
1868	48.5	1878	51.1	1888	62.2	1898	57.5	1908	53.2
1869	48.6	1879	57.6	1889	58.4	1899	60.1	1909	54.1
1870	52.0	1880	54.8	1890	57.4	1900	57.6	1910	53.8
1871	55.6	1881	56.2	1891	54.9	1901	56.4	1911	55.1
1872	45.4	1882	56.9	1892	62.5	1902	58.6	1912	52.2
1873	42.5	1883	55.1	1893	60.4	1903	61.7	1913	51.5
1874	46.1	1884	55.7	1894	69.5	1904	61.2	1914	36.7
1875	41.7	1885	58.6	1895	52.6	1905	57.3	1915	32.9
1876	45.1	1886	58.3	1896	46.1	1906	52.3	1916	31.0

* Sources: Gaffield Collection, Scraps, III, 134; *Glass Industry*, p. 335.

The important query — precisely how did continued protection affect the development of the domestic industry and especially its rate of technological change — remains to be considered. First,

[22] *Ibid.*, pp. 287–288, 327, 367–368; Tariff Commission, *Flat Glass*, p. 5.

it should be noted, domestic production expanded more rapidly than imports and hence supplied a larger and larger portion of the domestic market. This may best be shown (as in Table 21) by

TABLE 21

DOMESTIC PRODUCTION IN RELATION TO DOMESTIC CONSUMPTION, 1880–1921 *

	Containers	Pressed and Blown Glassware	Window Glass	Plate Glass
1880	96%	91%	78%	33%
1890	90	94	75	79
1914	99	100	95	98
1919	106	112	118	114
1921	105	110	92	98

* Sources: For the values of annual output in 1880, Weeks, *Report on Manufacture of Glass;* in 1890, *Eleventh Census*, Vol. VI, Part 3; in 1914, 1919, and 1921, *Biennial Census of Manufactures, 1921.* For the value of annual exports in 1880, *Statistical Abstract of the United States, 1886;* in 1890, *Statistical Abstract of the United States, 1899;* in 1914, 1919, and 1921, *Statistical Abstract of the United States, 1921.* For the value of annual imports in 1880, Weeks, *Report on Manufacture of Glass;* in 1890, *Statistical Abstract, 1899;* in 1914, 1919, and 1921, *Statistical Abstract, 1921.*

Exports and imports in 1880 and 1890 were for the fiscal years ending on June 30 and in 1914, 1919, 1921, for the calendar years. All exports for 1880 and 1890 (except almost $9,000 for window glass in 1890) were considered as pressed and blown glassware, since other kinds of glass had not yet entered the export list in appreciable quantities. Chemical and cut or engraved glassware were not included in the export (except in 1914), import, and production figures for 1914, 1919, and 1921. Glass containers both filled and empty were included in the total import figures.

comparing the value of domestic output for selected years with the value of new supplies of glass made available for domestic consumption.[23] The percentages for 1919 indicate that the United States in that year was exporting larger quantities of all kinds of glass than she was importing. The dislocation of foreign industry during the first World War was in part responsible for this situation, especially in regard to window and plate glass. But the fact that exports of containers and of blown and pressed ware in 1921 remained larger than imports significantly reflects

[23] The "new supplies available for domestic consumption" means the value of domestic output plus the value of imports minus the value of exports. Hence, a percentage greater than 100 indicates that the United States exported more glass than she imported.

the impact of technological change. Exports of window and plate glass, on the other hand, dropped precipitously in the years following the war, and imports again shot upward.[24] Both the high tariff (before 1913) and the war, therefore, had the effect of reducing foreign competition and of preserving more and more of the domestic market for American producers.

Does it follow from this that there was a direct causal relationship between the tariff and the technological changes brought about after 1890? After carefully studying the tariff's effects on the glass industry, Pearce Davis concluded that it would be unreasonable to impute to the tariff any of the major cost-reducing innovations adopted by American producers.[25] By destroying or minimizing competition on the foreign front, the tariff probably diminished by so much the incentive to improve production methods. Without the tariff, it is true, there would not have been as many glass factories in existence, and the plate-glass industry in particular would probably have experienced greater difficulties than it actually did during its infancy. The fact that protection encouraged and maintained domestic glass production at a certain level is not in itself a completely rational justification of the tariff. Were this not so, one might argue that any tariff would be desirable because it would cause new industries to appear at home if the rates were sufficiently high. Established producers were not always the ones who perfected the new machines. Several of the most epochal inventions stemmed from groups which were not previously engaged in making the lines of glassware most affected by their inventions.[26] Davis has suggested one possible indirect

[24] Although the dislocation of exchange rates was probably responsible for much of this reversal, it is not possible to assess quantitatively the influence of "exchange dumping." According to the sources cited in Table 21, the exports and imports of window and plate glass, in thousands of dollars, were:

	Plate Glass		*Window Glass*	
Calendar Year	*Exports*	*Imports*	*Exports*	*Imports*
1914	$ 194	$ 489	$ 405	$1,317
1919	4,211	3	6,493	138
1921	1,335	2,070	455	2,639

[25] "American Glass Industry," *passim*.

[26] The Toledo Group, for example, brought to perfection the bottle and sheet-drawing machines when its members were engaged in manufacturing other kinds of glass.

relationship. Since the tariff helped to maintain the relatively high wage and price schedules in the industry, it may have "hastened the search for, and development of, mechanical invention." [27] Even if this could be substantiated by anything other than inference, the argument would be far different in nature from the usual beneficence anticipated from tariffs by protectionists.

The tariff, without question, maintained domestic prices at a higher level than they would otherwise have attained. Until 1903 the restriction of imports seems to have supported monopoly profits and monopoly wages in the window-glass industry,[28] and the controls exercised after 1912 by distributors, manufacturers, and wage earners would scarcely have proved effective without protection against foreign producers. Everywhere the tariff facilitated the organization of workers in more or less closed, restrictive unions; and the rates on plate glass obviously were high enough to permit the Pittsburgh Plate Glass Company to assume the role of price leader in the domestic market and, consequently, to reap monopoly profits. In so far as the tariff stimulated noncompetitive controls, consumers not only had to buy from high-cost producers but also to pay tribute to monopolistic factions. Consequently, only part of the savings effected through the use of machinery was passed on to buyers.

2. Changes in Price and Consumer Demand

One way of gauging roughly the results of technological change upon selling price is to compare the course of glassware prices with that of wholesale commodity prices generally. Following 1880, commodities sold at wholesale in the United States, on the whole, declined in price until 1897 and thereafter increased; glassware prices fell faster than the average during the first period and continued to drop (after some recovery in 1897–1903) until 1912. During the war years 1915–1920 they rose less rapidly than whole-

[27] *Ibid.*, p. 336.

[28] Cf. *ibid.*, p. 298. "The tariff protection enjoyed by the American window glass industry from 1890 to 1903 was as heavy and probably heavier than during any previous period." — *Ibid.*, pp. 288–289. And it was precisely during this period that the "bilateral" monopoly was strongest.

sale prices generally. These various trends are easily discernible from the data presented in Chart V.

The index numbers for glass, unfortunately, do not include the prices of bottles and jars. If continuous quotations for such items had been available, the trend would have been less markedly downward in the years 1880–1897,[29] and the index numbers during the war years would not have increased so rapidly.[30] The lag of glassware prices behind those for other *manufactured* products after 1913 was somewhat greater than the general wholesale commodity series in Chart V would suggest.[31] If the window-glass, plate-glass, and machine-bottle-glass interests had not each acted in concert to maintain prices in their respective fields, the price of glass relative to the general price level would have fallen still more. Likewise, if the demand for glass products had not been increasing, the decline in price following each improvement in technological processes would have doubtless been greater.

The direct effect of changes in consumer demand upon prices during any historical period, to say the least, is difficult to isolate. According to economic theory, an increase in demand means that buyers become more anxious than formerly to purchase the product

[29] During this period the price of polished plate glass (supplying two of the five price relatives in 1880–1884, two of the six in 1885–1889, and two of the seven employed thereafter) decreased greatly because of constant but unspectacular improvements in production methods. The average of the relatives for the two sizes of plate included in the index numbers was 205 in 1880 and only 74 in 1897.

[30] The average unit values of four kinds of glass (obtained by dividing the value of output for the industry as a whole by the quantity of glass made) yield the following indices, with 1889 taken as the base.

	Bottles and Jars	Pressed and Blown Glassware	Window Glass	Plate Glass	Wholesale Prices of Manufactured Commodities
1889	100	100	100	100	100
1904	100	108	96	96	110
1909	105	109	68	84	125
1914	97	91	87	80	129
1919	152	138	223	192	273
1921	?	?	184	217	188

Productivity of Labor in the Glass Industry, p. 19. Although these figures should not be taken as limning the trend of glassware prices, they nevertheless lend support to the conclusion based primarily upon internal qualitative evidence that container prices fluctuated less violently than did those for window and plate glass.

[31] U. S. B. L. S., *Wholesale Prices*, Bull. 415, p. 31. The indices constructed by the Bureau of Labor Statistics from seventy series of prices for manufactured products rose about eight points higher than the general commodity index in 1918–1920.

at existing prices.[32] This will ordinarily occasion an increase in consumption and also in prices, but whether in the long run the price rise will be temporary or permanent or even give way to a

CHART V

PRICES OF GLASSWARE AND GENERAL WHOLESALE COMMODITIES, 1880 — 1920 *

(Base = 1901 — 1905)

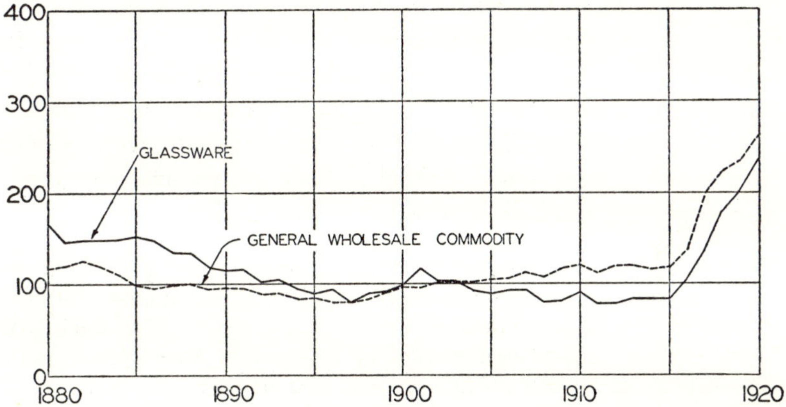

* Sources: The "glassware" series for 1890–1920 is composed of seven commodities (polished plate glass from three to five square feet in size, polished plate glass from five to ten square feet in size, grade A window glass, grade B window glass, four-inch nappies, one-half gallon pitchers, and one-third pint tumblers) taken from the prices published by the U. S. B. L. S., *Wholesale Prices*, Bull. 320, pp. 202–203, 206–207, 226–227. For 1880–1889 the prices found in the Aldrich Report, II, 233, 241–242, 276–278, for all these items except the poorer quality of window glass are utilized by splicing them with the respective series of the Bureau of Labor Statistics.

The wholesale commodity series is that given by Warren and Pearson, *Prices*, pp. 25–26, with each index number divided by the average for 1901–1905.

decrease will depend upon the economic and institutional character of the industry. Consumption of a product may also be increased through a reduction in its selling price (i.e., through an increase in supply made possible, say, by a lowering of production costs due to technological improvement) even though total demand has not increased. In the first case, the increase in demand may be said to

[32] It would be more accurate to state that the buyers become more anxious and ready to buy increased quantities of the product at existing and all other possible prices.

have "caused" the change in price and consumption; in the second, the decrease in price may be considered as directly responsible for the increase in consumption. If it were possible, therefore, to ascertain whether the increase in sales over some time period was due to increases in demand or to decreases in prices, one could then attempt to disassociate the effects of changes in consumption upon the growth of an industry from the effects which expansion of the industry had upon consumption.[33]

The lowering of the price of glassware made possible in part by the mechanical revolution, of course, occasioned some increase in consumption. But there is reason to believe that the demand for containers and window glass (if not for other kinds also) was relatively inelastic,[34] and hence that the increase in sales which would have resulted if consumer demand had not also changed would not have been very marked. According to Barnett,[35]

> The cost of the glass container for most products is only a small part of the total cost, and a reduction in the price of the container therefore stimulates very little the demand for the article . . .[36] Also, the extent to which glass is substituted as a container for tin or paper appears to be dependent more on slow changes in taste than on cost . . . In the main the great extension in the use of bottles and jars from 1899 to 1904 was due either to an increase in the consumption of the article, in no way

[33] Basically the problem involves determining whether or not the initiating cause was an increase in the supply curve or an increase in the demand curve. In either case the elasticities of demand and of supply would have important bearing upon the resulting increase in sales and change in price.

[34] Since the demand for window glass depended largely upon the amount of new building and since the cost of windowpanes relative to the total costs of construction was probably not over 1 or 2 per cent, the volume of window glass bought responded little to changes in its price. Barnett, on similar grounds, concluded that the demand for glass containers was also relatively inelastic. — *Machinery and Labor*, pp. 70–71. Polished plate glass for large display windows has few if any substitutes, and hence its demand seems to be inelastic — at least in the upper price ranges. At lower prices, however, the elasticity probably increases because it becomes economically and esthetically advisable to substitute it for window glass, which is an inferior product. Because of the lack of adequate substitutes, the demand for many kinds of blown glassware (such as tubing, lamp chimneys, and incandescent light bulbs) tends to be relatively inelastic.

[35] *Ibid.*, pp. 70–71.

[36] This, of course, refers to the total market demand and not to that of a particular firm. The Owens Bottle Company, for example, could have increased its sales enormously had it lowered prices below the established level commensurately with the decrease on costs of production occasioned by the automatic machine.

affected by a change in the price of the container, as was the case with beer bottles, or to changes in taste which demanded glass-packed goods for sanitary reasons.

The presumption is, therefore, that the increased number and purchasing power of consumers and their changing habits and tastes affected the course of development followed by the American glass industry.

Total population in the United States expanded from approximately 50,000,000 persons in 1880 to around 107,000,000 in 1920.[37] Concomitantly, the estimated national income, adjusted by changes in the general price level, rose from $15,442,000,000 to $60,615,000,000.[38] The per capita real income hence increased by about 81 per cent. Since not only population but also the material standard of living was mounting rapidly, American industry generally found encouragement in the prospect of ever-widening markets.

Between 1879 and 1899 the consumption of glass containers may have increased four fold and from 1899 to 1919 approximately three fold — or around twelve fold for the four decades.[39] With the growth in household preservation of foodstuffs, the number of fruit jars purchased annually increased about ten fold from 1879 to 1919.[40] As commercial canning and preserving of fruits and vegetables expanded from an industry with an output worth less

[37] National Industrial Conference Board, *Studies in Enterprise*, p. 28.

[38] Martin, *National Income in U. S., 1799–1938*, pp. 6–7.

[39] Statistics of the number of bottles and jars made do not appear in the census reports for 1879 and 1889, and the classification of shops according to output changed in 1899. The estimated increase for 1879–1899, accordingly, was based upon a comparison of the value of products made by green glassware houses (turning out bottles and jars chiefly but not exclusively) in 1879 with the value of containers produced in 1899, after allowances were made for changes in price and in volume of imports. Domestic production grew from around 8,000,000 gross in 1899 to over 24,000,000 gross in 1917 and to about 22,000,000 gross in 1920. One must recognized, however, that domestic production and domestic consumption were not necessarily the same. The United States imported more glass containers (nearly all of which were filled) and exported less in 1899 than in 1920. On the other hand, the stocks of unsold bottles and jars were appreciable at the turn of the century and either negligible or nonexistent in 1919–1920. Since figures on domestic consumption are lacking, one can estimate the increase only from changes in domestic production.

[40] *Twelfth Census*, Part III, 974; *Fourteenth Census*, X, 835.

than $25,000,000 in 1879 to one whose annual product was valued at almost $150,000,000 in 1914 and $400,000,000 in 1919,[41] the demand for glass containers for bottling "food preparations," relishes, pickles, preserves, and sauces steadily increased. Glass factories made at least five times as much packers' and preservers' ware in 1919 as they did in 1899.[42] Statistics on the output in 1880 are not available, but it is reasonable to assume that the quantity made in that year was considerably less than in 1899.[43] The number of glass bottles and jars used for this purpose, therefore, probably increased more than ten fold between 1880 and 1920. Apparently distributors began to use glass containers for delivering fluid milk to their customers in the late 1880's [44] because the first milk bottles made in the United States were produced by the Warren Glass Works at Uniontown, Pennsylvania, in 1887.[45] Thirty-two years later American factories made almost 900,000 gross.[46]

The vast majority of all containers blown in 1879 had been (1) prescription and druggists' ware, (2) "patent and proprietary" bottles for such things as home remedies and patent medicines, or (3) liquor and whiskey bottles and flasks. From 1899 to 1919 the value of druggists' preparations, patent medicines, and perfumery and cosmetic products prepared in the United States increased from about $89,000,000 to $386,000,000; and concomitantly the output of all glass containers used in merchandising such things grew from 3,720,000 gross to over 10,000,000 gross.[47] The distilled spirits industry expanded from one whose annual product was valued at $41,000,000 in 1880 to one with over $200,000,000 output in 1914. As one state after another legislated against the man-

[41] *Census of Manufactures, 1905*, Part III, 400; *Census of Manufactures, 1914*, II, 365; *Fourteenth Census*, X, 65. Most of the increase from 1914 to 1919 reflected the rise in prices rather than the increase in physical output.

[42] *Twelfth Census*, Part III, 993; *Fourteenth Census*, X, 835.

[43] The census reporter in 1899 stated that "the manufacture of bottles and jars for the packing and preserving industries, exclusive of the enormous production of fruit jars, has steadily advanced during the past decade, owing to the remarkable growth of the above interests and the increasing recognition of glass as the ideal container." — *Twelfth Census*, Part III, 976.

[44] *Ibid.*, Part III, 975.

[45] *National Glass Budget*, XXX, April 10, 1915, p. 3.

[46] *Fourteenth Census*, X, 835.

[47] *Ibid.; Twelfth Census*, Part III, 993; *Biennial Census of Manufactures, 1921*, p. 746.

ufacture and sale of spiritous liquors, and after the Eighteenth Amendment to the Federal Constitution was adopted in January, 1919, the industry rapidly declined in size. Its output for that year was valued at slightly less than $32,000,000.[48] The manufacture of flasks and whiskey bottles, reflecting the changes in demand occasioned by these trends, increased from about 1,000,000 gross in 1899 to almost 2,700,000 gross in 1914 and then fell to less than 1,000,000 gross in 1919.[49] Those bottle factories specializing in "whiskies" naturally bore the brunt of this decline. One of the two firms sharing the liquor rights on the Owens machine [50] closed down its works employing about seven hundred persons at Cincinnati, and its owners agreed to reorganize and sell controlling interest in its two plants to the Owens Bottle Company.[51] An important manufacturer stated in May, 1919, that "until we find some way of replacing this immense volume of production, wiped out in a flash [by prohibition], the glass bottle trade in the United States is going to face the greatest emergency possible in its history." [52]

Expansion in the mineral and soda-water business provided considerable relief. As compared with 1879 the value of soft drinks sold in the United States had increased around twenty-five or thirty fold by 1919 and 1921.[53] As a result, the output of over 4,000,000 gross of beer, mineral water, and soda-water bottles in 1919 was approximately three times greater than the number made by American factories in 1899. And this expansion materialized despite the fact that the beer industry, like the distilled spirits industry, suffered greatly from the effects of prohibition.[54]

The Owens Bottle Company and its licensees benefited not only from the general increase in the demand for glass containers during the first two decades of the twentieth century but also from the

[48] *Biennial Census of Manufactures, 1921*, pp. 684–687.

[49] *Twelfth Census*, Part III, 993; *Fourteenth Census*, X, 835.

[50] That is, the Charles Boldt Glass Company. The Illinois Glass Company also enjoyed the right to make whiskey bottles on the Owens machine. See above, p. 106.

[51] See above, p. 112.

[52] *National Glass Budget*, XXXV, May 10, 1919, p. 1.

[53] *Biennial Census of Manufactures, 1921*, pp. 684–687.

[54] The value of malted liquors legally made and sold in the United States was about 15 per cent lower in 1919 (almost two-thirds lower in 1921) than in 1914, whereas the value of mineral and soda waters sold in 1919 and in 1921 was more than double their 1914 value.—*Ibid.*

increased emphasis placed by buyers upon uniformity in weight, measure, and quality. The passage of pure food and drug legislation and the increasing popularity of filling machines among packers placed a premium upon bottles which conformed to this standard.[55] Those made on the Owens machine were much more uniform than those fashioned by hand or semiautomatic devices because exactly the same quantity of molten glass was gathered each time the suction cups of the machine dipped into the revolving pot. Gatherers working with hand blowers or feeding the semiautomatic could only approximate such precision, however practised and skilled were their eyes and hands. Many buyers, consequently, specified Owens bottles in their orders because of their superiority in quality to any others,[56] and both unions and hand manufacturers opposed the passage of every food and drug law which reduced the tolerance in weight and content for bottles and jars.[57]

One trend in the evolution of market demand, however, was not very favorably inclined toward the Owens machine. The desire of manufacturers to differentiate their product — be it perfume, horse radish, or whiskey — caused them to demand bottles of distinctive shape and size. The Owens machine lent itself especially well to making large quantities of containers of standard size and shape, but its economic advantage over hand and semiautomatic processes in small-lot production was not especially conspicuous. M. J. Owens was aware of this. During the closing years of his life he tried to persuade packers to redesign their bottles and order only those of more standard form. The fact that such a campaign made little headway helps to explain the continued existence of large numbers of hand factories, for these increasingly specialized in

[55] Cf. Barnett, *Machinery and Labor*, p. 87; Owens-Illinois Glass Company, Annual Reports of Owens Bottle Company, September 30, 1912; *National Glass Budget*, XXV, *passim*. Milk-bottling machines, operated by one man with the help of two others who fed the bottles, were in use by 1909 (*National Glass Budget*, XXV, July 3, 1909), and it is interesting to note that M. J. Owens himself made a catchup-filling machine in 1913.

[56] Cf. the report made by the president of the American Flint Glass Workers' Union in the April issue of *The American Flint* and reproduced in the *National Glass Budget*, XXV, April 23, 1910.

[57] The issues of the *National Glass Budget* around 1907–1912 are replete with brief articles on the subject.

filling small orders for unusual sizes and oddly shaped bottles.[58]

Until 1900 the demand for window and plate glass was almost wholly derived from the volume of new residential and business construction and from repairs and replacements in older buildings. After that year the automobile industry created an additional market which became ever more important with the passage of years and with the subsequent shift in popularity from open or touring models to enclosed automobiles.[59] The Libbey-Owens Sheet Glass Company especially benefited from the growing demand for relatively thick glass for automobile side lights and windshields. Its sheet-drawing process produced a glass of more even thickness, greater strength and flatness, and finer surface lustre than did blowing by hand or by cylinder machines.[60] In 1920 all but two of its machines at the large Charleston plant were making thick glass for automobile manufacturers.[61] Since the total amount of building space tends, over a long period of time, to maintain a constant proportion to the total population,[62] one may use the figures pertaining to population growth as roughly limning the increase in new construction. Between 1880 and 1920 the number of people living in the continental United States approximately doubled, with about half of the increase occurring before 1900. This means that there was roughly twice as much building space (not new construction) in 1920 as in 1880. Fortunately indices measuring the amount of new building construction also are available, and these are presented graphically in Chart VI.

Building activity in 1880 was just beginning to recover from the marked downward slump of the preceding decade,[63] and until the

[58] Cf. Barnett, *Machinery and Labor*, pp. 90–93.

[59] In 1929 the automobile industry took about half the plate glass made in the United States; and even before the first World War, when the open car was the most prevalent type, the industry's development was one of the most important factors conditioning the expansion of plate-glass production. — Tariff Commission, *Flat Glass*, pp. 82–83, 93.

[60] Cf. Libbey-Owens-Ford Glass Company, Libbey-Owens Sheet Glass Company Minutes, II, 27.

[61] *National Glass Budget*, XXXV, March 27, 1920, p. 2.

[62] See Newman, "Building Industry and Business Cycles," *Studies in Business Administration*, Vol. V, No. 4 (1935), p. 32.

[63] See Warren and Pearson, *World Prices and the Building Industry*, Figure 21, p. 112.

first World War the volume of annual construction increased with only temporary retardations. From 1880 through 1892 the trend was markedly upward (approximating a four fold increase); from 1893 through 1900, the amount of new building declined,[64] and from 1901 through 1916 renewed expansion was evident. On the whole, the volume of annual construction was almost four times as great during the years 1905–1916 as it had been during 1880–1882.[65] The resulting expansion in the demand for window glass [66] until 1893 and the absence of any very marked break in the market through 1903 help to account for the success attending the efforts of workers and manufacturers to administer wages, prices, and output. Since the Lubbers cylinder machine was introduced at a time (1903–1909) when building activity was great, most hand manufacturers did not perish in the resulting competition. The Colburn sheet-drawing machine, however, became commercially successful during the very years when the building industry was hit hardest by war-time shortages and restrictions. Nevertheless, there were at least two factors which tended to cushion the shock which such a revolutionary labor-saving invention would ordinarily have had. In the first place, the war cut off foreign supplies of window glass and gave American producers virtually complete freedom to meet the domestic demand (although greatly reduced) and also provided them an opportunity to supply large segments of demand in foreign countries. In the second place, the expanding automobile industry compensated in part for the decreased demand for windowpanes in building construction.

Manufacturers of plate glass also profited from the active state of the building industry in the years 1880–1916. As a matter of fact, they seem to have benefited more from the increase in demand for building glass than did producers of ordinary window lights

[64] The per capita trend was unquestionably downward. — *Ibid.*

[65] Obviously, the fact that the increase in annual construction was around four fold whereas population had only doubled over the four decades does not violate the statement previously made, that the volume of building space per capita tends to remain constant over a long period of time. Violation would exist only if total new construction over the four decades did not approximately equal the total amount of floor space available in 1880.

[66] There is no way of measuring the demand for building glass arising from minor repairs and replacements.

because there was a tendency for constructors to substitute polished plate squares for ordinary windowpanes in commercial and residential buildings in American cities. This tendency had become pronounced well before the beginning of the twentieth cen-

CHART VI

AVERAGE MONTHLY INDEX OF BUILDING CONSTRUCTION IN THE UNITED STATES, 1880 — 1920 *

(Base = 1913)

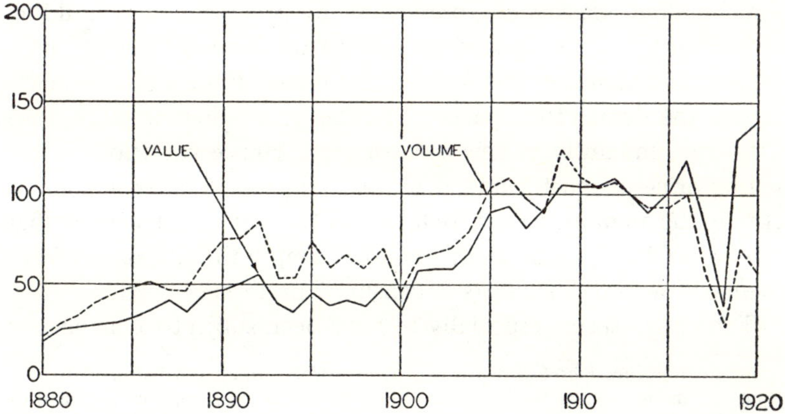

*Sources: The "index of value" was obtained from data supplied by Long, *Building Cycles*, pp. 214–219; the "index of volume," from data presented by Newman, "Building Industry and Business Cycles," *Studies in Business Administration*, Vol. V, No. 4 (1935), pp. 63–68. The figures used by Newman came from different sources than those employed by Long, but the value of building permits in various cities formed the basis for both studies. Newman divided his indices of permits by appropriate costs-of-production indices (hence, the "index of volume"); Long decided against such a procedure "because of distrust of all building cost data" (*Building Cycles*, p. 101). Long's series has been divided by his index for 1913 to facilitate comparisons with Newman's.

tury.[67] Around 1915–1916 the United States was said to consume approximately one-half the polished plate glass produced throughout the world.[68] From slightly more than one million square feet of polished plate in 1880, American factories enlarged their output to over 9 million square feet in 1889, 17.6 million in 1900, 44.9 million in 1910, and 69.6 million in 1916.[69]

[67] Gaffield Collection, Scraps, III, 134 ff.

[68] *Glass Industry*, p. 236.

[69] Weeks, *Report on Manufacture of Glass*, p. 11; *Twelfth Census*, Part III, 961;

Because of the great variety of products made, it is difficult to analyze the effects which changes in consumer demand had upon the pressed and blown glassware industry. One may reasonably assume in view of the relative availability of substitutable commodities that the demand for such items as lamps and lamp shades, tableware of all kinds, and ornamental household pieces was more elastic than the demand for building glass or containers. The former group of items were sold to the ultimate consumers either directly or through jobbers and distributors, whereas containers and building glass were purchased by processors and building contractors.[70] The demand for glassware used by private consumers, furthermore, was not a "joint demand" in the same sense as was the demand for packers' bottles or window or plate glass used predominantly in new construction. Hence variations in net selling price of most kinds of blown and pressed glass influenced greatly the amounts that were taken off the market. The most that can be profitably done at this point is simply to indicate a few of the more outstanding shifts in consumer preferences.

Tableware seems especially to have been subject to changes in

---

Tariff Commission, *Flat Glass*, p. 92. The American output of window glass grew from 93.2 million square feet in 1879 to 188.4 million in 1889, 217.1 million in 1899, 401 million in 1914, and 368.9 million in 1919.—*Ibid.*, p. 42. For both window and plate glass, as for containers, it is impossible to ascertain to what extent the increase in annual production (and sales) was due to shifts in consumer demand or to enlarged purchases as prices were lowered as a result of technological change.

[70] It is interesting to note that selling expenses in 1916 amounted to the following percentages of the total sales value of goods produced:

A. Building glass	
1. Hand-made window glass	1.46 per cent
2. Machine-made window glass	2.03
3. Plate glass	.25
4. Wire and opalescent glass	3.72
B. Containers	
1. Hand-made bottles	1.40
2. Automatic machine-made bottles	3.08
3. Semiautomatic machine-made bottles	4.63
4. Jars	3.30
C. Pressed and Blown Glassware	
1. Blown tableware	7.09
2. Pressed and blown tableware	6.66
3. Lighting goods	6.72
4. Lamp chimneys	4.52
5. Miscellaneous	3.46

Glass Industry, pp. 104–105.

styles and fads. The majority of establishments around 1900 had developed the custom of introducing many new designs and patterns at the opening of each year; if a pattern proved popular, production tended to be extremely profitable. But usually the popularity lasted only one season, for public favor frequently turned to new designs. During the first decades of the twentieth century, the general trend seems to have marked a shift from brilliant heavy-cut ware (and especially from imitations) to finer, more delicate pieces of blown glass.[71] Colored pieces, made in the Bohemian manner, rose in popularity, and American firms began to dispute with foreign producers for their market.[72] Jelly glasses and tumblers were employed increasingly in the packing trade, and thin-blown tumblers for household, restaurant, and other uses found an ever widening market.[73] Even more significant developments occurred in the lighting field. As gas jets invaded the market previously held by oil lamps, the demand for lamp chimneys and lantern globes ceased to increase,[74] but the market for shades, globes, and other fixtures used with gas lighting expanded rapidly.[75] Then the advent of incandescent electric lamps presaged additional shifts in the demand occasioned by lighting needs. Neither bulbs nor glass tubing was listed separately in the census figures until 1919, and for that year over 18,000,000 dozen bulbs and 18,000,000 pounds of tubing were reported as having been produced. The Libbey Glass Company profited from the growing

[71] Total sales of cut ware, however, increased absolutely because of a marked drop in cost of production. — Cf. Henderson, "Glass-Making," *Popular Science Monthly*, XXXVII (1890), 164.

[72] See *Twelfth Census*, Part III, 979.

[73] *Ibid.*, p. 980. It is interesting to note that the output of different kinds of pressed and blown ware increased as follows:

	1904	1914	1919
Tableware (per 100 pieces)	1,283,974	1,554,056	2,457,938
Jellies, tumblers, and goblets (per doz.)	7,346,214	18,030,243	25,182,133
Blown tumblers, stemware, and bar goods (per doz.)	6,282,606	11,377,310	9,668,638

Fourteenth Census, X, 835. The growth after 1914 may have reflected the curtailment of imports rather than an increase in the total domestic demand.

[74] American factories made about 400,000 dozen chimneys less in 1919 than in 1904, and their output of lantern globes showed only a small increase. — *Fourteenth Census*, X, 835.

[75] The output of such items increased more than twelve fold in volume between 1904 and 1919. — *Ibid.*

demand for glassware arising from the new electrical lighting industry. The reader may recall that this company not only expanded greatly its output of bulbs and tubing but also developed mechanical means of making such items automatically.

One may conclude from the foregoing discussion that in nearly every branch of the industry increases in consumer demand favored the Toledo entrepreneurs. The bottle trade was expanding rapidly at the turn of the century, thereby facilitating the introduction of the Owens bottle machine in 1903 and allowing its promoters to establish themselves securely before competition with hand and semiautomatic producers became intense. The fact that containers of uniform weight and measure grew in popularity with commercial food packers and preservers enhanced the relative value of Owens-made bottles as compared with other kinds. The increasing demand for window glass prior to 1916 had made it easier for both hand and machine-cylinder manufacturers to adopt "live-and-let-live" policies. The window-glass market, consequently, was not in a highly competitive state when the Libbey-Owens Sheet Glass Company introduced its Colburn machine. And at the same time, the first World War had cut off almost completely foreign imports of window glass and had created markets abroad for the American product. It is unlikely that the entrepreneurs at Toledo would have fared as well as they actually did had they entered the window-glass field before 1910. When they brought out their semiautomatic and automatic machines for making bulbs, tumblers, and tubing, they also entered an expanding market. The result was that in all cases the effect of their activities upon prices, employment, and competitive reprisals was less chaotic than it would have been in a stagnant or declining market.

3. Basic Materials and Shifts in the Industry's Location

As in earlier decades,[76] the cost of fuel and other basic materials used by glassmakers in the decades 1880–1920 amounted to about

[76] See above, p. 39.

one-third the value of the finished product.[77] With the exception of refractory clay, potash, and certain chemicals used in relatively small quantities to whiten or color the glass, most of the materials were drawn from domestic sources. Sand of superior quality was available in many places scattered throughout the glassmaking belt; soda ash was being made in very large quantities according to the Solvay process; and producers in few countries had access to as cheap fuel as did American manufacturers. Although they were working several clay beds in the United States, glassmakers continued to import German and English clays to mix with their domestic variety. The outbreak of the European conflict, of course, temporarily deprived them of foreign supplies.[78] As early as 1890 most of the potash used in making window glass came from Germany, and its price naturally soared during the war years. Even manufacturers confessed, however, that the net disrupting influences of the war were more beneficial than harmful. Even though certain lines of production suffered from inability to import important ingredients, the sudden increase in the demand for its products resulting from the cessation of glass imports stimulated the industry as a whole "to a remarkable degree." [79] This would scarcely have been possible had the industry been relying predominantly upon foreign sources of supply for its materials.

As glassmaking continued to grow in importance, the provisioning of factories with the materials they needed became increasingly a separate business. Most glassmen no longer supplied themselves with sand and soda ash but bought these from specialized dealers and firms. Even pots and furnace bricks were ordered from reputable houses rather than made in a department of the glass factory as had formerly been the case. Some of the larger firms, it is true,

[77] Based upon ratios of the costs of fuel and raw materials to the value of finished product as given in the various census reports.

[78] The only important domestic shortage caused by the war seems to have been coal. The directors of the Owens Bottle Company in September, 1917, told the managers of their various plants to acquire and store away as much coal as possible. — Owens-Illinois Glass Company, Owens Bottle Company Minutes, VI, 152–153. And the Charleston plant of the Libbey-Owens Sheet Glass Company was operated at about half capacity in 1918 in compliance with the government's request to conserve fuel. — Libbey-Owens-Ford Glass Company, Libbey-Owens Sheet Glass Company Minutes, I, 255–261.

[79] Tariff Commission, *Glass Industry as Affected by War*, p. 9.

formed subsidiary companies to provide themselves with fuel or sand, but even the Owens Bottle Company wrote off its Toledo-Owens Glass Sand Company as a failure and made no attempt to mine its own coal.[80]

For purposes of the present inquiry, the primarily important question relating to fuel and other materials is whether they were a positive (as contrasted with a passive or negative) factor in the industry's expansion. If their price, relative to the price of glass, declined during the period, one might justifiably conclude that developments in the materials market encouraged the expansion of glassmaking. Sufficiently reliable and complete price data exist for making the comparison. The index numbers for the price of glassware and of fuel and other materials, therefore, are pictorially presented in Chart VII. The fuel-materials series has one outstanding weakness. It does not reveal the effects of reduced fuel costs resulting from natural-gas booms in various areas because prices for this combustible are not included.[81]

From 1880 until 1897 the price of glassware decreased more rapidly than the cost of its materials and fuel (other than natural gas); and from 1901 through 1915, while materials increased or remained more or less stable in price, the trend of the glassware index was unmistakably downward. Furthermore, during the critical war years, the ingredients and fuel used by manufacturers increased more sharply in price than did the glass they made. All of this means that the price margin between glassware and its basic materials lessened almost constantly throughout the period under observation. In addition, a comparison of the index numbers for materials and fuel with the index numbers for general wholesale commodities (shown on Chart V) suggests that developments in the materials field as a whole were not revolutionary. The two price series did not diverge by more than one could reasonably

[80] See above, p. 114. It did develop, however, a profitable natural-gas company.

[81] Natural-gas prices varied greatly from region to region, and the opening up of a new field benefited only those plants close enough to utilize its supply. It is significant, however, that the census reports for 1890 and 1900 show that all material costs amounted to about 30 per cent of the value of glass made whereas those for 1880, 1910, and 1920 reveal the percentage to have been 38, 35, and 35, respectively. The great gas booms in the Middle West occurred between 1880 and 1910.

explain in terms of the inadequacies of the statistical measures employed. Hence the conclusion may be drawn that the expansion of glassmaking between 1880 and 1920 did not rest primarily upon the increased availability or cheapness of the materials which figure in the indices.

Since fuel was the only material for which glassmakers could avoid transportation costs by building their factories near impor-

CHART VII

PRICES OF GLASSWARE AND BASIC MATERIALS, 1880 — 1920 *
(Base = 1901 — 1905)

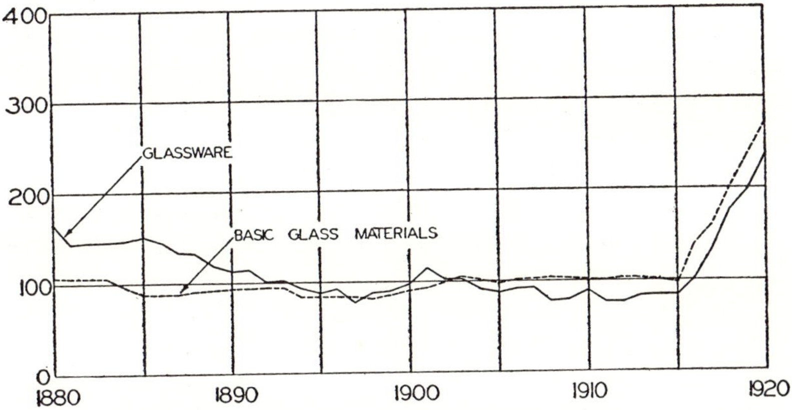

* The commodities used in the materials series are lime (1880–1920), potash (1880–1914), soda ash (1880–1920), salt cake (1893–1920), red lead (1880–1920), egg-size anthracite (1880–1920), two grades of bituminous coal (1880–1920), and crude petroleum (1880–1920). The prices for lime, anthracite, two grades of bituminous coal, and crude petroleum given in the Aldrich Report (Vols. II, 174, 178, 179, 227; IV, 1836) for 1880–1889 were spliced with the prices of the same commodities for 1890–1920 reported by the U. S. B. L. S., *Wholesale Prices*, Bull. 320. The wholesale prices of potash, soda ash, and red lead in Philadelphia which were generously supplied by Miriam Hussey and Anne Bezanson for 1880–1892 have been spliced with the prices of the same commodities for 1893–1920 found in the "prices current" published in the monthly issues of *Drugs, Oils and Paints*, VIII–XXXVI (1893–1921). The composition and sources of data used in the glassware series are described in the footnote to Chart V.

The prices of potash, an imported commodity, are not included in the indices for 1915–1920 because of the marked vitiation which otherwise would have resulted. The price of potash in 1916, for example, was as much as thirteen times its average price in 1901–1905. If potash prices had been included in the index, the indices for 1915, 1916, 1917, 1918, 1919, and 1920 would have been 125, 272, 263, 273, 219, and 281, instead of 98, 140, 180, 200, 197, and 271, respectively.

tant sources of supply,[82] it exercised an influence upon the industry's location which was out of proportion to its importance as an item in the cost of production.[83] According to the census, the outlay for various kinds of fuel amounted to 5.7 per cent of the total value of glass produced in 1889 and 1899, 7.8 per cent in 1904, 8.2 per cent in 1909, and 8.9 per cent in 1914.[84] Shifts in the industry's location, in terms of both number of establishments and value of output, are revealed in the absolute and percentage figures given in Table 22. In 1919 the New England area, once so important as a glassmaking center,[85] could boast of only one furnace, and its output was too inconsequential to tabulate separately in the census reports for that year. Although New York and New Jersey together continued to produce glass valued at about $25,000,000, their importance relative to the industry as a whole fell by 50 per cent between 1879 and 1919. Even Pennsylvania, the most important producing state, experienced a marked decrease in its margin of superiority; and West Virginia, Ohio, Indiana, and Illinois became increasingly important glassmaking areas. California, also, rose in significance, with only one factory in 1899, three in 1909, and seven in 1919.

The opening up of new gas fields naturally offered an incentive for someone to construct and operate a glass furnace in the vicinity. Technologically, natural gas is the fuel *par excellence* for glassmaking; and wherever new wells were brought in, it was also the cheapest fuel.[86] The natural-gas booms in Ohio during the 1880's, in Indiana during the 1890's, and in West Virginia during the first two decades of the twentieth century captured the imagination of entrepreneurs, and they built many new works in these

[82] See above, pp. 41 ff.

[83] Prior to 1880 also fuel had largely determined the locational pattern. A time series correlating fuel and glassware prices would be enlightening, but it is not possible to construct a very meaningful index of fuel prices. Although natural gas was at times a very important glassmaking fuel, there are no usable price series for it available.

[84] Fuel outlays were 19 per cent of total materials costs in 1889 and 1899 and approximately 23.5 per cent in 1904, 1909, and 1914. The reason for the increased importance of fuel as a cost element was the dwindling supplies of cheap natural gas after 1900.

[85] See above, pp. 44–45.

[86] It was both expensive and relatively difficult to store natural gas for future use or to pipe it to distant markets.

states. Between 1909 and 1919, when oil and gas were abundantly brought to the earth's surface in Oklahoma, the number of glass factories in that state rose from one to sixteen. Oftentimes, however, gas wells lost their pressure and "dried up" almost as quickly and as unexpectedly as they had come into existence, thus leaving the neighboring glass furnaces stranded. In the wake of the gas boom's collapse, the number of furnaces in Ohio dropped from 59 in 1889 to 28 ten years later and in Indiana from 110 in 1899 to 44 in 1909. Some men, like E. D. Libbey who had moved his factory from New England to Ohio because of the cheapness of natural gas in the West, were soon faced with depletion of the wells and had to turn to producer gas or to petroleum if they were to keep their fires lighted. Coal and oil both were relatively abundant in most areas where natural gas had constituted the original attraction; and technological improvements in the construction of furnaces, the production of gas from soft coal, and the utilization of petroleum in direct firing kept many factories from shutting down. As a result of these technological advances and the discovery of new fuel-producing areas, the glass industry showed a tendency to expand beyond the confines established in earlier days by the necessity of direct firing with coal or wood. In 1920 it was much easier for a manufacturer to find a location which was close to both fuel sources and major glass markets than had been the case in 1880 and earlier.

4. Conclusion

On the whole, the four decades beginning in 1880 were marked by lusty growth in business activity generally. The recession in 1883–1884 left few scars upon American industry; and the important depression following the panic of 1893 showed signs of having run its course by the end of 1897. From then until 1920, a period of almost unparalleled prosperity reigned, with only brief reactions in 1903, 1907, and 1914.[87] Although glassmaking felt the effects of a general decline in business activity during each of these periods (especially, in 1894–1895),[88] its progress was not seriously

[87] Cf. Wright, *Economic History of U. S.*, pp. 872 ff.

[88] Owens-Illinois Glass Company, Annual Reports of Owens Bottle Company, September 30, 1915.

TABLE 22

LOCATION OF THE AMERICAN GLASS INDUSTRY, 1879 — 1919
(Based on Census Data)

A. Number of Establishments

States (by regions)	Number of Establishments					Per cent of Total				
	1879 *	1889	1899	1909	1919	1879 *	1889	1899	1909	1919
Massachusetts	11	6	5	1	1	5.2	2.0	1.4	0.2	0.2
Maryland	8	11	7	7	8	3.8	3.7	2.0	1.9	2.1
New Jersey	27	34	26	23	21	12.8	11.6	7.3	6.3	5.7
New York	32	30	27	24	19	15.2	10.2	7.6	6.6	5.1
Virginia	—	—	—	5	4	—	—	—	1.4	1.1
Illinois	7	13	6	11	14	3.3	4.4	1.7	3.0	3.8
Indiana	4	21	110	44	35	1.9	7.1	31.0	12.1	9.4
Missouri	6	5	3	4	4	2.8	1.7	0.8	1.1	1.1
Ohio	20	59	28	45	44	9.5	20.1	7.9	12.4	11.9
Pennsylvania ‡	78	99	119	112	102	37.0	33.7	33.5	30.9	27.5
West Virginia	4	7	16	51	77	1.9	2.4	4.5	14.1	20.7
All Others	14	9	8	36	42	6.6	3.1	2.3	10.0	11.4
Total	211	294	355	363	371	100.0	100.0	100.0	100.0	100.0

B. Value of Output

States (by regions)	Value of Output					Per cent of Total				
	1879	1889	1899	1909 **	1919	1879	1889	1899	1909	1919
Massachusetts	$ 854,345	$ 431,437	$ 418,458	†	†	4.0%	1.0%	0.7%	†	†
Maryland	587,000	1,256,697	557,895	1,038,000	4,031,841	2.8	3.1	1.0	1.1	1.5
New Jersey	2,810,170	5,218,152	5,093,822	6,961,000	13,694,677	13.3	12.7	9.0	7.6	5.2
New York	2,420,796	2,723,019	2,756,978	4,509,000	12,995,741	11.4	6.6	4.9	4.9	5.0
Virginia	—	—	—	682,000	831,455	—	—	—	0.7	0.3
Illinois	901,343	2,372,011	2,834,398	5,047,000	18,244,717	4.3	5.8	5.0	5.5	7.0
Indiana	790,781	2,995,409	14,757,883	11,593,000	30,106,652	3.7	7.3	26.1	12.6	11.5
Missouri	919,827	1,215,329	765,564	1,993,000	6,927,042	4.3	3.0	1.4	2.2	2.7
Ohio	1,549,320	5,649,182	4,547,083	14,358,000	35,240,877	7.3	13.8	8.1	15.6	13.5
Pennsylvania ‡	8,720,584	17,179,137	22,011,130	32,818,000	80,479,752	41.4	41.8	38.9	35.6	30.7
West Virginia	748,500	945,234	1,871,795	7,779,000	42,730,338	3.5	2.3	3.3	8.4	16.3
All Others	851,905	1,065,397	924,706	5,317,000	16,500,470	4.0	2.6	1.6	5.8	6.3
Total	$21,154,571	$41,051,004	$56,539,712	$92,095,000	$261,884,080	100.0	100.0	100.0	100.0	100.0

* The figures for this year are for the total number of establishments whether active, idle, or under construction. Only 169 were making glass at the time.

** Figures are rounded to the nearest thousand.

† Value of output was not listed separately but is included in the value for "all others."

‡ After 1879 the census reports do not break the states down into counties; hence is was impossible to divide Pennsylvania into an eastern and a western section as was done in Table 2. Almost without exception, however, the new factories established in Pennsylvania after 1879 were built in her western counties.

arrested at any time. The increase in the number of factories and in the value of their combined output, as shown in Table 22, continued unabated. The outbreak of the European conflict in 1914 brought in its train several months of business uncertainty, which, however, proved to be the prelude to one of the most intense periods of production the industry had ever known. Imports of glass from Europe declined sharply, and exports soared.[89]

General business conditions in the country after 1880 (and especially after 1897, when the entrepreneurial group at Toledo was most active), therefore, were influential factors in the technological and industrial expansion of glassmaking. The argument of this and the immediately preceding chapter has been that certain additional factors of a less general nature also contributed their share in creating a favorable conjuncture. High tariffs gave domestic manufacturers continued protection from foreign producers, and as a result innovators may have felt a little more reckless in their schemes than they would have otherwise. Then, too, the domestic demand for most kinds of glass was increasing as population expanded, the standard of living rose, and consumer preference shifted to things made of glass. The natural-gas boom in the Middle West temporarily lowered fuel costs and wrought changes in the industry's locational pattern. But variations in the price of other materials do not appear to have influenced the expansion realized between 1880 and 1920. The tactics employed by organized labor were unsuccessful in halting mechanical revolution, and the "co-operation" of organizations embracing hand manufacturers or skilled workers with those who controlled the machines permitted the latter to amass larger profits than probably would have been possible had unbridled competition resulted in price-cutting and greater increases in output.

[89] From 1880 through 1893 annual imports were from 7 to 9.7 times greater in value than exports, and from 1898 through 1914, 1.6–3.4 times greater. In 1915, however, exports were 21 per cent greater than imports in value and in 1918 and 1919 they were over 1,100 per cent greater. — *Twelfth Census; Thirteenth Census; Census of Manufactures, 1914; Statistical Abstract, 1917* and *1921*. Glassware exports in 1920 were roughly 40 times their value in 1880, or, after allowance for changes in glass prices, about 28 times greater.

CHAPTER X

THE TOLEDO ENTREPRENEURS

The various factors listed and evaluated in the three preceding chapters as having conditioned the technological and industrial growth of glassmaking at Toledo do not provide a complete explanation of that development. Three examples may throw light upon the problem which remains unsolved. Although the tariff on imports — by effectively reducing foreign competition — protected the relatively high wages of glassworkers, sheltered inefficient high-cost producers, and helped maintain domestic glassware prices at a level above foreign prices, it was hardly responsible for the revolutionary trends initiated after 1890. In so far as the tariff increased the profit opportunities awaiting those who succeeded in developing cost-reducing machines, it helped to create a favorable conjuncture. Then, too, it may have been an influential factor in the industry's expansion during the nineteenth century.[1] This is important because, if glassmaking in America had been in an infant state around 1890, its subsequent development would have been different. Despite these considerations, one must remember that high tariff walls had sheltered domestic manufacturers throughout most of the century. Why, then, did not the acceleration of mechanical innovation and growth in output appear much earlier? If one takes the position that high tariffs retard technological change, then why did the revolution occur at a time when tariff barriers were unusually forbidding?

It has also been suggested in previous chapters that the relative costliness of skilled labor was a factor which apparently encour-

[1] It is hazardous to be dogmatic on this point. The plate-glass industry was a child of the tariff, but one must remember that other branches of glassmaking made their appearance in America before import duties were raised to truly protective levels. In addition, because overland transport costs would have been prohibitive, factories located in the Allegheny Mountains or on their western side probably could have competed effectively with foreign producers in inland markets had there been no tariff at all. There is some analytical basis, furthermore, for supposing that the tariff may have retarded the industry's growth by facilitating the formation and maintenance of monopolistic controls.

aged entrepreneurs to adopt labor-saving devices. Throughout the nineteenth century, however, there was a pronounced tendency for the wage differential enjoyed by skilled glassmen to decline. Hence, considering the wages factor alone, the incentive for manufacturers to develop and install glassmaking machinery must have been stronger before the Civil War than later. High wage rates, therefore, fail to provide an adequate explanation of what transpired after 1890.

The third example concerns the state of technology itself. The argument advanced in the first part of Chapter VII was that certain basic technological innovations — for example, the continuous tank and the gas furnace — were prerequisite for the mechanization of the glass-blowing process. The fact that these improvements materialized only after 1860 helps to account for the delayed appearance of bottle-blowing machines. But why the machines appeared at Toledo rather than at some other glassmaking center has yet to be resolved. Perhaps the answer is the favorable environment created there by that elusive factor which economic theorists call entrepreneurship.

Considerable uncertainty exists about just what functions are involved in the entrepreneurial role. The early classical economists frequently failed to draw a sharp distinction between the capitalist on the one hand and the entrepreneur on the other. Alfred Marshall stressed the managerial duties of business leaders on both the technological and the pecuniary plane, and several writers have explained that entrepreneurs are primarily concerned with bringing the other productive agents together into a "going concern." Many textbooks emphasize that risk is inescapable in modern industry and that it is the primary function of entrepreneurs to bear risk. Schumpeter has offered a somewhat different conception. To him the entrepreneur is ideally the innovator — the one who brings about new combinations of the means of production.[2] According

[2] See, especially, Schumpeter, *Theory of Economic Development*, pp. 74 ff. Schumpeter distinguishes five broad categories of "new combinations": (1) new products, (2) new methods of production, (3) new markets, (4) new sources of supply of materials, and (5) new forms of industrial organization. — *Ibid.*, p. 66. J. B. Clark, earlier, had stressed practically the same functions as did Schumpeter, though less explicitly. — *Distribution of Wealth*, especially Chap. XXV.

to Thorstein Veblen's characterization, the leaders of American business enterprise in the late nineteenth and early twentieth centuries became decreasingly "captains of industry," concerned with technological and organizational efficiency, and increasingly "captains of finance," concerned primarily with pecuniary management.[3]

Frank H. Knight [4] has made what appears to the present writer to be a most helpful suggestion. The entrepreneur is basically the personal element in the "economic equation;" it is he who makes decisions and chooses between alternative possibilities in a world of uncertainty. According to Arthur H. Cole,[5] his functions may be tentatively classified as (1) instituting innovations (technological and managerial); (2) adjusting (when and as necessary) to external conditions, whether business, economic, political, or social; and (3) maintaining an efficient operating unit. None of these is sufficient by itself, and all of them require the moving force of executive decision.

The very nature of decision-making renders entrepreneurship in industry an elusive factor to evaluate. Why Libbey, Owens, and their associates — rather than their competitors — seized the opportunities for profit afforded by the state of glass technology in 1890 and pioneered with new inventions probably can never be definitively answered. One can do little more than state the basic decisions they made and attempt to explain why they made them. For this, reliance must be placed upon inferences drawn from available records of stockholders' and directors' meetings and from personal correspondence, on the one hand, and upon reminiscences of the few people who personally knew the principal characters involved, on the other.[6] To summarize and appraise

[3] *Absentee Ownership*, especially Chap. VI.

[4] See, especially, "Profit and Entrepreneurial Functions," *The Tasks of Econ. Hist.*, II (1942), 126–132.

[5] The analysis summarized here appeared in his correspondence with me and in an unpublished memorandum prepared in February, 1946. He has suggested the same ideas in slightly different form in his Presidential Address before the Economic History Association, "An Approach to the Study of Entrepreneurship: A Tribute to Edwin F. Gay," *The Tasks of Economic History*, VI (1946). Unfortunately, this address was delivered at a date too late to be of much use to me.

[6] The men interviewed at Toledo may be classified in three main groups: (1) those who believe that primary credit should be given to E. D. Libbey, (2) those

the role which E. D. Libbey and M. J. Owens played must involve a considerable amount of information which has already been presented in earlier chapters. Hence, the following pages are a sort of summary and conclusion.

1. Members of the Toledo Group

Edward Drummond Libbey never made or blew glass; but he was actively associated with the industry from 1874, when he was twenty years of age, until his death in 1925. In fact, he was born into a glass-manufacturing family. His father, William L. Libbey, had entered a Boston firm engaged in manufacturing and importing glass, and in the 1860's operated his own factory, first in South Boston and then in New Bedford. Around 1870 or 1872, William L. Libbey became general manager of the New England Glass Company at Cambridge. His son, educated at Kent Hill's Academy in Maine,[7] joined the same company a short time later as clerk and quickly rose to the position of bookkeeper. The elder Libbey leased the New England Glass Company in 1878, formed a partnership with his son two years later, and died in 1883. E. D. Libbey then found himself at the head of a factory which had reached its apogee some thirty years earlier and which had encountered increasing difficulties since 1870.[8] By acting boldly on at least two occasions, he was able to salvage enough of his inheritance to form the financial basis for the industrial empire which he was later instrumental in creating.

While left temporarily in complete charge of the Cambridge factory just before his father's death, E. D. Libbey had to devise a way to meet an important financial note then falling due. He discovered a large stock of peculiarly colored tableware which had been thrown into a storeroom and written off the company's books as a loss because the glass lacked the rich ruby color it was sup-

---

who are still loyal to the memory of M. J. Owens, and (3) those who think that neither man deserves more credit than is usually paid one who accidentally discovers a rich vein of gold. Enough points of view were obtained, however, to offset most of the bias. There is little to be gained from acknowledging the personal source of much of the information presented in this chapter. Hence, source references are much less common in this chapter than in foregoing ones.

[7] Owens-Illinois Glass Company, "Lives of E. D. Libbey and M. J. Owens," p. 2.

[8] See above, pp. 87 ff.

posed to have. Young Libbey packed some representative pieces in his sample case, boarded the train for New York City, and called upon one of his father's best customers — the famous Tiffany jewelry firm. When he saw the samples and heard Libbey's enthusiasm and artful salesmanship, the buyer for Tiffany offered to purchase the entire stock. This "Amberina Glass," a name apparently coined by Libbey himself, immediately captured the public's fancy and yielded enough profits to save the New England Glass Company from bankruptcy.[9]

Unfortunately, the reprieve was only temporary. For several decades New England had been losing her advantage over other areas in the production of glass, and it had become increasingly evident that her industry would have to yield to the competition of factories located in cheap-fuel areas of the Middle West.[10] Whereas his father had been unable to find enough courage or financial resources to move the plant westward, E. D. Libbey did so in 1888.[11] After relighting the furnace at Toledo, Ohio, he canvassed the area for men to supplement the working force which had accompanied him from New England. Among the first ones employed was a skilled flint-glass worker named Mike Owens.

Michael Joseph Owens' parents had left Ireland in the early 1840's, and his father had entered the coal mines in that part of the country which later became West Virginia (Mason County). Ten years after his birth on January 1, 1859, Mike found employment in Wheeling with J. H. Hobbs, Brockunier, and Company — one of the most successful and progressive flint-glass houses in the country. He quickly worked his way upward through the various jobs customarily held by boys in glass factories of that day and actually began to blow glass when only fifteen. He soon became proficient in the art. He joined a local debating society, to which he later attributed a large portion of his success,[12] and became a leader in the local branch of the American Flint Glass Workers' Union. He helped organize the workers' co-operative Union Flint

[9] See above, pp. 90–91. The version presented here is the one Libbey himself recounted to his associates at a later date.

[10] See above, pp. 91–92.

[11] See above, pp. 92–93.

[12] Cady, "Builders of Business," *System*, June, 1919.

Glass Company at Martin's Ferry, Ohio, in 1882,[13] and he aggressively solicited funds to maintain the *Ohio Valley Boycotter*, the union's regional publication.[14] By 1888 he was generally recognized as one of the most forceful members of the national union.

When the union decided to send agitators to Massachusetts in that year for the purpose of persuading workers at the New England Glass Company and the Boston and Sandwich Glass Company to strike for higher wages and a reduction in the daily output, Owens was among those selected for the task. Hence he actively participated in stirring up the labor trouble which occasioned Libbey's decision to move westward.[15] Upon his return to Wheeling, Owens attended his union's eleventh annual convention at Canton, Ohio, and staunchly defended its various policies for controlling output.[16] He insisted that all workers who refused to co-operate in restricting output should be summarily expelled from the union.[17] With all the prejudices and viewpoints of a typical labor unionist, he entered E. D. Libbey's employ at Toledo in August, 1888, as a "servicer" or second-ranking blower of a small group engaged in making glass shades for oil lamps.

How this worker, endowed with most of the character traits commonly associated with men of Irish extraction, first gained the attention of his employer is uncertain. According to some contemporaries, Owens strode into Libbey's office one day and announced that he was quitting work because the plant superintendent was not properly disciplining his workers. He further predicted that the company would be bankrupt within a few years. Libbey agreed and then asked Owens if he thought he could restore order in the furnace room. When the latter emphatically affirmed that he could, Libbey offered him the superintendency. Owens accepted,

[13] Owens-Illinois Glass Company, Owens' Private Files.

[14] According to Thomas W. Rowe, later president of the union and one of Owens' close friends. Tom Owens, Mike's brother who is still living in Toledo, claims that Mike was actually editor of the paper for a short time.

[15] See above, pp. 91–92.

[16] Such as refusal to work during the two hottest months of each summer and limits upon the amount of ware each worker could make every five hours.

[17] Years later, after his own inventive efforts had smashed the union's restrictions on output and had aligned him with the capitalist-employing class, the editor of the *National Glass Budget* (XXVI, May 21, 1910) reminded him that he had once been as "unprogressive as the common herd" of glass blowers.

but only upon the condition that neither Libbey nor any other member of the "front-office" force would interfere in the least with his duties. He then proceeded to discharge the entire laboring force and to rehire only those men whom he considered both capable and manageable. According to other contemporaries, Owens constantly crammed a box which his employer had set up for employee suggestions about how conditions and production in the plant could be improved. Libbey consequently could no longer overlook the author of so many scrawled notes.[18] Whatever the actual course of events may have been, Owens rose from the ranks of workers primarily because he possessed remarkable imagination, ingenuity, initiative, and will power. All who knew him consider it only "natural" that he always came to dominate any group of which he was a member.

The "facts" appear to have been as follows: After he had been with Libbey three months at Toledo, he was appointed foreman of the blowing department, and two years later he became superintendent of the factory.[19] He had charge of the workmen blowing bulbs at the branch shop located at Findlay, Ohio, in 1891–1892 [20] and superintended Libbey's glass exhibit at Chicago's Columbian Exposition in 1893.[21] Upon his return to Toledo in the fall of that year, Owens had already formulated rather definite ideas about a machine which he thought would be able to blow glass bulbs in paste-molds. He later testified in court [22] that he originally had planned to withdraw from the Libbey Glass Company and to seek financial support from some glassmen he knew in Wheeling. E. D. Libbey heard of his plans and promised to provide adequate financial backing for his experiments if he would remain in Toledo. Together they formed the Toledo Glass Company in 1895.

The incorporation of this new company entailed two significant

[18] William H. Boshart, a later associate of the men, claims that this was the version given him by Libbey himself.

[19] According to Owens' testimony presented in a lawsuit brought by the Owens Bottle Company against the Libbey Glass Company in 1921. A copy of the court proceedings which I examined is now preserved in an old safe in the basement of the Libbey Glass factory at Toledo.

[20] See above, pp. 93–94.

[21] See above, pp. 94–96

[22] During the Owens Bottle-Libbey Glass lawsuit.

changes in personal relationships. In the first place, casting aside their relationship as employer and employee, E. D. Libbey and M. J. Owens became business associates. And in the second place, a schism began to develop between Libbey and his former associates. Solon O. Richardson, Jr., Jefferson D. Robinson, and William F. Donovan — three salesmen who had accompanied him from Cambridge — had become important stockholders and directors of the Libbey Glass Company. But they had never been able to get along with Mike Owens. Furthermore, they were experienced in only the tableware and bulb-and-tube branch of glassmaking; and by nature they were inclined to conservatism and cautiousness in business affairs. They had not been enthusiastic when Libbey proposed participation in the Columbian Exposition,[23] and they were reluctant to venture time and money in any of Owens' experiments. They remembered the failures which had beset other manufacturers and inventors who had sought to perfect glassmaking machinery.[24] Consequently, Libbey felt well advised to find other associates for the new enterprise which he and Owens planned to start. Owens withdrew from active participation in the Libbey Glass Company in 1896, and Libbey himself arranged to devote most of his time to the Toledo Glass Company. Richardson, Robinson, and Donovan were left to operate the Libbey Glass Company [25] although Libbey remained a director, officer, and large stockholder.[26] These three men directed the company's affairs from 1895 through 1919 and constituted what was called in Chapters IV and V the "Libbey Glass Faction."

Perhaps the most capable man whom Libbey interested in the

[23] See above, p. 94.

[24] See above, pp. 178 ff.

[25] Richardson and Robinson, on February 11, 1895, signed ten-year contracts with Libbey whereby they agreed to act as general manager and assistant general manager, respectively. Subsequently, the contracts were modified by new ones which stipulated increased salaries and enlarged duties. — Libbey Glass Company, Libbey Glass Company Minutes, I, 81–95, 126–130, 155–166, 222–233.

[26] Libbey continued to be a director and held the office of president or vice-president through 1919. He was the largest single stockholder until at least 1915. After 1898, however, he failed to attend board meetings with increasing frequency, and his official status became more and more an honorary one. This does not mean that his influence in shaping major company policies ceased altogether but only that he took no part in routine administration.

Toledo Glass Company was Clarence Brown, a brilliant Toledo lawyer who had been his legal advisor in business matters. F. L. Geddes, Brown's law partner, and William S. Walbridge, Libbey's brother-in-law who had not been eminently successful in the furniture business at Cambridge, also participated in the new company. These five men — Libbey, Owens, Brown, Geddes, and Walbridge — composed the nucleus of the "Toledo Glass Faction" and complemented one another to an unusual degree. Libbey always supplied most of the venture capital needed, supported promotional activities in various fields, and formed many valuable contacts with business men outside Toledo. Owens offered his invaluable experience gained while actually working molten glass, formulated ways whereby machine processes could supplant the old handicraft methods, and provided a liaison between the executive and the production branches of the various companies. Brown invested some of his own funds and, what was more important, offered sound advice on a variety of problems. Geddes handled most of the legal details involved in the business and thereby helped formulate its policies in two very important fields: patents and licensing. Walbridge had charge of the more or less routine matters pertaining to administration.

Owens possessed a dominating personality, lacked sufficient business acumen at times, and was ever impatient with anything which impeded mechanical experimentation. These characteristics, if left unbridled, obviously could have embroiled him and his associates in business difficulties and perhaps in financial bankruptcy. Yet they were also the traits which could and did lead the group to success in its pioneering ventures. Libbey's problem was to control and direct Owens' enthusiasm and impatience without destroying or thwarting them. To this end he frequently used his other associates — Walbridge, Geddes, and to a lesser extent Brown — as counterbalances, and thereby avoided clashing with Owens personally.[27]

All five men were directors of the Toledo Glass Company, the

[27] Most of the men now living who knew the five principal characters have independently suggested that upon innumerable occasions Libbey "used" Walbridge, Geddes, and Brown to hold Owens in check.

Owens Bottle Company, and the Owens European Bottle Machine Company simultaneously and, except for Walbridge, were also directors of the Libbey-Owens Sheet Glass Company.[28] During the majority of years prior to 1920 they filled the most important offices of these four companies [29] and always retained large enough blocks of voting stock in each to prevent others from acquiring control. At one time or another some of them were either directors or officials, or both, in the Macbeth-Evans Glass Company, the Charles Boldt Glass Company, the Whitney Glass Works (before it consolidated with Owens Bottle), and other independent glass firms. They also held positions in various subsidiary companies controlled by Owens Bottle.

The schism among E. D. Libbey's associates appearing in 1895 grew more pronounced.[30] Although those in the Libbey Glass Faction long remained on more or less friendly terms with the men belonging to the Toledo Glass Faction, they retained very few mutual business ties. Perhaps the strongest tie was personal. Libbey, as a major stockholder, director, and officer in both the Toledo Glass and the Libbey Glass companies, managed to maintain considerable harmony between the divergent interests of the two entrepreneurial groups. A second tie was provided by the system of cross-licensing, which permitted each company to use the inventions developed by the other in manufacturing its own particular line of glassware. The Westlake Machine Company,

[28] Brown died in 1918; Owens severed his connection with Toledo Glass in January, 1917; and Geddes resigned as director of Libbey-Owens in 1918.

[29] On October 16, 1903, the entire group resigned as directors and officers of Toledo Glass, and a panel of "straw men" was installed in their place. Two days later, as directors and officers of the Owens Bottle Company, they signed the contract whereby Toledo Glass sold Owens Bottle exclusive American rights on the Owens machine. On February 7, 1905, they resumed their former functions with Toledo Glass.

[30] In 1921 the Owens Bottle Company brought suit against the Libbey Glass Company on the ground that the Danner machine, developed by the latter, produced tubing which subsequently was sometimes made into containers (test tubes and vials). In accordance with prior contracts, it was argued, Owens Bottle should receive an exclusive royalty-free license in its use. After a heated contest during which personal animosities came to light, the court decided in favor of Libbey Glass. — Libbey Glass Company, Documents relating to licenses, etc., Safe No. 1. Several contemporaries, moreover, have affirmed that shortly before Owens' death in 1923 and Libbey's death in 1925, individuals in the two groups were scarcely on speaking terms.

engaged in experimental work with automatic bulb-blowing machinery,[31] may be considered a third tie, since it was in some respects a joint enterprise. Libbey, Walbridge, Donovan, Richardson, Robinson, and Owens all held stock in the company; and the first three were among its directors and officers. The Buckeye Clay Pot Company definitely was jointly owned. On the other hand, none of the Libbey Glass group — except E. D. Libbey himself — held stock or positions with the Toledo Glass Company, the Owens Bottle Company, or the Owens European Bottle Machine Company. After these companies proved successful, Robinson and Richardson regretted their failure to support Owens in 1895 and finally persuaded Libbey to let them buy a few shares in the Libbey-Owens Sheet Glass Company, incorporated in 1916.

Only stock in Owens Bottle and Libbey-Owens came to be widely held. The other companies remained "closed corporations" at least until 1920. The members of four families[32] retained all the shares in Libbey Glass, and fourteen individuals owned all the stock of the Westlake Machine Company in 1917. Toledo Glass and Owens European shares rarely appeared on the market, and at no time before 1920 were there more than fifty stockholders in either.[33] The Owens Bottle Company, on the other hand, had 70 different stockholders in 1904–1905, 458 in 1914, 712 in 1916, and 1,609 one year later;[34] more than 1,200 people held shares in Libbey-Owens by 1920. The stock of both of these companies was actively bought and sold on the New York Stock Exchange after 1916.

2. Managerial Difficulties

Managerial difficulties never loomed very large in the Libbey Glass Company, the Westlake Machine Company, the Toledo

[31] See above, pp. 130 ff.

[32] The Libbeys, Donovans, Richardsons, and Robinsons. Until 1913 there were only seven shareholders; in 1919 there were seventeen.

[33] The stock list for Toledo Glass included 39 names in 1920; and the owners of Owens European numbered 46 in 1919. — Libbey-Owens-Ford Glass Company, Toledo Glass Company Minutes, IV; Owens European Bottle Company Minutes, III, 231–232.

[34] The great increase in the number of stockholders in 1916–1917 was due to a rapid rise in the company's capitalization (which nearly trebled from September

Glass Company, or the Owens European Bottle Machine Company. This was due chiefly to two facts. First, none of these enterprises was extremely large in terms of output, capital investment, and number of plants; nor did any one of them have an unwieldly number of stockholders or directors. Secondly, most of them had rather limited, carefully defined objectives which changed little over the years. Robinson, Richardson, and Donovan, who managed the Libbey Glass Company after 1895, were congenial because they were all conservative and because they had acquired considerable experience in the tableware and bulb-and-tube fields. Furthermore, their offices were located at their firm's only factory. The Westlake Machine Company had set out with a very definite objective — to develop the automatic bulb machine and to license other producers in its use. It took ten years to achieve this goal. The Toledo Glass Company was primarily a machine-developing and licensing agency. It employed few men, and its directors (usually numbering only five) had a unity of interest which allowed them to formulate and execute policies without any great difficulty. It emerged from periods of relative inactivity only when it undertook a new experimental project such as developing the Owens bottle machine or the Colburn window-glass machine. The Owens European Bottle Machine Company completed within two years the job for which it had been created — namely, to sell all European rights on the bottle machine. After 1907 its directors had only to receive the annual payments from the European purchaser, distribute them to its own stockholders as liquidating dividends, and render decisions on routine matters.

The Owens Bottle Company, on the other hand, began to encounter managerial difficulties as soon as it started to make bottles in large quantities. The reader may recall that it had been organized in 1903 primarily to license domestic producers who might wish to employ the bottle machine. Not one of its small group of directors had ever had experience in the bottle field or with managing several factories located in different places. In 1911, when

30, 1915, to September 30, 1917), a four-for-one common-stock split in 1916, and the listing of its securities on the New York Stock Exchange in the same year.

three plants were in operation at Toledo, West Toledo, and Fairmont, West Virginia, E. D. Libbey recommended to the board of directors that it elect a new president to replace him because he himself was unable to keep in constant touch with all departments of the business.[35] He also suggested that the company needed an enlarged number of directors in view of its expanding activities. The board refused to consider Libbey's first suggestion at that time, and only reluctantly elected Clarence Brown president in June, 1917, after Libbey had once again tried to resign.[36] It did, however, accept his second suggestion in November, 1914.[37] Upon its recommendation, the stockholders gradually increased the number of directors until there were fifteen in 1917 compared with seven in 1903.[38] Most of the additional members were drawn from executive positions in the various companies which Owens Bottle was then absorbing.

The enlarged board, however, proved too unwieldy to handle many problems expeditiously; and, in addition, several of the new members had divergent interests. Owens frequently complained that some directors had become more concerned about the rate of next year's dividends than about improvements in production and marketing techniques. Probably in order to keep the reins of control in the hands of the original entrepreneurial group, an executive committee — composed of Owens, Brown, Libbey, Walbridge, William H. Boshart, and John D. Biggers [39] — was established in December, 1916. This committee met every week, kept minutes of its proceedings, and handled much of the company's business formerly conducted by the board. Until some time after

[35] Owens-Illinois Glass Company, Owens Bottle Company Minutes, II, 86.

[36] In his letter to the board, dated June 23, 1917, Libbey wrote, "For some years my attention and energies have been devoted, and I purpose increasingly devoting them, to activities which I deem even more useful than those primarily concerned with business profits . . ." — *Ibid.*, VI, 139. After Brown became president, Libbey remained chairman of the board and about one year later resumed the presidency upon Brown's death.

[37] *Ibid.*, IV, 63–64.

[38] *Ibid.*, V, 147.

[39] Owens, Brown, Libbey, and Walbridge had controlled the company from its beginning; and they had brought in Biggers and Boshart at an early date and had schooled them in their ways. Boshart and Biggers were members of the Toledo Group's "second generation" and helped to carry on after the older generation passed away.

1920 it formulated most policies, and there is no evidence that the directors or stockholders ever rejected its recommendations.

The executive committee was part of an extensive plan, adopted in October, 1916, to integrate management and control. The physical plant and volume of business of the Owens Bottle Company had expanded rapidly after 1908; [40] but the loose, personal administrative framework which it had had in 1903 had undergone little change. Although additional men had been added to the managerial group from time to time, they acted more in the manner of personal assistants to either Libbey or Owens than as business associates with equal influence and responsibility. By 1916 the company's affairs were threatening to get out of hand. Following the report of an efficiency engineering firm from New York,[41] the various executive and administrative departments were grouped under three vice-presidencies, and the duties of officials and department managers were specifically defined in order "to let each individual know the scope of his duties and responsibilities and thus avoid complications which might otherwise result." [42] The new plan represented improvement over conditions existing earlier but failed to eradicate all trouble.

The difficulties of management appear to have stemmed from two sources: the tendency of some members of the original entrepreneurial group to divide their interest and loyalty between two large-scale enterprises in quite distinct branches of glassmaking, and the inability of Owens to adjust himself to an institutionalized pattern of control. In regard to the first point, it should be recalled that the Toledo Glass Company in 1912 had purchased Irving W. Colburn's patents on certain window-glass machinery and, after making his process practicable, had incorporated the Libbey-Owens Sheet Glass Company in 1916.[43] The entrepreneurial group which had directed the destiny of the Owens Bottle Company since its foundation hence assumed the responsibility of another great company. No member of that group, moreover, had ever had any previous experience with the technical problems

[40] See above, pp. 109 ff.
[41] Owens-Illinois Glass Company, Owens Bottle Company Minutes, V, 5, 46–48.
[42] *Ibid.*, Executive Committee Minutes, I, 5 ff.
[43] See above, p. 127.

involved in producing window glass or with the wholesale or retail outlets in the field. At the outset, the Libbey-Owens Company suffered much more from this situation than did Owens Bottle, but the latter occasionally was handicapped by the fact that some of its key executives were devoting more of their attention to establishing the new company on a sound operating basis than to formulating and enforcing various policies for the old company.

M. J. Owens became general manager of Libbey-Owens in June, 1917, and appointed William J. Jacobs — who, for want of a better term, may be considered his protégé — superintendent of the large factory built near Charleston. Neither man was able to set the furnaces to melting the proper kind of glass for window-panes nor to eliminate breakage on the drawing machine. Conditions rapidly deteriorated in the plant. Libbey visited Charleston in December and immediately recommended a change of management.[44] Clarence Brown urged, on January 10, 1918, that the company employ "a General Manager or a President who would devote his entire time and ability to the affairs of the Company, with full authority over all, subject only to the Board of Directors."[45] Libbey was the man selected. After 1918 he devoted most of his attention to Libbey-Owens, discharged several men at the factory, and replaced them with others who were better acquainted with the sheet-glass business. Together with James C. Blair,[46] who succeeded Owens as general manager, he effected considerable improvement in the management. Consequently, Libbey showed less and less concern about the Owens Bottle Company, and his inactivity increased the relative strength of the new

[44] Owens resigned at once, and Jacobs went back to Toledo. — Libbey-Owens-Ford Glass Company, Libbey-Owens Sheet Glass Company Minutes, I, *passim*.

[45] *Ibid.*, I, 170.

[46] Blair had entered the executive offices of the Owens Bottle Company sometime earlier as M. J. Owens' "understudy." He was one of the few associates who enjoyed Owens' complete confidence and until the latter's death acted in many matters as Owens' agent. After 1919 Blair was simultaneously general manager of Owens Bottle and Libbey-Owens, but according to his own admission, M. J. Owens remained "the power behind his throne" in both companies. Libbey had informed Owens that he would enjoy free access to Libbey-Owens' factory but that he should never give any orders there directly to anyone and should discuss all matters with Blair. Libbey then suggested to Blair that he always consult Owens on matters pertaining to company policy.

financial interests on that company's board. As will be suggested shortly, this was partially responsible for the strained relations arising at that time between Owens and Libbey.

The second source of managerial difficulties in the Owens Bottle Company mentioned earlier was the inability of Mike Owens to adjust himself to the institutionalized pattern of control which became increasingly desirable as the company's chief objective shifted from technological improvement and licensing to large-scale production. Owens had risen from the ranks of labor, and for almost two decades after his entry into the entrepreneurial group in 1895 he had exercised authoritarian control over matters pertaining to experimentation, management, and production. Until 1908, management and production problems had never become serious because the group employed only a small personnel and had not yet seriously undertaken the manufacture of bottles. Even in 1915 the Owens Bottle Company had only four plants in operation. But the following year it controlled eleven.[47] Inevitably, questions relating to personnel and management suddenly grew more complex; and "outsiders" were elected to the board of directors who did not understand and appreciate Owens' peculiarities and hence were not always in sympathy with his proposals. Neither his nature nor his experience had conditioned him for the necessity of curbing his will and actions in order to conform to those of men whom he could not dominate. Essentially a driver of men rather than a leader, Owens favored the pyramiding of control and responsibility into one person's hands. In a directors' meeting of the Owens Bottle Company on December 10, 1918, he insisted that additional centralization of authority was badly needed—especially since its president, Libbey, was devoting so much of his time to Libbey-Owens' affairs, vacationing in California, and collecting art objects in Europe and elsewhere.[48] The lack of such centralization was responsible, he contended, for recent failures to carry out certain proposals which had been made and whole-

[47] See above, pp. 110 ff.

[48] Again, on March 20, 1923, in a letter addressed to John H. McNerney, Owens returned to this theme: "I fear Mr. L. is to [sic] far from the field to be worth much to us on new things."—Owens-Illinois Glass Company, Owens' Private Files.

heartedly endorsed.[49] The following year he resigned as general manager and advised that "the company's Code of Regulations should be so amended that the Board of Directors should elect only the president of the company and that the president should appoint its general manager, who should [then] appoint all other officers." [50] His resignation was accepted but his advice rejected.

This incident was part of an explosive situation which threatened to disrupt the original entrepreneurial group. Owens learned in 1917 that he had a weak heart, and this disquieting knowledge made him even more intractable than normally. In addition, he came to envy the material wealth and social position enjoyed by Libbey, who had increased the polish and culture which he had brought with him from Boston. Owens ineptly managed his personal fortune and, in spasms of childish annoyance at officials or stockholders in various companies, frequently sold the securities he held in them at disadvantageous prices. Nearly always he lived to regret his action.[51] There is also reason to believe that he was deeply hurt when Libbey, resigning as president of Owens Bottle in 1917, recommended as his successor Clarence Brown instead of Owens. Then, too, he felt that Libbey was trying to gain the limelight by becoming the chief figure in the Libbey-Owens Sheet Glass Company, whereas he himself considered the bottle enterprise more important.[52] He must have been keenly aware of the fact that Libbey enjoyed to a far greater degree than he did the confidence and esteem of most of their other associates.

The first tangible evidence of the approaching crisis appeared when the question arose as to what part Owens was to play in the prospective sheet-glass company. On February 19, 1916, Libbey

[49] *Ibid.*, Owens Bottle Company Minutes, VI, 280. [50] *Ibid.*, VII, 97.

[51] His friends have said that Mike told them that the only time he made bad financial decisions was when he was mad at someone. He never knew what to do with his money beyond satisfying his personal desires for expensive clothing and good food and meeting the expenses piled up by his immediate family. In his later years, however, he came to recognize that especially in America private fortunes impart considerable social esteem to their possessors.

[52] He may have felt as he did about the Owens Bottle Company because it handled the machines which had made him famous and which also bore his name. Owens was intensely egotistical, which may have been nothing more than a protective covering for an inferiority complex. He invariably referred to his associate as "Mr. Libbey" although the latter always addressed him as "Mike."

wrote Brown, "I believe that *with proper restrictions, properly prescribed*, Mr. Owens would make an ideal manager and, as he is really the only available person at the present time, I am inclined to recommend his appointment to that position."[53] Even Libbey's half-hearted approval of a plan to place his associate in a position of authority and responsibility was not shared by all the others. Four months later Brown reported to Libbey the latest consensus of opinion on the subject.

Concerning the temporary organization of the Libbey-Owens Company, the situation is this. It had been suggested that I should be named President and Fowle[54] vice-President, while Owens was not even to be made a director. It occurred to me that this would not be wise and certainly not be agreeable to Owens. There seemed to be no sufficient reason why Fowle or I should be put, even in temporary authority over Owens, nor any justification for not putting Owens on the Board of Directors, in view of his past connections with the enterprise and his large financial interest in the future of the Company . . . I think Owens' cooperation in the enterprise is very important and have no doubt, when it comes to a permanent organization, you will agree with me that he should be placed in a position of authority.[55]

The final decision was that Owens should be a director and also general manager.[56] But this did not entirely clear the atmosphere.

Owens, in his capacity as president and director of the Toledo Glass Company, moved on June 22, 1916, that Toledo Glass be dissolved (since it "has fulfilled its mission") and that the stock which it held in Libbey-Owens ($500,000 in preferred and over $1,500,000 in common) be distributed proportionately among its own stockholders. No one but Owens voted in favor of the motion. Libbey and Walbridge immediately offered to resign as Toledo

[53] Owens-Illinois Glass Company, Brown's Private Files. Italics not in the original.

[54] Arthur E. Fowle was one of Libbey's relatives. Owens liked him even less than he did Walbridge.

[55] *Ibid.*

[56] Libbey-Owens-Ford Glass Company, Libbey-Owens Sheet Glass Company Minutes, I, *passim*. On June 4, 1917, he signed a contract with the Toledo Glass Company to act as Libbey-Owens' general manager for five years.—*Ibid.*, Toledo Glass Company Minutes, IV, 201–203.

Glass directors; [57] but at the end of what must have been a tumultuous session, no definite action was taken.[58] Owens resigned as president and director in January, 1917, and at the stockholders' meeting held the following month, unsuccessfully moved once more that the company be dissolved.[59] He then proceeded to sell what shares he still held in the company. Eleven months later he also resigned as general manager of the Libbey-Owens Sheet Glass Company because, he later said, he "was dissatisfied with the way the affairs of that company were being conducted." [60] At the same time he was quarreling with Clarence Brown over the management of the Owens Bottle Company; and, as a result, Brown offered to resign as its president. Libbey, at that time vacationing on his estate at Ojai, California, wired Brown to withdraw his resignation as a personal favor to him.[61] Brown obliged and even succeeded in placating Owens. Only at this point did the strained atmosphere existing for one and one-half years begin to dissipate.

Although Owens filled fewer offices in the various companies after 1918 than earlier, he continued to exercise considerable influence over managerial problems by working through the men who did fill them. This arrangement and two important financial agreements appear to have been chiefly responsible for reinstituting and maintaining harmonious relationships. In 1917 the Toledo Glass Company consented to pay M. J. Owens $200,000 in cash and Libbey-Owens common stock for seventy-five shares of its own stock which Owens had "loaned" or returned to the company in 1897.[62] And in 1922–1923 the Owens Bottle Company con-

[57] By July 18 Walbridge had sold all his shares in the company and did not repurchase any until after Owens had ceased to be a stockholder.—*Ibid.*, IV, 224–225

[58] *Ibid.*, IV, 159–160.

[59] *Ibid.*, IV, 188–189.

[60] Owens-Illinois Glass Company, Owens' Private Files. See also above, p. 128.

[61] *Ibid.*, McNerney's Private Files.

[62] *Ibid.*, Owens' Private Files; Libbey-Owens-Ford Glass Company, Toledo Glass Company Minutes, IV, 192, 194–195, 201–203, 220–221, 233, 244. The terms of remuneration were later modified in January, 1919. The 75 shares of stock loaned Toledo Glass was merely the excuse for paying Owens the $200,000. Both he and Libbey had received equal portions of half the stock in Toledo Glass when this company had been formed in 1895. Two years later, when the company needed additional working capital, both men agreed to return 75 shares of the original 375 received by each. According to the records, Libbey never claimed and never received reimbursement as did Owens.

sented to pay him (or his estate) for the next seventeen years a one-cent royalty on each gross of bottles made on any of his machines. The maximum (and actual) royalty was fixed at $100,000 a year.[63]

3. Internal Policies

Thus far the discussion has concerned mainly the personality relationships within the entrepreneurial group. In no economic problem encountered in the real world should the personal element be completely ignored, and certainly not in any problem involving entrepreneurship. The strength and weaknesses of individuals probably loom as large in the business world as they do in other realms. Diplomacy, friendship, jealousy, rational and irrational animosities, and other such manifestations of human character affect the management of economic enterprises in much the same way as they condition the development of other institutions which are basically noneconomic. Pecuniary motives, it has long been recognized by many economic theorists, may be outweighed, or at least tempered, by personal pride and desire for power and social esteem.[64] The entrepreneurial group at Toledo was no exception. As its pecuniary success became more and more substantial, other incentives gained in relative strength. An examination of the way the men formulated certain internal and external policies may illustrate the point.

M. J. Owens appears to have exercised the greatest influence on the group's internal policies. Chapters IV and V have already related how Owens assumed the initiative and persuaded Libbey and his other associates to transform their bottle company from a

[63] Owens-Illinois Glass Company, Owens' Private Files.

[64] Precisely this point has been made by Marshall E. Dimock in his recent book, *The Executive in Action*, pp. 252–253. "Up to a certain point, rising executives seek increased income first of all; it gives them the comforts and luxuries they want, plus a sense of security. But eventually, varying somewhat with the individual, this financial urge is superseded by a desire for power and position — the chance to make and change policy, and to do things one's own way. This in turn gives way to a desire for recognition, for prestige as a result of work well done. Finally, this egotistical yearning is appeased and top business executives seek an opportunity to work at something where public spirit and altruism are primarily involved . . ." (I am indebted to Dr. Arthur H. Cole for this reference.)

machine-licensing agency to a concern primarily engaged in manufacturing glass containers; how Owens continually emphasized the importance of building adequate plants and engaging in mass production; and how he pushed the Toledo Glass Company into the flat-glass field. Furthermore, he was the group's fountainhead for ideas about technological improvements.

It is surprisingly difficult to discover the precise nature of his genius or, perhaps better still, his greatness. He was not equipped by education with what many people would consider the rudimentary requirements of an inventor.[65] He could neither speak nor write grammatically, and he did not master decimal fractions until shortly before his death. Furthermore, he could neither draw nor read blueprints. For years he maintained an office at the company's engineering and experimental shop in Toledo and spent most of his time there when he was not visiting factories in other cities. Whenever he had an idea to communicate to his draftsmen or engineers, he called them together in conference and drew rough sketches of the device he had in mind on a blackboard. In other words, he provided them with the crude idea, and they tried to translate it into manageable engineering terms. Several of them have independently affirmed that Owens was always shocked by the size and complexity of the machines for which he had supplied the basic principles.

Perhaps this, more than anything else, directs attention to those abilities which constituted his greatness. Other inventors like Irving W. Colburn had fertile imaginations but lacked sufficient understanding and experience in glassmaking to realize success. Most practical glassmen, on the other hand, were slaves of their art. Why Owens shook off the intellectual fetters of craftsmanship while thousands of other contemporary members of his fraternity did not probably can never be resolved. The "facts" are that he understood many of the properties of molten glass and his restless, imaginative mind conceived ways whereby it could be successfully handled by machines. His subordinates had to overcome nearly all the technical difficulties involved in any of his schemes and may have suggested many of the important features later incorporated

[65] There is no indication that any other member of his family shared his talents.

in his machines.[66] One of them presumably pointed out that six instead of five arms were preferable on the bottle machine from a strictly engineering standpoint, and another claims to have suggested the basic idea for the revolving pot. One may conclude, of course, that Owens deserved fame, if for no other reason, simply because he selected men competent enough to carry out his hazy plans. But this appraisal would be misleading. Owens never succeeded in overriding his personal likes and dislikes when it was a question of employing or retaining men,[67] and there is evidence that he was sometimes jealous of his subordinates.[68] This psychological weakness, as a matter of fact, magnifies rather than diminishes Owens' greatness in *technical* matters.[69]

[66] Perhaps it was to be expected that some who worked with him came to feel that they contributed more to the combined effort than Owens or posterity has conceded. In whatever is essentially group endeavor, it is hazardous to attempt to attribute individual ideas to the various men. But, when four of the men involved independently concur in the same judgment, as they do in regard to the Owens' bottle machine, their views can not be entirely overlooked.

[67] As a matter of fact, it seems almost sheer chance that he engaged and retained such competent help. Personal animosity developed between him and William Emil Bock, who had performed essential service on the early bottle machines; and as a result, Bock resigned around 1910. Richard La France, who later assumed the post vacated by Bock, began working for Owens while still attending high school and could scarcely have shown at that time any indication of his engineering abilities. Charles W. Schwenzfeier, another important figure around the research shop, owed his connection with the group to Bock and was always in jeopardy of being discharged because of his continuous disagreement with Owens. Realistically speaking, he remained "in spite of" rather than "because of" Owens.

[68] Especially, Bock, Schwenzfeier, and Seth Henshaw, who at one time was manager of Libbey-Owens' plant at Charleston. The parallelism between Owens and James Watt in many particulars will not escape all readers. They both were fortunate in finding a congenial business man to finance their efforts; Watt apparently became jealous of William Murdock, as Owens did of his underlings; and "there was, in fact, a trace of intellectual arrogance in Watt's character, which increased with time." — Marshall, *James Watt*, pp. 169 ff., especially. The degree of Owens' intellectual arrogance is treated on pp. 293 ff.

[69] Parallels, like analogies, are frequently misleading and even dangerous. Yet the similarity of Thomas A. Edison and M. J. Owens is so striking that Professor Earl J. Hamilton has suggested to me that it deserves special mention. According to M. A. Rosanoff, "Edison in his Laboratory," *Harper's Magazine*, Vol. 165 (1932), 402–417, "Edison seldom worked with his own hands. He had a mechanical man who did all the manipulating, while the master did the experimenting in his head" (p. 403). Like Owens, Edison refused to allow cost to interfere with any experiment, borrowed ideas freely from others, evinced unusual patience and perseverance in his laboratory, and found great difficulty in understanding theoretical chemistry and physics. "No wonder," Rosanoff wrote, "conventional academic people would get the impression that his only gift was a shrewd capacity for finding trained men to

In addition to formulating most of the basic ideas, Owens also had the conviction and perseverance necessary to carry each project through to completion. He threw many machines upon the scrap pile before he obtained one which measured up to his standards. E. D. Libbey consistently supported him in his demands for financial assistance, no matter how uncertain and how distant the anticipated results.[70] One or two other members of the original entrepreneurial group, however, were inclined to view the mounting pile of scrap metal with alarm and occasionally suggested to Owens that perhaps he could manage with a smaller research budget. Since they were unable to influence Libbey on this matter, Owens fortunately was unhampered in developing his bottle machine, in modifying and improving it constantly,[71] and in spending over four years perfecting Colburn's sheet-drawing machine.[72] In all matters pertaining to technology and experimentation Owens continued until his death in 1923 to have the dominant, if not the final, word. It was he who persuaded Libbey and the Toledo Glass Company to buy up Colburn's patents and enter the window-glass field, and it was also he who urged the Owens Bottle Company to acquire the Graham bottle machine and the Homer Brooke feeder as a hedge against competition.[73]

Owens, on the other hand, was reputedly responsible for some miserable blunders. Two of his acquaintances have affirmed that he bluntly told one inventor hired by the Libbey Glass Company

---

make inventions for him." — *Ibid.*, p. 411. But his real greatness lay in his brilliant thinking.

[70] The records reveal only one occasion when Libbey failed to support a research program initiated by Owens. This occurred in March, 1918, after Owens had spent several months and around $12,000 at the Libbey-Owens' plant trying to make optical glass and a "wood block bending device" for the window-glass machine. Without a doubt, this incident played a part in the broader, more serious dissension which was then threatening their long-standing association.

[71] At the hour of his death he was trying to persuade the board of Owens Bottle to equip the Charleston plant completely with his latest twenty-arm machine.

[72] It may be conservatively estimated that the Toledo Glass Company and its offshoots spent between $2,000,000 and $3,000,000 on experiments during Owens' lifetime and that the Libbey Glass group spent around $1,000,000 in developing the automatic bulb and tumbler machines and the Danner tube machine. In 1912 there were at least twenty persons employed in Owens Bottle's engineering department, and in 1914, at least fourteen with five additional men engaged as draftsmen in the mold-design department. — Owens-Illinois Glass Company, Soubier's Private Files.

[73] See above, pp. 111, 162 ff., 182–183.

and the Westlake Machine Company that it was impossible to construct a completely automatic paste-mold machine for tumblers and light bulbs. Yet that same inventor did it. Again, in 1919, one of his plant managers finally convinced him by successful demonstrations that a worker could change the molds on an Owens bottle machine within five minutes without stopping its rotary motion — a feat which Owens had violently argued could never be performed.[74] He also ridiculed the suggestion of another employee that a revolving rod would greatly increase the efficiency of devices like Homer Brooke's [75] for feeding gobs of molten glass to forming machines. Four years later, however, the Hartford-Fairmont Company, Owens Bottle's great rival, introduced a glass feeder embodying essentially the same principle. Another individual in 1922 suggested that Owens attempt to buy the rights to an annealing lehr which purportedly consumed about one-tenth as much gas as did his own lehr; he, too, met with rebuff and was told that the Owens model was the most efficient lehr on the market.[76]

The worst blunder of all concerned the glass-feeding devices developed by the inventor Peiler employed by Hartford-Fairmont.[77] In either 1915 or 1916 the inventors offered to sell their device and patent rights to the Owens Bottle Company for about $100,000 or $200,000. Owens superficially examined their process and advised his company not to accept their offer. Less than a decade later the company would have gladly welcomed an opportunity to purchase the rights for many times the original sum stipulated. Perhaps one can explain (without excusing) Owens' shortsightedness in terms of a rather common human weakness — the inability to consider fairly the merits of suggestions made by subordinates and the tendency to forget that one's own outstanding achievements may lose their uniqueness with the passage of time and may even be surpassed by other men's accomplishments.[78]

[74] This accomplishment saved much time formerly lost whenever a machine began to make a different kind of bottle. The same factory manager also demonstrated that molds with multiple cavities were feasible and that amber glass could not be melted in as deep a tank as clear glass because heat penetrated amber very poorly.

[75] See above, p. 182.

[76] Experience has proved conclusively that he, not Owens, was right.

[77] See above, pp. 183 ff. Hartford-Fairmont became Hartford-Empire in 1922.

[78] According to James Harvey Robinson (*The Mind in the Making*, *passim*), this

The Toledo Group placed primary reliance upon Owens not only in matters regarding mechanical development but also, especially during its early period, in affairs relating to factory management. His practical experience and his ability to fraternize with glassmakers rendered him adept at locating and eradicating production troubles at the various plants operated by the Owens Bottle Company or its licensees. His associate, Walbridge, around 1905 wrote the president of the Thatcher Manufacturing Company as follows:

> Bear in mind this: Owens is brusque in many ways, but as a factory manager he has few equals and his advice is sought by a great many and as an example he is consulting superintendent of the Libbey Glass Company, employing some thousand hands, and through his advice their factory end is in most excellent shape. He may suggest radical changes but I should heed all he says and I believe [he] can get you out of the hole.[79]

Even the *National Glass Budget* paused in its vitriolic attack upon Owens' extravagant egotism and intellectual arrogance (which it considered unwarranted) to pay due tribute. "Having thus advanced himself from a position in the trade of no greater comparative importance than a pimple on a pickle to that of being the admitted greatest glass-house manager this country ever produced," the editor wrote, "he is entitled to unstinted credit, not an iota of which would *The Budget* deny him." [80] But in his capacity as general manager and managerial consultant for licensee firms, Owens made so many costly mistakes that one might well question whether the *Budget*'s appraisal was not too extravagant. As in the field of technological research, his errors usually resulted from impulsive judgments, vanity, and an unwillingness to take advice from subordinates.

---

weakness — expressed in more universal and philosophical terms — is the greatest single impediment to social progress. Men have first to disabuse themselves of all prejudices and tendencies toward rationalization, he contends, if they are to create the "unprecedented attitude of mind" necessary for coping with "unprecedented conditions" and for utilizing "unprecedented knowledge."

[79] Owens-Illinois Glass Company, Walbridge Letters, I, 417.

[80] *National Glass Budget*, XXVI, May 21, 1910. See also the editorials appearing in the April 23, 1910, and May 14, 1910, issues.

It was he who decided to build the bottle factory at Fairmont, West Virginia, in 1910 next to the steep slope of a mountain hemming in a narrow valley. Wind currents sweeping down the mountain interfered with the furnace drafts, and annual spring floods frequently caused the creek to rise and extinguish all fires at the factory.[81] He also selected Clarksburg, West Virginia, as the site for another plant in 1912 although railroad connections were extremely poor.[82] Without authorization from Owens Bottle's board and at an expense of approximately $100,000, he laid additional foundations there in 1917–1918 with the view of enlarging the plant. The directors decided against the project, and the factory at Clarksburg remains to this day a relatively useless possession of the company.[83] A man who acted as plant manager, first at the West Toledo factory and then at Glassboro, New Jersey, is authority for the statement that the layout, made under Owens' supervision, of gas producers, storage bins, furnaces, and machines in both places was very inefficient.[84]

The experience of the Toledo Glass Company and the Owens European Bottle Machine Company with the demonstration factory established at Manchester, England, throws additional light upon M. J. Owens' limitations. Heeding the advice of foreigners who were mere acquaintances, and unmindful of the warnings voiced by his American subordinates, he committed two grave errors. First, he caused a newly built furnace to explode by insisting that the workers light it full blast without going through the

[81] On December 20, 1916, the plant was absolutely idle because the natural-gas supply in the area had been depleted.—Owens-Illinois Glass Company, Executive Committee Minutes, I, 28.

[82] How he came to select Clarksburg provides an illuminating anecdote. The fact that natural gas was then cheap in that area first captured Owens' attention. Other associates, however, suggested a site in northern New Jersey, close to New York City. Owens along with a few directors of the Owens Bottle Company set out to visit the New Jersey site by automobile on a bleak, wintry day. They had to leave their car on the highway and trudge a few miles across marshland in order to reach the exact place. Owens lost his patience and reputedly exclaimed, "Hell! I'm cold, and I've already decided on Clarksburg anyway." The group never saw the suggested site in New Jersey.

[83] The factory made an extremely poor showing its first few years. See above, p. 110.

[84] It cost between $200,000 and $300,000 to rectify the mistakes at Glassboro's new factory.

usual slow process of gradually heating the crown and sides. And, secondly, he choked up another furnace and stopped production by mixing clay with the usual glass ingredients on precisely the day that representatives arrived to inspect the furnace and machines in behalf of a combination of prospective buyers. As long as he remained in Manchester, Owens made the factory manager vary the glass formulae two and three times a week, apparently unaware that the same formula would have to be followed for two or more weeks before the results could be fully determined from the bottles blown.[85] The plant there began to show an operating profit, instead of losses, soon after Owens returned to Toledo.[86]

Throughout his lifetime, when production difficulties developed at any factory, Owens frequently prescribed modification in the glass formulae even though he was not at the plant to investigate whether other factors were responsible for the trouble.[87] Together with some of his associates, he regarded himself as an expert on the proper proportions of glassmaking ingredients. In December, 1913, the directors of the Owens Bottle Company considered hiring an "expert chemist and batch mixer"; [88] Owens delayed doing so until two or three years later. Then he hired an experienced batch mixer rather than a university-trained chemist.[89] As long as he lived, he held firmly to the belief that in order for any

[85] Charles W. Schwenzfeier, now living in Toledo, has preserved the Manchester "batch book," which contains dates and entries made in Owens' own handwriting.

[86] See above, p. 121.

[87] William J. Jacobs claims that while he was at the Libbey-Owens' plant in 1916–1917 Mike made him change the formulae again and again, and that this was partially responsible for the trouble at the factory which brought about the change in management described earlier in this chapter.

[88] Owens-Illinois Glass Company, Owens Bottle Company Minutes, III, 160. The Bureau of Foreign and Domestic Commerce stated as a result of its investigation of the industry in 1916 that "not more than one chemist was found in about every 20 plants visited." — *Glass Industry*, p. 316.

[89] Elbert E. Fisher could analyze accurately the composition of glass and the purity of its materials, but he never tried to develop new kinds of glass. The company on February 15, 1915, authorized the construction of a small laboratory at its Toledo plant for his use. — Owens-Illinois Glass Company, Owens Bottle Company Minutes, IV, 121. And the board of the Libbey-Owens Sheet Glass Company on December 9, 1918, authorized the employment of a competent chemist "to work in experimentation and analysis . . . for the purpose of determining the nature and quality of the raw materials used and of the glass produced in the Company's factory." — Libbey-Owens-Ford Glass Company, Libbey-Owens Sheet Glass Company Minutes, I, 231.

man to be of much help in any position around a glass factory he had, first of all, to be a practical glassmaker.

Owens never curbed his pride to the point where he could cheerfully support any subordinate who seemed to know more than he or who attracted much public attention. According to the testimony of several individuals who worked under him, he apparently felt that to admit that an employee (or even a business associate) knew more than he about any practical angle involved in glassmaking was to admit his own general incompetence. Several capable men resigned or were discharged because they had been a little too successful in their accomplishments or because they had been too outspoken upon occasion. He sometimes used spies in each plant in order to keep himself informed about what the men in authority there were doing.[90]

E. D. Libbey also kept himself fully informed about business affairs during protracted absences from Toledo, but he usually obtained his information from daily or weekly letters written by the plant managers themselves. Moreover, Libbey followed the policy of compensating essential employees with blocks of company stock in addition to their regular salaries, for, as he once said, "When I find any of my associates or employees contributing to my success, I tie them to me and make their interest mine, as my interests are theirs." [91] Libbey seems to have mitigated to some extent the influence which Owens exerted over subordinate members of the managerial staff.

Owens maintained more amicable relations with ordinary glassmen employed by his companies than he did with supervisors and factory managers. According to a one-time president of the American Flint Glass Workers' Union, Mike remained sympathetic with the ideals of organized labor. If this was the case, his sympathy must have been vague. Workers in his mold shops and at the Libbey Glass Company continued to belong to unions, but he consistently refused to allow those who worked his machines to organize.[92] When the machines had been first introduced in differ-

[90] One of his henchmen, still living in Toledo, was kind enough to explain the various tasks of espionage he was asked to perform at different times.

[91] Owens-Illinois Glass Company, "Lives of E. D. Libbey and M. J. Owens," p. 7.

[92] See above, Chap. VIII.

ent plants, skilled bottle blowers had tried to sabotage their efficiency. Hence the Owens Bottle Company considered it safer (and cheaper) to employ unskilled labor. Then, too, Mike's belief in integrated control denied the possibility that his men could owe allegiance to both the plant manager and the union's leaders and yet perform their tasks as efficiently as they should. Occasionally in the early days he would take off his coat, roll up his sleeves, and show an inexperienced man the proper way to manipulate glass. As the chasm separating him from his workers widened, he lost much of his personal contact with employees; but he always commanded their respect whenever he visited the furnace rooms. His physical bearing and dress demanded deference, while his intimate acquaintance with practical problems and glassmaking jargon invited confidence.

4. External Policies

The Toledo entrepreneurs exercised their greatest influence upon glassmaking as a whole primarily through their activities in the technological field. Earlier chapters [93] have described, first, the processes and machines developed by others after M. J. Owens and E. D. Libbey introduced their models and, secondly, the degree to which the industry had become mechanized by 1920. During the latter half of the nineteenth century several men in both Europe and America were trying to devise ways of blowing glass which would lessen labor costs and reduce the industry's dependence upon skilled workmen. Not one of them, however, succeeded in solving the basic problem of how molten glass could be manipulated by mechanical contrivances without destroying those features which make glass the most desirable commodity for certain household and industrial purposes. Owens' great contribution in 1895 consisted in showing how this problem could be resolved. In 1903 he further demonstrated that glass could be gathered from the furnace prior to blowing in such a way that skilled gatherers no longer were indispensable. Still later he, Colburn, and others successfully pioneered in pulling a continuous sheet of glass without human or mechanical blowing at all. These and similar ac-

[93] See, especially, pp. 179 ff.

complishments yielded the competitive advantage which assured the Toledo enterprises of success. As usually happens, however, this advantage dwindled and probably would have disappeared after 1920 had not institutional controls intruded and prevented the rise of effective competition.[94]

Invention is frequently contagious. Once Owens pointed the way, others followed. The success achieved by the small group of engineers and inventors supported by the Toledo Glass Company yielded substantial profits and fired the imagination of even the conservative heads of the Libbey Glass Company. They, together with E. D. Libbey, organized the Westlake Machine Company and set men to work on an automatic bulb machine. Later, the Libbey Glass group financed the experiments of Edward Danner with his tube-drawing process and August Kadow's work on his automatic tumbler machine. In addition, it is interesting to note that the independent O'Neill Machine Company, located in Toledo, was responsible for some of the best semiautomatic bottle machines,[95] that Irving W. Colburn's father and brother at one time were under contract with the Toledo Glass Company, and that Irving himself lived in Toledo at about the time that he became interested in glassmaking machinery.[96] All of these facts suggest that spatial proximity may accelerate the spread of interest in technological change from one group to another.

Eventually the activity at Toledo radiated throughout the industry. Many individuals and companies tried to develop and improve paste-mold and bottle-blowing machines in order to regain their relative position in the market. None showed much progress

[94] Reference here is to such controls as patents and the agreement concluded between Owens Bottle and Hartford-Empire for their mutual benefit.

[95] *National Glass Budget*, XXXIII, August 18, 1917.

[96] H. J. and L. H. Colburn were mechanical engineers at Toledo and were employed by the Toledo Glass Company in 1896–1898 as consultants for glassmaking machinery. — Libbey-Owens-Ford Glass Company, Toledo Glass Company Minutes, I, 46, 103 ff. Irving himself left Fitchburg, Massachusetts, where he had an electrical and mechanical engineering shop, and moved to Toledo in August, 1898. — *Ibid.*, Colburn Records, Vol. A, 45. A year later he lived in Philadelphia for about eleven months and then moved back to Toledo. — *National Glass Budget*, XV, September 23, 1899; XVI, August 11, 1900. He may never have worked for the Toledo Glass Company before 1912, but he probably became familiar with the experiments which it was carrying on.

in perfecting a completely automatic machine to compete with Owens' until Hartford-Fairmont's experiments with gob-feeding devices gave promise of success in the years 1914–1915.[97]

Mechanization would have eventually invaded glassmaking had Owens and Libbey never formed their association, but their accomplishments stimulated others. The actual course of events, as described in preceding chapters, conformed to what is probably a common pattern. Owens' specific inventions gave a decided advantage to the group which supported him; this situation in turn incited competitors to develop and adopt other innovations; and thus the industry advanced in productive efficiency all along the line, like a battle line with bulges appearing continuously but at different points along the front. Society ultimately benefited because glass prices gradually fell.

The Toledo entrepreneurs in general and E. D. Libbey in particular never favored aggressive price competition. They preferred to take most of the reductions in cost occasioned by their machines as profits rather than attempt to expand output by slashing selling prices. They were fearful of initiating general price wars and of completely disrupting trade relations even though they realized that they would not be among those producers who consequently would have been forced out of business. On January 19, 1903, Libbey explained his philosophy about co-operation and competition in a letter addressed to an official in the Indiana Match Company.[98] "In nearly all of my large industrial concerns, with which I am connected," he wrote, "it has been my policy to always go out of my way to bring about closer relationships with my competitors, and in nearly every case it has proven not only a benefit to me but to those with whom I have been fortunate enough to make business connections." In November of the same year, he wrote a certain glass manufacturer who was competing with the Libbey Glass Company in the cut-glass market:[99] "We do believe . . . that in order to make the most money for all, now is the time when

[97] See above, pp. 183 ff.

[98] Owens-Illinois Glass Company, Libbey Letters, I, 166. Libbey had been a director and heavy investor in the Toledo Match Company prior to the incorporation of the Owens Bottle Company.

[99] *Ibid.*, II, 103.

we should get together upon some satisfactory basis whereby the blank business [100] may not be demoralized, which certainly will be the case if something is not done by all for our mutual benefit." One more excerpt from his correspondence will suffice to depict Libbey's views on competition. On December 28, 1907, he wrote Hermann Heye, one of the European manufacturers interested in acquiring exclusive rights to the Owens bottle machine, "I still feel, as I did, when I mentioned the subject to you the first time I had the pleasure of meeting you in Manchester, that it might be to the mutual advantage of all bottle manufacturers in the world to have some kind of an organization whereby the product of the Owens machine might be controlled to the advantage of all . . ." [101]

Throughout the period covered by this study the Toledo Group followed price and licensing policies which implemented such a philosophy. The Libbey Glass Company, after 1894, co-operated with the Corning Glass Works in dividing up the American bulb market and regulating prices.[102] The Toledo Glass Company between 1895 and 1899 disposed of the *exclusive* bulb, tumbler, and chimney rights on its Owens' paste-mold machine to the Libbey Glass Company, the Rochester Tumbler Company, and the Macbeth-Evans Glass Company, respectively.[103] Soon after its incorporation in 1903 the Owens Bottle Company issued several *exclusive* licenses to other American manufacturers and retained the right to administer prices set by nonexclusive licensees.[104] Libbey Glass in 1904 signed a contract with the H. C. Fry Glass Company whereby both agreed to maintain prices and divide the market for all cut glassware made under one of Owens' patents.[105] Working through the Owens European Bottle Machine Company, E. D. Libbey in 1906–1907 tried to form an association of European bottle manufacturers which would buy the *exclusive* rights

[100] Glass "blanks" were pieces of tableware which were later cut or engraved either in the factory or in shops of independent workmen. M. J. Owens had recently perfected a novel and cheap way of pressing patterns on blanks to guide the cutters. See above, pp. 133 ff., 150.

[101] Owens-Illinois Glass Company, Europäischer Verband Lawsuit Files, *Hearings*, p. 637.

[102] See above, pp. 135 ff.

[103] See above, pp. 97 ff.

[104] See above, pp. 106 ff.

[105] See above, p. 135.

on the Owens machine for Europe. His efforts successfully materialized in the Europäischer Verband des Flaschenfabriken Gesellschaft mit beschränkter Haftung.[106] As part of the purchase agreement, each party agreed not to sell bottles in the other's territory. M. J. Owens in 1912 unsuccessfully tried to use his influence with organized labor to keep the bottle union from cutting wages because he feared this would cause independent hand producers to upset existing prices.[107] And in July, 1920, the executive committee of Owens Bottle forwarded certain suggestions to the National Bottle Manufacturers' Association. It endorsed the principle that all firms should decrease output in times of depression and thereby maintain "fair prices" instead of cutting prices to cost or below cost levels in order to maintain output; it professed a willingness to furnish the Association with cost data for its own plants and applauded all efforts to persuade other companies to do likewise after adopting "sound" cost-accounting systems; and it admitted that it already adhered to a one-price, open-price policy and urged others to follow its lead.[108] The Libbey-Owens Sheet Glass Company also tried to lessen competition. It sold *exclusive* Japanese, Canadian, and European rights on the Colburn machine to three different foreign companies[109] and contracted with the Johnston Brokerage Agency to market most of its domestic output in much the same way that the Agency sold the output of other firms.[110]

In nearly every case, therefore, Libbey and his associates guided their enterprises into those channels of administered prices and monopolistic controls which had persisted since organized labor and employers' associations in the industry had discovered that they could obtain mutual benefits by working together. It should be remembered, however, that when Libbey, Owens, Brown, Walbridge, and Geddes first came together in the Toledo Glass Company, they never intended to produce glass. They started out as developers and promoters of glassmaking machinery. It is true

[106] See above, pp. 122 ff.
[107] See above, p. 212.
[108] Owens-Illinois Glass Company, Executive Committee Minutes, II, 396–397.
[109] See above, pp. 128–129.
[110] See above, p. 234.

that they incorporated the American Lamp Chimney Company in 1898 and began to make bottles at Manchester, England, in 1907. But these excursions into the manufacturing field were designed merely to demonstrate the practicableness of their machines and to intimidate established producers into purchasing their rights.[111] The same motive lay behind the incorporation of the Northwestern Ohio Bottle Company in 1904.[112] Four or five years later M. J. Owens finally convinced Libbey that the Owens Bottle Company should embark upon a large-scale production program in addition to its licensing activities.[113] New plants were built and old ones were acquired by taking over independent firms. By 1917 Owens Bottle was operating fourteen factories, either owned outright or controlled.

Even this expansion did not satisfy Libbey. On August 7, 1917, he confided his ambition to Clarence Brown in a letter:

> Our most favorable channel through which to expand is, in my opinion, the acquisition of the other large U. S. companies — viz. *Ball Brothers*, *Hazel*, and *Illinois* . . . [These] if secured upon a fair basis . . . would give our company not only supreme control of all branches of the bottle trade, but to its stockholders, a total of 6 to 7 million dollars earnings, which if proven true would exceed my previous year's estimate of $5,000,000 net profit in all branches of the bottle business under the controlling influence of the great and wonderful invention of ours. I hope to see this a fact soon and I shall then feel that it was worth living for — to be one part of the mechanism that gives its stockholders, its original inventors, and promoters a keen satisfaction in its success physically and financially and, greater than *all*, [to] be instrumental in giving to the world cheaper bottles when the time [for the consolidation?] may come.[114]

[111] A spokesman for the British group of manufacturers who participated in the Europäischer Verband said, "We had to accept the Owens machine. They came to us, and held a bludgeon at our heads. In effect they said, 'You have got to buy the machine or we shall sweep the deck.' " — *National Glass Budget*, XXVI, August 27, 1910.

[112] See above, pp. 103 ff.

[113] See above, pp. 109 ff. It was also Owens who convinced Libbey that the group should undertake the development of the Colburn machine. See above, pp. 126 ff. Since they were so successful in manufacturing bottles, the Toledo entrepreneurs did not hesitate to make sheet glass themselves once they had perfected the machine.

[114] Owens-Illinois Glass Company, McNerney's Private Files. The three companies specifically mentioned — Ball Brothers (fruit jars), Hazel-Atlas (packers' ware), and Illinois Glass (prescription, proprietary, and whiskey ware) — were the largest independent firms in the bottle field.

Almost immediately Owens Bottle began to investigate the possibility of carrying out the plan. Negotiations, however, were dropped in 1918 because the interested parties could not agree upon terms.

Those men now living in Toledo who participated in these negotiations affirm that fear of violating the antitrust laws was in no way responsible for the failure to carry out Libbey's proposal.[115] Disagreement over personality and financial matters, according to them, was the real obstacle. Yet the records of the Owens Bottle Company reveal that its officials upon several occasions tried to avoid trouble with the government. In 1911 they revoked all price-control clauses in contracts with their licensees;[116] and on June 27, 1916, they voted not to exercise their option to purchase control over five or six glass companies because, they said, they did not wish to become embroiled with the Sherman and Clayton acts and with the Supreme Court's "rule of reason."[117] For the same reason and in the same year, Walbridge, Brown, and John Biggers resigned as directors of the Whitney Glass Works, which their company controlled.[118] Furthermore, the executive committee of the Owens Bottle Company decided on January 9, 1917, not to purchase the Libbey Glass and Westlake Machine companies,[119] and in October, 1920, not to buy controlling interest in the Maryland Glass Company[120] lest its action be construed as unreasonably restraining interstate commerce.

Perhaps the desire for industrial power and bigness, which may account for Libbey's plans to expand in 1917, was a "natural" outgrowth of the astounding success achieved initially by the

[115] Apparently, Libbey wanted to purchase all the assets of the three companies and consolidate them with those of his own company. This procedure would not have run afoul of the antitrust laws as they were then being administered and interpreted by the courts.

[116] See above, p. 108.

[117] Owens-Illinois Glass Company, Owens Bottle Company Minutes, V, 231–232.

[118] *Ibid.*, V, 281.

[119] *Ibid.*, Executive Committee Minutes, I, 37. Since Libbey Glass and Westlake were engaged in different lines of production than was Owens Bottle, it was feared that the company would encounter difficulties in justifying its purchase if the government raised the question.

[120] *Ibid.*, II, 391, 413.

Toledo Group. Libbey, because of his pleasing personality and cultivated diplomacy, had been able to make many valuable contacts with important manufacturers and capitalists both in the United States and abroad. He greatly influenced the selection of possible licensees [121] and was always able to discover sources of additional capital whenever the original entrepreneurs were unable or unwilling to provide the necessary amounts. He brought Adolphus Busch, the American brewer, and Julius C. Prince, the managing director of a British mineral-water company, into the Owens European Bottle Machine Company; [122] and primarily through Prince, Libbey was able to interest important bottle manufacturers on the Continent in the exclusive rights to the Owens machine. Even earlier he had obtained outside venture capital for the Owens Bottle Company [123] and then had persuaded a group of independent manufacturers to merge, thereby creating one of Owens Bottle's first licensees.[124]

He himself provided the largest and most reliable source of venture capital. His personal funds and promissory notes financed the Libbey Glass Company during its early years, made possible its participation in the Columbian Exposition at Chicago,[125] met all deficits accumulated by the Toledo Glass Company while Owens was experimenting with his bottle machine,[126] and provided

[121] In 1909 he persuaded Owens Bottle to license the Whitney Glass Works in spite of M. J. Owens' strong opposition. For personal reasons Owens preferred the Whitall-Tatum Company at Millville, New Jersey. There is some evidence that Owens tried to sabotage Whitney so that the license could then be offered to his favorite.

[122] See above, pp. 119 ff.

[123] In 1903 he asked friends and lawyers in various cities to submit confidential reports on the character of men whom he might possibly interest in his prospective company. — Owens-Illinois Glass Company, Libbey Letters, I and II, *passim*. He realized that the company would be a somewhat speculative venture, and he did not want any of his Toledo friends to risk their money simply because his name was connected with the enterprise. He asked that the information which he sent men outside Toledo be regarded as confidential and at all costs kept from falling into the hands of Toledoans. — *Ibid.*, II, 1, especially.

[124] See above, p. 104.

[125] See above, pp. 92 ff. He risked over $200,000 in the model factory exhibited by the Libbey Glass Company and gained rich rewards from its spectacular success.

[126] According to an executive colleague now living in Toledo, when the possibility of success seemed dim and when the company's funds were exhausted, Owens told Libbey that he honestly thought it unwise for Libbey to risk any more of his personal funds in the endeavor. Libbey then replied that he would say when no more

most of the working capital for the Owens Bottle Company.[127]

Bankers and underwriters played no part in financing these ventures. Several associates who are still living have independently affirmed that bankers in Toledo and, especially, those in New York considered Libbey a poor risk once he began supporting Owens' experimental schemes. They felt that he relied too much upon intuitive judgment because he rarely could support a request for financial assistance with the sort of data and market information which bankers usually desire.[128] The coldly polite reception which New York houses extended him during the early period (say, from 1895 to 1910) irritated Libbey. He must have enjoyed immensely their changed attitude after he had become a recognized success, for they actually chided him for taking his banking business elsewhere. He finally utilized their services in 1916 when Owens Bottle greatly expanded its security issues in order to acquire the stock of other bottle companies and when the Libbey-Owens Sheet Glass Company was incorporated with an initial capitalization of $6,000,000.

5. Conclusion

M. J. Owens and E. D. Libbey formed the nucleus and guiding spirit of the entrepreneurial group at Toledo. They had been drawn together because they shared an interest in improving glass technology and because they respected and trusted each other. As far as personality and both natural and acquired abilities were concerned, they were quite unlike. Whereas Owens was uncultured, brusque, domineering, strong-willed, but highly skilled in working glass, Libbey was suave, diplomatic, relatively rich, and skilled in the art of managing other business men. Owens in his formal education never got beyond the grammar grades, and even Libbey stopped short of college after attending an academy in

funds were available and that until then Owens should continue his experimentations without worrying about the risk involved.

[127] See above, pp. 102 ff. Later, the various companies helped one another with intercompany loans.

[128] As Veblen probably would have said, conservative business men and bankers viewed Owens and Libbey as "irresponsible, fanciful project-makers, footless adventurers, fit only to work out innovations that were of doubtful expediency in a business way." — *Absentee Ownership*, 109.

Maine. Both were venturesome and inclined to act intuitively, but Libbey was guided less than his associate by personal likes and dislikes.

Owens' inventive bent would probably have led him to develop many useful improvements in the industry had he never met Libbey. Tom Owens, his brother, and others associated with him over a long period of time have attested that he was forever trying to invent something.[129] It is doubtful, however, whether he would have accomplished what he actually did if he had worked with someone else. Libbey and Owens made an excellent, harmonious team for what they originally set out to do. Everyone who knew Owens agrees that he was intractable and could not brook supervision which interfered with his work. This is scarcely surprising, for genius usually is intractable and withers away or becomes frustrated if subjected to controls ineptly administered. Libbey was one of the few people who understood Mike and had the ability to direct his energies into their most fruitful channels. Libbey patiently supplied the necessary funds and encouragement and, with only one or two exceptions, refrained from antagonizing Owens by interfering in matters over which the latter had responsibility and control. Had Libbey been parsimonious and outwardly anxious about his associate's progress in solving technological problems, insurmountable conflicts of personalities would have arisen. Certain personal and managerial difficulties actually did appear but only after the group's major interests had shifted from developing and licensing inventions to manufacturing bottles and window glass.

At the outset, E. D. Libbey had become the head of a moribund factory in Cambridge, Massachusetts, and had saved it from bankruptcy. By developing and marketing new tableware patterns,[130] he managed to cover operating costs and thereby keep the furnaces lighted until labor troubles made the high fuel costs in the East insupportable. He then moved to Toledo and continued

[129] At one time he tried to make gas from oil, and upon another occasion he temporarily became interested in folding streetcar doors and steps. So far as I know, however, he made no successful invention outside the glass field.

[130] His experience with "amberina" glass and the Tiffany firm in New York has already been recounted. See above, pp. 90–91, 274–275.

to manufacture glassware of excellent quality. M. J. Owens, as factory superintendent, helped him to build an efficient working force there and also at a branch bulb factory in neighboring Findlay. Once the company added bulbs and tubing for incandescent lamps to its tableware line, financial success began to materialize. Libbey's subsequent participation in the Columbian Exposition at Chicago not only gave evidence of his daring in business affairs but also securely entrenched his firm in the consumer market.

He and the other Toledo entrepreneurs began another phase of activity in 1894–1895, when they decided to form the Toledo Glass Company for the purpose of developing Owens' glassmaking machinery. That their basic contributions were technological in nature cannot be too strongly emphasized. Their machines in the paste-mold, bottle, and window-glass fields were truly revolutionary and laid the foundation for successful operation of the other companies they formed. Until 1916 the group remained small in number and unified in purpose. Its industrial structure was neither large nor complex. Hence personnel and production problems did not become too complicated for Libbey, Owens, and their associates to handle. The Owens Bottle Company did not seriously enter the manufacturing field on a large scale until 1911; and from 1912 to 1916 the Toledo Glass Company was still primarily concerned with mechanical improvements, for it had just purchased the Colburn patents on sheet-drawing. In view of their objectives at the time, one must admit that the entrepreneurs achieved a high degree of success. There may be some question about the wisdom of their licensing and pricing policies,[131] but this does not diminish their reputation as outstanding revolutionaries in the field of technological change.

As they moved from one project to another, as the number of their companies multiplied, as their private fortunes mounted, and as manufacturers throughout the world grudgingly accorded them respect and esteem, Libbey and Owens discovered that their mutual interests were at last diverging. After 1916 the primary concern of the Toledo Group centered increasingly around problems con-

[131] See above, pp. 302 ff.

nected with the production and sale of glassware. Not only did the Owens Bottle Company grow rapidly in size and complexity, but the group assumed the responsibility of another great enterprise — the Libbey-Owens Sheet Glass Company — in an entirely different branch of the industry. M. J. Owens was unable to adjust himself to the institutionalized pattern of management which now became imperative; and furthermore, his personal objectives and interests in technical improvements did not change with those of his associates. During this third phase of activity, on the other hand, Libbey seems to have been more conservative and less venturesome than he had been earlier. Once the enterprises were "going concerns" and so complex that more responsibility had to be delegated to other executives and subordinates, he lost his vital interest in them and diverted his energies to the arts and noneconomic pursuits. The new crop of associates initiated into the group after 1916 were definitely "captains of business and finance" and not technological innovators. A few even speculated in their companies' securities. Owens, in this fashion, increasingly found himself hemmed in by conflicting interests; and it was from this situation that the executive, managerial, and personality difficulties of 1916–1923 arose.[132]

Some men associated with the companies today believe that during the 1920's the Libbey and Owens network definitely outgrew its managerial and entrepreneurial framework and that relative stagnation actually commenced. After 1920 the promotional, pioneering work was practically at an end. Libbey celebrated the sixty-sixth anniversary of his birth in that year, and Owens was only five years his junior. It should not be surprising, therefore, that like most men Libbey — and to a lesser extent Owens — had lost much of their former enthusiasm and resiliency. They had already attained both prestige and fortune. Altogether the companies which they created and directed possessed assets in 1920 valued at between $50,000,000 and $60,000,000. Libbey had

[132] Many people may see in this history of the Toledo Group's activities at least a partial illustration of Thorstein Veblen's analysis of the increasing division of labor between those engaged in technological activities and those engaged in the pecuniary control and direction of productive activities. See, especially, *Absentee Ownership*, pp. 106–113.

arrived at Toledo in 1888 with a private fortune of only about $100,000, and Owens at that time possessed practically no material wealth. Yet both were millionaires when they died. And large fortunes often have the tendency to make people lethargic to change and "naturally" conservative.

APPENDIX A

THE GENERAL TREND OF GLASSWORKERS' WAGES, 1880–1920

Chapter VIII discussed the wages paid certain grades of glass labor along with other aspects of the policies formulated by unions and associations of producers. It was there suggested that the high wage rates maintained by skilled workmen greatly enhanced the manufacturer's incentive either to develop or to adopt mechanical means for making glass. Little statistical evidence, however, was presented which would indicate how important an element wages were in total production costs, how the earnings of glassmakers compared with those of men employed in other industries, and how the advent of machines affected the real income of glass-factory employees. It may prove helpful, therefore, to append a note summarizing the trend of wages for the industry as a whole from 1880 through 1920.

Of the 334 industries reported by the *Census of Manufactures* for 1914, glassmaking ranked thirteenth in the percentage of labor costs to the value of output produced.[1] The Bureau of Foreign and Domestic Commerce in 1917 published the results of its investigation into the cost of making glass in the United States, and the percentages of labor costs to the total sales value of glass made (which it calculated from data obtained from 213 establishments) are reproduced in Table 23.

A comparison of these percentages with similar ones presented in Chapter II reveals that the relative importance of labor as a production cost had not decreased *markedly* since 1860–1880 except in the manufacture of bottles on the Owens machine and of jars on various wide-mouth machines.[2] This conclusion may at first sight appear rather startling since several revolutionary changes in the techniques of glassmaking occurred in the interim. There are several reasons, however, why the percentages for 1916 fail to reflect clearly the labor economies made possible by the new technology described in Chapters VI and VII. In the first place, the percentages given in the table are for the manufacturer's *total wage bill* and cover more than the wages paid those who operated the new machines or themselves worked directly with molten glass. In the second place, since the long-run trend of glass prices was

[1] *Glass Industry*, p. 49. The total value of glass made, in round numbers, was $123,000,000; and total wages amounted to almost two-fifths this sum.

[2] See above, pp. 30–31. Factory records of the Owens Bottle Company reveal that labor accounted for 30.67 per cent of its manufacturing costs in August, 1923. — Owens-Illinois Glass Company, Owens' Private Files.

TABLE 23

PERCENTAGE OF LABOR COSTS TO THE TOTAL SALES VALUE OF GLASS MADE, 1916 *

Nature of Product	Percentage
Window Glass (hand)	56.80
Window Glass (cylinder machine)	47.74
Plate Glass	33.76
Wire and Opalescent Glass	24.03
Bottles (hand)	53.00
Bottles (Owens automatic machine)	25.04
Bottles (semiautomatic machines)	48.04
Jars	28.91
Tableware, Blown	53.68
Tableware, Blown and Pressed	48.01
Lighting Goods	40.15
Lamp Chimneys	52.40
Miscellaneous	44.85
All Kinds	40.26

* Source: *Glass Industry*, pp. 104–105. The remuneration to superintendents and foremen is included in the labor costs; the salaries of officials and office workers are excluded. The royalties paid on the Owens bottle machine, the semiautomatic bottle machines, and the jar machines have *not* been included in the percentages in this table. See *ibid.*, p. 107.

definitely downward after 1880, the percentage of total labor costs to the *sales value* of glass output cannot, by itself, be taken as a true indicator of the increased productivity of men engaged in forming glassware. And, in the third place, only jars and bottles made on the Owens machine were the result of fully automatic processes. All other categories of output listed for 1916 were made in plants following semiautomatic or handicraft methods. Ordinarily, technological improvements only slowly permeate an industry, and glassmaking was no exception. The full impact of the new technology, largely developed in 1890–1920, was not felt until after the first World War.

All available evidence suggests that prior to 1880 skilled glassworkers had received substantially higher wages than unskilled workmen and many other skilled craftsmen in America.[3] The differential had been lessening both before and after the Civil War, and during the four decades following 1880 the same trend continued. From Paul H. Douglas' summary of the annual earnings of laborers in various industries from 1890 through 1926,[4] it appears that the average earnings of all employees in the glass industry were not consistently much greater than the average annual wages paid by all industries grouped together. As a matter of

[3] See above, pp. 32 ff.

[4] *Real Wages in U. S., 1890–1926*, pp. 299, 391.

fact, the annual earnings of glassworkers were greater than the general average for only twelve of the thirty-one years in 1890–1920. The data presented by Douglas and reproduced in Table 24, however, should be used with extreme caution in comparing the income of *skilled* glassmen with the earnings of other industrial workers. He based his averages

TABLE 24

DOUGLAS' ANNUAL MONEY AND REAL EARNINGS OF GLASSWORKERS AND INDUSTRIAL WORKERS GENERALLY *

Year	Annual Earnings		Relative Earnings (1890–1892 = 100)		Real Earnings ** (1890–1892 = 100)	
	Glassworkers	All Industrial Workers	Glassworkers	All Industrial Workers	Glassworkers	All Industrial Workers
1890	$479	$486	98	99	96	97
1891	478	489	98	100	99	101
1892	513	497	105	101	105	101
1893	491	483	100	98	102	100
1894	451	457	92	93	97	98
1895	462	478	94	97	99	102
1896	461	472	94	96	97	99
1897	493	474	101	97	103	99
1898	513	482	105	98	107	100
1899	513	493	105	100	105	100
1900	521	503	106	102	102	98
1901	524	518	107	105	101	99
1902	547	530	112	108	103	99
1903	524	548	107	112	94	98
1904	583	546	119	111	105	98
1905	582	559	119	114	105	101
1906	719	571	147	116	126	99
1907	686	593	140	121	113	98
1908	559	564	114	115	96	97
1909	570	594	116	121	97	102
1910	592	627	121	128	97	102
1911	581	624	119	127	92	98
1912	574	638	117	130	90	100
1913	603	666	123	136	92	101
1914	563	673	115	137	85	101
1915	495	672	101	137	76	103
1916	527	748	108	152	74	104
1917	668	866	136	176	78	101
1918	919	1,088	188	222	88	104
1919	1,129	1,245	230	254	95	105
1920	1,393	1,459	284	297	101	106

* Source: Douglas, *Real Wages in U. S., 1890–1926*, pp. 60, 299, 391.

** The indices of real earnings were obtained by dividing the indices of relative earnings by the appropriate "cost-of-living" indices (reduced from "1890–99=100" to "1890–92 = 100") found in *ibid.*, p. 60.

upon total payrolls and numbers employed. Since employees with degrees of skill varying from the very least to the greatest were crudely lumped together, the earnings of skilled glassmen were much higher than his figures would suggest. The annual reports of the National Window Glass Workers' Union, for example, reveal that its blowers averaged $961 for the 1916–1917 "blast" and $1,126 for the 1917–1918 "blast"; its blowers, gatherers, flatteners and cutters — all of whom were quite skilled — together averaged $884 and $1,008 for the same two "blasts." [5]

Available information on the hourly earnings of different classes of workers in the industry offers a rather reliable measure of the *wage* differentials (as distinct from *annual income* differentials) enjoyed by craftsmen. According to *Wages and Hours of Labor* (the Nineteenth Annual Report of the U. S. Commissioner of Labor), the average hourly wage for seven classes of skilled glassworkers [6] in 1890 was almost three times the wages paid five classes of unskilled men [7] employed in the same industry.[8] Data published by the Ohio Bureau of Labor Statistics for 1910 and 1911 and by the U. S. Bureau of Labor Statistics for 1919 indicate that this differential in wage rates decreased as competition from glassmaking machinery became increasingly severe.[9] Index numbers based upon the hourly wages received by glassworkers in 1890–1911 and 1919 suggest that the differential margin enjoyed by those in the skilled group tended to narrow in 1893–1895, to widen again in 1896–1903, to decrease temporarily in 1904, to increase again in 1905–1907, and to diminish a great deal in 1908 and 1911. These movements clearly reflect fluctuations in the bargaining strength of unions and in the competitiveness of trade conditions as outlined in Chapter VIII. The relative margin in favor of craftsmen was less in 1919 than at any other time since 1890, with the exception of 1895 and 1911.[10]

[5] Tariff Commission, *Glass Industry as Affected by War*, p. 48. The "blast" was the annual production period.

[6] That is, (1) flint-glass blowers, (2) green-glass blowers, (3) window-glass blowers, (4) window-glass cutters, (5) window-glass flatteners, (6) flint-glass gatherers, and (7) window-glass gatherers.

[7] That is, (1) batch mixers, (2) male laborers, (3) male lehrsmen, (4) male packers, and (5) male teasers (furnace men).

[8] *Wages and Hours of Labor*, pp. 93–97.

[9] The hourly earnings of city glass blowers, village glass blowers, city cutters, city flatteners, city gatherers, and village gatherers in Ohio factories averaged in 1910 about 2.7 times (and in 1911, 1.7 [sic] times) the earnings of batch mixers, laborers, packers, lehrsmen, and teasers employed in glass plants located in cities (Ohio Bureau of Labor Statistics, 35th and 36th *Annual Reports*, 1912–1913). The average hourly earnings in 1919 of blowers of lighting goods and tableware, window-glass cutters and flatteners, and glass gatherers were about 2.5 times the wages paid batch mixers, teasers, laborers, lehrsmen, and packers. — *Monthly Labor Review*, X (1920), 102.

[10] The index numbers which have been calculated for skilled and unskilled glass-

Unfortunately there is no series of hourly or daily wages paid glassworkers which runs continuously from 1880 through 1920. Data from four separate sources have been employed in the construction of the indices which are presented in Table 25, along with index numbers for general wholesale commodity prices and general wages. The daily wages for window-glass blowers and gatherers and for five classes of green-glass bottle blowers given in the Aldrich Report for New Jersey have been used for 1880–1892;[11] the hourly earnings of thirteen classes of skilled and unskilled workers given by the U. S. Department of Com-

---

workers with 1890–1892 as the base period are as follows:

Year	Skilled Glassworkers	Unskilled Glassworkers	Ratio of Skilled to Unskilled Glassworkers' Index
1890	101	100	101
1891	99	100	99
1892	101	100	101
1893	98	100	98
1894	97	100	97
1895	86	100	86
1896	93	99	94
1897	96	99	97
1898	101	98	103
1899	105	99	106
1900	112	99	113
1901	110	101	109
1902	123	101	122
1903	128	106	121
1904	111	107	104
1905	120	107	112
1906	130	109	119
1907	137	113	121
1908	109	112	97
1909	115	107	107
1910	132	122	108
1911	107	144	74
....	...	...	...
1919	224	242	93

One should consult the text and Table 25 of this appendix for the sources of data employed and the classes of workers included in these indices. It is interesting to note, furthermore, that the average weekly wages of blowers, batch mixers, and common laborers in hand shops making window glass increased by a much smaller percentage between 1914 and 1917 than did the wages of similar classes of workers in machine window-glass plants or of unskilled or semiskilled workers in plate-glass factories. — Tariff Commission, *Glass Industry as Affected by War*, pp. 25, 27.

[11] *Wholesale Prices, Wages, and Transportation*, IV, 1,566, 1,572. Bottle blowers were classified according to the kind and size of containers made. They and window-glass blowers are the only ones entering into the indices following 1880 whose wages were also included in the indices for 1840–1880, shown in Chart I.

TABLE 25

GENERAL WAGES, GLASSWORKERS' WAGES, AND WHOLESALE PRICES IN THE UNITED STATES, 1880–1920 *

(Base = 1890–1892)

Year	General Wages	Glassworkers' Wages	Wholesale Prices
1880	87	84	125
1881	90	100	129
1882	91	100	135
1883	93	99	126
1884	93	99	116
1885	93	99	106
1886	93	100	103
1887	97	96	106
1888	97	100	108
1889	99	100	101
1890	100	100	103
1891	100	99	103
1892	100	100	95
1893	100	99	98
1894	97	98	88
1895	99	92	89
1896	100	95	85
1897	100	97	85
1898	100	100	89
1899	101	103	96
1900	106	107	103
1901	107	107	101
1902	112	114	108
1903	116	120	109
1904	116	109	109
1905	119	115	110
1906	123	122	113
1907	129	128	119
1908	129	111	115
1909	130	112	124
1910	135	128	129
1911	138	123	119
1912	141	*	126
1913	145	*	128
1914	148	*	124
1915	149	*	126
1916	161	*	156
1917	186	*	215
1918	235	*	239
1919	267	234	253
1920	339	*	283

* Source: The general wage series is from U. S. B. L. S. *History of Wages, Colonial Times to 1928.* In order to facilitate the comparison, the index numbers constructed

merce and Labor have been used for 1890–1907;[12] the hourly earnings of twelve classes of skilled and unskilled workers given by the Bureau of Labor Statistics of Ohio have been spliced with those of the U. S. Department of Commerce and Labor and used for 1908–1911;[13] and the average hourly earnings of nine classes of workers presented by the U. S.

---

with 1913 as the base have been divided by 69, their average for 1890–1892. The glassworkers' index was compiled from data for 1880–1892 found in the Aldrich Report, IV, 1566, 1572; from wage relatives for 1890–1907 found in *Wages and Hours in Manufacturing Industries, 1890–1907*, pp. 40–41, 85–87; from wages for 1902–1911 reported in the *Annual Reports* of the Ohio Bureau of Labor Statistics, XXVII–XXXVI (1904–1913), and from wages for 1919 published in the *Monthly Labor Review*, X (1920), 102. The wholesale price series is that prepared by Warren and Pearson, *Prices*, with the indices divided by 80, their average for 1890–1892.

The seeming incongruity between this table and Chart I on p. 35 (where the 1880 indices for general wages, glassworkers' wages, and wholesale prices were 100, 102, and 106, respectively) is due to the fact that the general relationship of overall wages, glassworkers' wages, and wholesale prices in 1878–1880 (the base period for Chart I) was different from their general relationship in 1890–1892. Unfortunately, there is no defensible way of splicing glassworkers' wages in 1841–1880 with the series used for 1880–1920.

---

[12] *Wages and Hours of Labor in Manufacturing Industries, 1890–1907*, pp. 40–41, 85–87. The skilled workers included flint-glass blowers, green-glass blowers, window-glass blowers, window-glass cutters, window-glass flatteners, flint-glass gaffers, flint-glass gatherers, and window-glass gatherers; the unskilled workers (all males) included batch mixers, laborers, lehrsmen, packers, and teasers. From the wage relatives for 1890–1907 and the actual hourly wages for 1906 and 1907, the wages were interpolated for 1890–1905. Since there was no justifiable way of splicing the data given by the Aldrich Report with these figures for 1890–1907, the only overlapping years (i.e., 1890–1892) had to be taken for the base period.

[13] Ohio Bureau of Labor Statistics, 27th through 36th *Annual Reports* (1904–1913). For each class of workers the average daily earnings were divided by the average length of the working day in order to obtain the average hourly wage. Then the average of the annual ratios of the U. S. Bureau of Labor hourly wages to the Ohio hourly wages for 1902–1907 was used for splicing the following Ohio series with the U. S. series:

Ohio series	U. S. series
City batchmen	batch mixers
City glass blowers	flint-glass blowers
City glassworkers	green-glass blowers
Village glass blowers	window-glass blowers
City window-glass cutters	window-glass cutters
City window-glass flatteners	window-glass flatteners
City glass gatherers	flint-glass gatherers
Village glass gatherers	window-glass gatherers
City laborers	male laborers
City lehrsmen	male lehrsmen
City packers	male packers
City furnace men	male teasers

CHART VIII

ANNUAL REAL EARNINGS OF INDUSTRIAL WORKERS AND GLASSWORKERS, 1890–1920 *

(Base = 1890–1892)

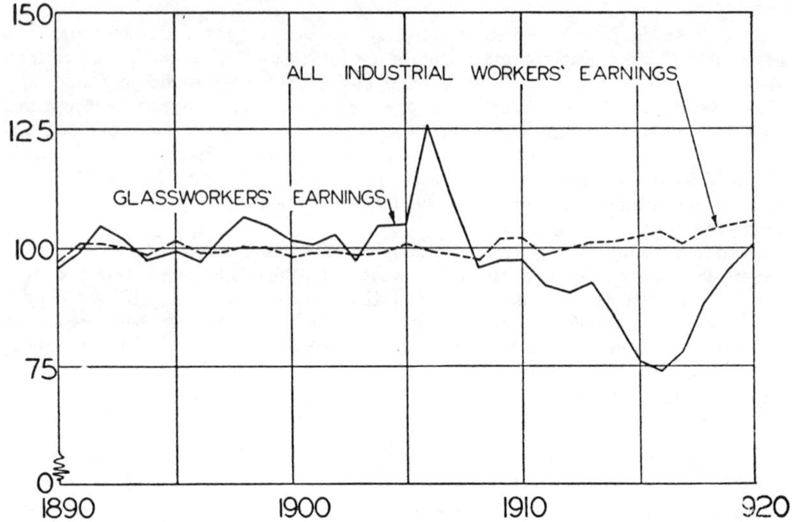

* Source: See Table 24.

Bureau of Labor Statistics have been employed for 1919.[14] The lack of homogeneity in the data constitutes a serious defect, but the resulting index numbers of hourly wage rates in the industry, it is believed, reflect the long-run trends rather satisfactorily.

Throughout the period general wage rates tended to increase more than glassworkers' wages, and thus the gap between the two narrowed.[15] Since wholesale prices declined rather steadily until 1898 while wages showed a strong tendency toward stability, the real hourly earnings of employees in the glass industry probably rose until near the end of the

[14] *Monthly Labor Review*, X (1920), 102. Out of all the categories of labor listed for 1919, only eight classes were considered comparable to the classes employed in the indices for 1890–1911. These were batch mixers, lighting and tableware blowers (considered as flint-glass blowers), window-glass cutters, window-glass flatteners, furnace men (teasers), gatherers (considered as flint-glass gatherers), laborers, lehrsmen, and packers.

[15] Of the twenty-nine industries for which the average working day and hourly wages were calculated for 1919, the glass industry ranked fifteenth as far as hourly wages paid male employees were concerned; and twenty industries maintained longer working days than did glass. — Douglas, *Theory of Wages*, p. 530.

nineteenth century. And until 1908 their real wages, on the whole, remained at a higher level than they had been in 1890–1892. Mainly as a result of mechanization, glassworkers' wages from 1908 through 1920 increased more sluggishly than wholesale prices and the cost of living.[16]

The introduction of glassmaking machinery further affected the general labor situation by decreasing the incentive of manufacturers to employ children. In 1880 slightly over 30 per cent of the employees in plants making blown and pressed glassware and 24 per cent of all workers in bottle shops were either boys under seventeen years of age or girls under sixteen.[17] At the same time, children composed less than 4 per cent of the laboring force in window and plate-glass factories. Many of those who helped gatherers and blowers or who carried ware to the annealing oven and later decorated or packed it were twelve years old or even younger.[18] In many bottle and tableware plants they worked long hours around furnaces housed in drafty sheds and hence were subjected to great variations in temperature.

The Owens bottle machine, especially after it was coupled with automatic take-outs and conveyors, used practically no child labor. In 1913 M. J. Owens received an unsolicited letter from the National Child Labor Committee which congratulated him warmly upon having accomplished more toward the elimination of child labor in bottle factories through the medium of his machine than it had been able to do through labor legislation.[19] Producers making jars and other containers on semiautomatic machines, however, still hired many boys.[20] Before 1920 most of the functions previously performed by children on these machines had been either eliminated or mechanized. The development of the rotary presses, the semiautomatic tumbler and lamp-chimney machines, and the semiautomatic and completely automatic bulb machines further reduced the incentive of glass manufacturers to employ young boys and girls. In 1880 children accounted for 23.4 per cent of the working force for the entire industry; in 1904, only 10.1 per cent; in 1919, 1.8 per cent.[21] The

[16] The U. S. Department of Labor estimated from the figures provided by the Census of Manufactures that the real earnings of persons engaged in glassmaking fell from 119.5 in 1899 and 121 in 1904 to 100 in 1914 and 94.6 in 1919. — *Productivity of Labor in the Glass Industry*, p. 22.

[17] Weeks, *Report on Manufacture of Glass*, Tables III and IV.

[18] Henderson, "Glass-Making," *Popular Science Monthly*, XXXVI (1889–1890), 168–169; Walbridge, *American Bottles*, pp. 96–99; Morrison, *Romance of Glass.*

[19] Owens-Illinois Glass Company, Annual Reports of Owens Bottle Company, September 30, 1913.

[20] In 1916, in fact, shops equipped with semiautomatic machines employed relatively more children than did those following the handicraft method. In both, however, child labor was much less important than it had been in bottle factories around 1880. — *Glass Industry*, p. 250.

[21] *Productivity of Labor in the Glass Industry*, p. 24.

introduction of machines, it is true, was not entirely responsible for this decline. The increasing number and improved enforcement of child labor law by states and also the tightening of apprenticeship rules by labor unions were also important. It is impossible to rate these three agencies as to their relative importance in eliminating child labor.[22] Available evidence, however, lends support to the conclusion drawn by investigators of the U. S. Bureau of Labor Statistics, who reported in 1927 that "child labor in the glass industry has now become almost a thing of the past, and credit for this is due in no small measure to Michael J. Owens [and other inventors of glass machinery] . . ."[23]

[22] It is possible that the increasing social concern over the employment of children in industry acted in some measure as a stimulus to the development of glassmaking machines.

[23] *Ibid.*, p. 25.

APPENDIX B

SOME NOTES ON TECHNOLOGY

1. Bottle and Jar Machines

The first attempts to make bottles or jars mechanically leaned heavily on the processes incorporated in side-lever pressing. William T. Gillinder, on December 5, 1865, received a patent (numbered 51,386) on a combination press-plunger and blowpipe which was supposed to press the top, lip, and handle of such things as water pitchers. The worker was then to remove the glass from the mold and blow the body of the pitcher to proper size while shaping it with hand tools. Although the contrivance saw little or no commercial use, Gillinder's was among the first in the line of inventions preceding successful bottle machines.[1] Atterbury, a glassmaker at Pittsburgh, in 1873 secured a patent on a device which was intended to press and then to blow a bulbous pitcher in one and the same mold. But his process, apparently, was never successful.[2] Then Philip Arbogast in 1881 secured patents on a double-mold method which ultimately led others to success. After a worker had gathered sufficient glass from the furnace and had dropped it into the blank mold, another individual lowered the plunger. The plunger pressed the glass up and around the top portion of the mold — thereby forming the jar's mouth and lip and also leaving an initial cavity in the lump of glass. It was through this cavity that compressed air blew the glass to full dimension once the blow mold had replaced the "blank" mold.[3]

The D. C. Ripley Glass Company of Pittsburgh acquired his rights in 1885 and actually pressed-and-blew a few large pieces of tableware. Since the company was unionized, and since the union fought the new process by fixing very small outputs and high wages for the operatives, Ripley decided to put aside the machines he had had built.[4] A glassmaker named Dos Taylor took the next step in 1892, when he adopted Arbogast's principles and began making wide-mouth ware in a nonunion plant at Huntington, West Virginia; [5] and one C. L. Flaccus, after secur-

[1] James Gillinder of Port Jervis, New York, one of William Gillinder's descendants, was kind enough to provide me with a copy of the original patent.

[2] I am indebted to my former colleague, Robert L. Bishop, for much of the information on early bottle-machine patents presented in this Appendix.

[3] A "blank" mold shaped an almost solid gob of viscous glass somewhat along the lines that the finished bottle would eventually have.

[4] Barnett, *Machinery and Labor*, pp. 73–74.

[5] *National Glass Budget*, XXVI, November 26, 1910.

ing a machine license from the United States Glass Company[6] in 1893, began to produce vaseline jars in his nonunion Enterprise Glass Company at Beaver Falls, Pennsylvania.[7] Hence the first jar machines did not really appear on the American market until 1892–1893.

Flaccus's jars at the outset had rough, sharp edges along their mouth and lip,[8] but before the beginning of the twentieth century many other manufacturers had enhanced the quality of their output after building improved machines. Charles E. Blue of the Wheeling (West Virginia) Mold and Foundry Company, for example, began work on a machine which pressed glass in a blank mold telescoped within the regular blowing mold.[9] The president of the Atlas Glass Company financed his experiments and in 1896 was rewarded with one of the most important of the early jar machines.[10] The Ball Brothers Glass Manufacturing Company paid $10,000 to the United States Glass Company for rights covered by the Arbogast patent and in 1898 successfully completed its own fruit-jar machine.[11] Other inventors and companies were also busy developing competing models,[12] with the result that most fruit jars and similar

[6] This company, incorporated on February 12, 1891, was a combination of some 15 tableware factories (see above, pp. 236 ff.). It had acquired the Arbogast patents by absorbing the Ripley company.

[7] Barnett, *Machinery and Labor*, p. 68; Davis, "American Glass Industry," p. 34. The C. L. Flaccus Glass Company of Tarentum, Pennsylvania, controlled the Enterprise Glass Company.

[8] The president of the Chesebrough Manufacturing Company, which produced petroleum jellies, testified in 1898 that the Flaccus vaseline jars had been of inferior quality.—*National Glass Budget*, XIV, July 30, 1898.

[9] He soon discovered that two English patents, the so-called Windmill and Rylands patents, covered the telescoped mold, but he secured permission from the patentees to use their principles. He also operated under a license issued by owners of the Arbogast and Ripley patents.

[10] *National Glass Budget*, XIV, June 18, 1898; XXV, July 24, 1909; XXVI, May 14, 1910. The Blue machine (variously called the Atlas and the Beatty-Brady) did not require the transference of the glass blank from one mold to another. After the glass was pressed, the blank mold fell out of the blow mold, a bottom was moved into place, and the glass was then blown to the dimensions of the blow mold.

[11] *National Glass Budget*, XIV, July 16, 1898.

[12] See above, pp. 178 ff. The various machines built before 1901 included the Butler (at the Redkey, Indiana, Glass Company), the Haley-Bridgewater (at the Cumberland Glass Company in Bridgeton, N. J., and at the Illinois Glass Company at Alton), the Pyle (at the Greenfield, Indiana, plant of Hollweg and Reese), and the Powers (at the Swayzee, Indiana, Glass Company and formerly at the Gilcrist Jar Company in New Jersey). Atlas, Flaccus, Ball, the Marion Fruit Jar and Bottle Company, the Rochester (N. Y.) Glass Company, and several more enterprises used other types of wide-mouth machines.

The Julius Proeger model, intended both for wide- and narrow-mouth containers, deserves special mention because, unlike the others, it did not *press* glass blanks. After a sufficient quantity of metal was cut off the punty and allowed to fall into an inverted blank mold, suction drew the glass down and around that

wide-mouth containers made in the country around 1901–1902 were pressed and blown on semiautomatics.

No manufacturer by that time had succeeded in producing narrow-mouth bottles for the market without the help of skilled blowers. Glass for bottles could not be pressed as in the first stage of jar-making because the aperture of the container was too small to allow workmen to insert and then to withdraw the plunger. Two Englishmen, Ashley and Arnold, had tried in 1886 to overcome this "bottleneck" by employing a bottomless mold which they proposed to invert in the preliminary stages. A worker could then drop a gather of glass in it and insert a plunger to press the glass around the *lower* portion of the mold and thereby form the bottle's mouth and neck. At this stage a bottom was slipped onto the mold, the mold righted, and the glass blank blown to its proper dimensions by admitting air through the bottle's fully-formed mouth. Their machine never attained the success requisite for commercial use. About twenty years later the first practicable semiautomatic for forming bottles (as contrasted with jars) appeared in America after having been acclaimed in England.[13]

Meanwhile M. J. Owens and the Toledo Glass Company had built a machine which not only gathered and blew glass (and hence was the first completely automatic in the field) but also made bottles as well as jars. Each machine, equipped with six or more "arms" with their own set of blank and blow molds, made several bottles simultaneously and gathered molten glass by sucking it up from a revolving pot. Glass flowed from the melting tank through a clay spout [14] into this pot, which was built just inside the furnace so that part of it extended beyond the furnace walls. The pot was shallow and slowly revolved on its axis, thereby

---

portion of the mold which formed the mouth and lip. Proeger had begun his experiments in 1898 at the Greensburg, Pennsylvania, Glass Company.—*National Glass Budget*, XIV, December 17, 1898. His machines never made glass successfully; yet the Owens Bottle Company felt compelled to purchase his patent rights in 1903 for a very large sum because they broadly related to the use of suction in forming bottles. See above, p. 102, n. 60.

[13] See above, p. 180.

[14] Owens' experience with this clay spout is interesting. He had wanted to inject the glass from the furnace into that side of the pot from which the gathering took place, apparently in order to preserve as much homogeneity and fluidity in the glass as possible. This would have necessitated a long clay spout. Time and again the spout broke, leaving only a short stub protruding. Associates claim that this almost caused Owens to despair of achieving success with his automatic gathering. Finally, a friend suggested that he try running the machine with the broken spout, and Owens discovered not only that success was possible but also that a short spout was definitely superior to a long one. When the glass flowed into the pot at the point farthest inside the melting tank, there was less likelihood that it would still contain bubbles when it eventually came underneath the dipping arm than if it entered the pot on the gathering side.

allowing the furnace temperature to keep the glass in it heated and in a workable state. The fact that the pot revolved offered another advantage in that successive gathers were not drawn from that portion of the glass which had been disturbed and made full of bubbles by the immediately preceding gather. As each arm of the machine came opposite the pot, the arm dipped the bottom of its blank mold slightly below the surface of the glass and sucked up a sufficient quantity to form a bottle or jar. A broad-bladed knife attached to each mold automatically severed the glass from that remaining in the pot and at the same time provided a bottom for the mold.

Once the glass was gathered in this fashion, the second phase of bottle-making began. It should be remarked at the outset that, unlike the bottles blown by hand, those made on the Owens machine had their mouth and lip formed first. The blank mold was attached to the so-called neck-ring, which in reality was a full-sized mold for shaping the bottle's top. When the machine sucked up the molten metal from the revolving pot, the glass filled both the blank mold and the neck-ring. In this way the mouth and lip of each bottle were completely formed before any other part. As the machine continued to rotate on its axis, the blank mold eventually opened and dropped about three feet, thereby leaving the gob of glass suspended from the neck-ring. There then occurred a short, preliminary blowing before the blow mold rose from below and enclosed the glass once again. Compressed air, passing through the bottle's completely formed mouth, inflated the still-viscous glass to the dimensions of this second mold. As the machine neared a complete revolution, the neck-ring separated, joined the blank mold once again, and assumed the position necessary for another gathering. The bottle, meanwhile, remained in the blow mold to harden as the mold dropped to its original position below the pot and the dipping arm. The bottle fell onto a conveying belt leading to the annealing oven when the blow mold finally opened and again rose to envelop the second gob of glass which had just been sucked up into the blank. Hence, each arm of the machine made one bottle about every one and one-sixth revolutions. For one-sixth of a revolution each arm carried a gob of glass in its blank mold and a blown bottle in its other mold. A six-arm machine, consequently, held seven bottles in different stages of completion at any instant of time; a ten-arm or fifteen-arm machine carried simultaneously about twelve or seventeen, respectively.

The first machines built by M. J. Owens did not conform precisely to the description just given. His earliest experimental device was like a bicycle pump or spray gun and was operated by hand.[15] Still it embodied the basic principles which were later incorporated in his commercial mod-

[15] See above, pp. 156 ff.

els.[16] In 1900 he and his assistants converted the hand "gun" into a small powerdriven, nonrotating machine equipped with only one set of molds. A year later they constructed another model which revolved on its axis but which came to a complete stop each time one of its five arms gathered glass. This machine proved to be unsatisfactory primarily because it sucked up glass from a stationary pot. Each of its arms, conse-

[16] Owens did not originate any one of these principles. Arbogast had employed the blank and blow molds alternately, and he and others had used the full-sized neck-ring for forming the container's mouth and lip during the pressing stage. Even gathering by suction — usually attributed to Owens — does not appear to have been wholly original. According to a memorandum prepared by L. D. Soubier of the Owens-Illinois Glass Company's patent department on April 17, 1943, and according to the information obtained from John Staiger, an old employee of the Libbey Glass Company, there was a device called the "tallyho" or the "sucker-upper" in use at the Libbey Glass plant around 1890–1895. The "sucker-upper" which I examined during one of my visits at the factory consisted of a long tubular cylinder with a cup-like opening at one end and a hand-operated piston rod at the opposite end. A worker inserted the open end of the cup in a pot of flint glass and filled the cup through suction created by pulling the piston back. He then quickly turned the cup over, cut away the stringy tail of glass, and attached a punty or working rod to the gob of glass gathered in this manner. The "sucker-upper" consequently permitted the worker to gather exactly the same quantity of glass over and over again and also to avoid striae which usually resulted whenever viscous glass was collected on the end of slowly revolving gathering rods. Perhaps Libbey's employees had carried this device with them when they moved from Cambridge to Toledo. No patent covering the "sucker-upper" has come to light; but if it was used by glassmen at the Libbey Glass Company around 1890–1895, surely M. J. Owens must have been acquainted with it.

An examination of pertinent patents also suggests that Owens was not entirely original. He filed a patent application for his simple bottlemaking "gun" on December 26, 1899, and received it on May 10, 1904. One Julius Proeger, however, had priority to claims "broad enough to cover any apparatus in which the glass is first sucked to shape, combined with means for subsequently blowing the molten glass." — Owens-Illinois Glass Company, Patents Department, Opinion of Whittemore. Herman Hilde, a German inventor, had secured a German patent in 1891 (filed in the United States in October, 1894) covering the broad principle of suction used in making glassware. Also, one John J. Power, another glassmaker, obtained a patent (application filed on March 1, 1901; issued on July 4, 1905) on a device almost identical with Owens'. Hence, near the turn of the century, four men — Hilde, Proeger, Owens, and Power — were trying to gather or shape glass by suction. Hilde and Proeger enjoyed priority over Owens in the matter of patents even though they were never able to make their process practicable; Power's patent probably would have been held invalid in court because of Owens' rights. Once the Owens Bottle Company began activity, the Toledo Group nevertheless successfully attempted to forestall any litigation by buying exclusive rights to the patents of all three men (see above, p. 102, n. 60). In all fairness to Owens it should be pointed out that there is no evidence whatsoever that when he first began his experiments in bottle-blowing, he was familiar with the work these other men were doing. Furthermore, he was the first to bring together all the various principles into a workable machine, and this alone was a sufficiently great accomplishment to warrant his fame.

quently, dipped its blank mold at approximately the same spot in the pot, and the bubbles and striae left in the glass by the immediately preceding gather could not be eradicated before the next arm moved into place. Between 1901 and late 1903 Owens and his engineers developed the revolving pot,[17] made the machine rotate continuously,[18] and added a sixth arm. These improvements resulted in the so-called "A-type" or first commercial machine. At later dates it underwent additional modification.

In the first place, the introduction of a fire-polishing process and, later, improved construction of the blow molds themselves smoothed the mold seams and rough edges left on the bottles' sides, neck, mouth, and lip.[19] These imperfections had materially lessened the saleability of the first bottles. In the second place, around 1908 the Owens Bottle Company introduced a new model which kept each bottle in its blow mold a longer time than had the first machines. Formerly, each arm of the machine had completed a bottle during five-sixths of the time that it took the machine to make a complete revolution.[20] This had not been long enough to allow the glass to harden before it was taken from the molds; and, consequently, the neck and sides of the bottles had often sagged out of shape.[21] On the new model the blow mold dropped below the blank mold while the latter was gathering a new charge of glass, and each bottle remained in the blow mold approximately two-thirds longer than before.[22] Hence the amount of misshapened ware decreased. A third modification — and one of the most important — was the introduction of the so-called "dip-head" in 1911.[23] Instead of the entire machine dipping each time one of its arms gathered glass from the revolving pot, only the head on the end of each arm dipped. This greatly reduced the amount of wear on all mechanical parts and also allowed the machine to rotate faster.[24]

[17] Owens' patent bearing the number 774,690, which was filed on December 22, 1903, and issued on November 8, 1904, covered the revolving pot. — Owens-Illinois Glass Company, "Revision of Sketch of M. J. Owens," p. 4.

[18] The "continuous-rotating" principle was covered in the Bock patent 870,664, filed on May 2, 1905, and issued on November 12, 1907. — *Ibid.*

[19] The *National Glass Budget* reported on November 12, 1904, that the Owens bottles then had fire-polished edges and that the mold marks formerly visible were no longer in evidence.

[20] Owens-Illinois Glass Company, Owens' Private Files.

[21] See, e.g., the letter addressed to the American Bottle Company by William S. Walbridge. — *Ibid.*, Walbridge Letters, I, 200.

[22] The blow mold was not emptied, therefore, until just before it was needed to receive a new gob of glass from the blank mold.

[23] Owens' patent which bore the number 766,768 and was issued on August 2, 1904, covered the dip-head principle, but the engineers who constructed the earlier machines had not made use of this device.

[24] Owens-Illinois Glass Company, Owens' Private Files. The older machines had

(Courtesy of Owens-Illinois)

FIG. 9. THE OWENS "AT" OR CARBOY MACHINE

Only the upper half of the machine is visible.

Beginning in 1909 the Owens Bottle Company put on the market several new models which were designed either to make special kinds of containers or to increase the rate of production. The first of these, installed in 1909 and called the "AD," had ten instead of six arms. Two other ten-arm machines, brought out in 1911–1912, were the "AR" and the "AN," both of which employed the dip-head principle. The former was specifically designed for making large bottles, and the latter was intended to produce smaller ware or "prescription" bottles. The eight-arm "AL" model, introduced in 1912, made gallon jugs. Owens' engineers later perfected two fifteen-arm machines — the "AQ" in 1914 for large bottles and the "AV" in 1917 for smaller ware. The largest of all was the "AT" or carboy machine,[25] upon which work had begun in December, 1911. It was operating successfully in 1915, and two years later the Illinois Glass Company obtained an exclusive license to use it. Finally, mention should be made of the twenty-arm machine under construction before the end of 1920. This machine had twenty sets of blank and blow molds, with two, three, or four cavities each. Consequently, each of the twenty arms could make two, three, or four bottles simultaneously.[26]

2. Window-Glass Machines

The first successful window-glass machine, developed by John H. Lubbers and the American Window Glass Company, blew cylinders. Workers first ladled molten glass from the melting tank into special pots which were heated from beneath and which held just enough glass for one cylinder. Directly above each of these pots hung the drawing apparatus. A flanged metal disk or "bait," on the end of a large blowpipe,

---

worked very well as long as twelve or fifteen bottles were made per minute; but when they were speeded up, the machines vibrated so much that they had to be clamped to the floor. The dip-head models completed one revolution about every twenty seconds. — Walbridge, *American Bottles*, p. 85.

[25] This machine, weighing over 100 tons, was approximately fifty feet high and thirty feet in diameter (see Fig. 9). The first six-arm machine had weighed around ten or eleven tons, and the fifteen-arm "AQ" weighed close to fifty tons.

[26] Whether the Owens Bottle Company or one of its licensees first developed and employed the plural mold is not clear. Charles W. Schwenzfeier claims that while plant manager at the Whitney Glass Works he started making two bottles at the same time on each arm of the Owens machine; but the Illinois Glass Company, according to some glassmen, was the company which originally developed the plural bottle mold. See, e.g., Owens-Illinois Glass Company, "History," p. 26. On the other hand, the directors of the Owens Bottle Company learned on March 19, 1912, that M. J. Owens was then working on a ten-arm machine which would gather and form two small bottles simultaneously in each mold. — *Ibid.*, Owens Bottle Company Minutes, II, 143. Whether one accords priority of claim to Schwenzfeier at Whitney, to the Illinois Glass Company, or to M. J. Owens himself, it is a fact that "two-cavity" molds were employed on some machines after 1912.

then descended into the pot. After the glass had hardened sufficiently in the flange, mechanical power slowly raised the bait between two guiding shafts. Compressed air, continuously flowing through the blowpipe, expanded the shoulders of the nascent cylinder to the desired diameter, which it thereafter maintained. The speed of drawing the bait upward, the temperature of the glass in the pot, and the rate at which air was admitted determined the thickness of the glass and the circumference of each cylinder. When the pot was about empty, the rate of drawing suddenly increased, and the cylinder broke loose at its base. Workmen then carefully lowered it onto a support so that others could cut away the cap and divide the cylinder into three or more sections. Each section was split, reheated, flattened out, and finally annealed.[27]

Window glass made in this way always distorted light somewhat because it was impossible to flatten cylinders without causing slight ripples on the glass's outside surface. This defect was gradually lessened over a period of years but never entirely eliminated. The promoters, however, eventually corrected other faults latent in the first machines. The glass collar in the flange of the metal bait frequently cracked because the glass and metal did not have the same coefficient of expansion and contraction. Hence cylinders often crashed to the floor before they were completely drawn.[28] Then, too, when compressed air passed through the blowpipe and came in contact with the hot glass, it expanded rapidly and caused undulations in the cylinder's walls.

Within the decade after the Lubbers' machine became a commercial success, the Toledo Group introduced their Colburn sheet-drawing process.[29] And Emil Fourcault in Belgium was busy perfecting his method for drawing a continuous sheet of glass vertically from the furnace. His "debiteuse" or floating block made of refractory clay, with a narrow slit down its length, rested in a shallow tub of glass connecting with the melting furnace. Downward pressure, constantly applied to the floating block, squeezed the glass up through the slit.[30] After the glass passed closely between two water coolers, its strength was increased to

[27] The machine that did the hoisting and blowing was not in itself very expensive. It cost only about $1,500. But many changes in the ordinary workroom and factory structure were necessary before the machinery could be installed. William L. Monro, general manager of the American Window Glass Machine Company, in 1918 estimated that a battery of "eight machines could be installed at prices prevailing prior to the war at something like $40,000." — Tariff Commission, *Glass Industry as Affected by War*, p. 75.

[28] One report states that the breakage on the Lubbers machine in 1904 was as high as 80 per cent. — *National Glass Budget*, XXXIV, November 23, 1918, p. 1.

[29] See above, pp. 127 ff.

[30] The effect of the downward pressure on the block was to keep the glass sheet as wide as the slit. Knurled rollers, similar to those employed in the Colburn machine, hence were unnecessary.

such a point that it did not tear or break when seized by several pairs of asbestos-covered rollers arranged in a vertical series. These rollers, carefully synchronized and enclosed in a fifteen-foot lehr, pulled the glass upward through the annealing lehr. The outstanding advantage of the Fourcault over the Colburn process was the fact that it was never necessary to bend the continuous sheet of glass; on the other hand, the Fourcault lehr was not nearly as long as the Colburn lehr and consequently did not anneal the glass as well.[31]

3. The Toledo Group's Paste-Mold Machines

In order to appreciate the contribution made by the Toledo Group in the paste-mold [32] field it will be necessary to describe the method followed in producing bulbs by hand. The first ones ever blown in the United States probably were made by heating glass tubing over a flame and then inflating the melted end. Workers at the Corning Glass Works, apparently in the early 1880's, started blowing bulbs on the end of their pipes and shaping them with hand tools.[33] Since there were no technical problems involved in shifting from this so-called "off-hand" process to one requiring paste-molds,[34] blowers soon began to form bulbs in water-soaked wooden molds or cast iron ones coated with a mixture of linseed oil and charcoal. A gatherer collected from a furnace pot a suitable quantity of glass on the end of his blowpipe, marvered it on a metal or wooden slab to give it an initial shape, and then handed the pipe with its attached gob of partly blown and shaped glass to a master blower. This individual, aided by a mold boy, enclosed the glass in a paste-mold at his feet and slowly turned the pipe as he blew. Another worker or "boy" took the pipe as soon as the blower stopped and cracked the bulb off its end; still another carried the bulb to the annealing oven; and a third scraped the pipe clean of all glass fragments.

Owens' semiautomatic bulb machine, introduced in 1894–1895, incorporated these principles. A skilled workman collected sufficient glass on the end of a blowpipe, inflated the gather slightly, and then fixed the pipe in an upright position on a revolving circular machine. A carbon-coated iron-mold automatically rose from a trough of water and closed around the suspended gob of glass. Next, as the glass slowly turned in the mold, compressed air flowed through the pipe. Just before the whole machine

[31] The Colburn lehr was about 200 feet in length.

[32] Paste-molds were usually coated with carbon and lubricated with water so that the glass object being blown could turn in the mold. Consequently, seams on the glass's surface were eliminated.

[33] *Glass Industry*, p. 18.

[34] Paste-molds had come into common use for other ware. It seems doubtful, therefore, whether bulbs were ever made in truly commercial quantities from tubing or by off-hand methods.

completed one revolution, the mold around the bulb opened, and a boy lifted out the pipe and its attached charge.[35] Owens' semiautomatic machines for tumblers and lamp chimneys differed little from the one which turned out light bulbs.

The services of skilled gatherers in such work did not become dispensable until more than twenty years later. In 1916 the Westlake Machine Company succeeded in combining suction gathering (employed on the bottle machine) with the paste-mold principle. Each Westlake machine was composed of several blowing arms, and each arm had a gathering ram [36] which darted into the furnace and sucked up sufficient glass from the melting tank. After a charge of glass was gathered in this manner, it was dropped into the shallow cup on the end of an inverted, slowly turning blowpipe. Next, the blowpipe (still revolving on its long axis) flipped over and retained its hold on the gob of glass which now dangled from its end. A carbon-coated iron-mold emerged from a trough of water, closed about the glass, and gave it final shape as compressed air rushed through the pipe. Finally, the mold opened, and an automatic device cracked the bulb off the blowpipe. These various steps were all part of a continuous process. As the entire machine rotated about its central shaft, each of its several arms (equipped with gathering ram, blowpipe, and paste-mold) carried a charge of glass through all the stages involved in producing the finished object.[37]

4. Pressing and Finishing Machines

Glass pressing had already become quasi-mechanical long before 1880.[38] A worker gathered molten glass on the end of his rod or punty

[35] According to a report made by attorneys to the Toledo Glass Company, one of the patents issued to M. J. Owens on a similar machine (patent #548,588, dated October 22, 1895) "shows and describes certain mechanisms for receiving a blowpipe to which is attached a portion of molten glass and for manipulating said blowpipe and glass in all necessary ways to produce from said glass a completely formed hollow glass article which is removed from the machine with said blow-pipe; the proper action of said machine necessarily comprising a mold (sectional or not) and means for rotating the said glass article in the mold or for holding the glass article while the mold revolves around it, and for moving the blow-pipe and mold in unison. The said Owens patent contains the first group of mechanism, so far as is known, which was ever contrived for doing the thing above specified." — Libbey-Owens-Ford Glass Company, Licenses and Correspondence of Toledo Glass Company, Vol. I.

[36] Instead of having a gathering ram for each set of molds, later models used only one gathering ram.

[37] Because it made paste-mold ware, the Westlake was a more complicated mechanism than the Owens. The glass in process had constantly to be twirled in order to preserve its symmetry and to avoid mold seams. But in a sense the Westlake was less "marvelous" than the Owens. The latter had really paved the way for the former.

[38] See above, pp. 18–19.

and then cut off an appropriate amount into a mold. Another individual at this stage lowered the plunger by means of a side lever and forced the glass around the interior surface of the mold. Since it was necessary to withdraw the plunger each time it was inserted in the mold, it was impossible to press pieces of glassware if their aperture was not as wide as or wider than their cross-section at any other horizontal plane. This fact, consequently, imposed limitations on the variety of objects which could be pressed.[39] Between 1880 and 1920 many different companies built various semiautomatic rotary presses — the first ones being operated by means of compressed air and later ones by electrical power. The application following 1916 of the flow-feeding and gob-feeding principles to the earlier models had only mildly revolutionary effects upon workers.[40] The side-lever press continued to be common, especially in the manufacture of "novelties" and in the filling of small orders. Such a wide range of glassware was pressed that almost any producer could continue operations with his old equipment simply by shifting to another kind of product if power-driven machines invaded his old field.

The first semiautomatic presses — as contrasted with side-lever presses — appeared on the American market during the latter half of the 1890's. The *National Glass Budget* (Vol. XIV) reported on July 2, 1898, that "for some reason that is yet veiled in mystery" the numerous presses introduced only a short time before had been withdrawn from the market. It suggested that power pressing was inferior to manual pressing because the machines were not sensitive enough to exert exactly the right amount of pressure required by variations in the temperature and in the quantity of glass gathered by hand. Two months later, however, the same trade journal eulogized the Hitchcock electric press as "that marvel of modern mechanical construction [which is] so sensitized as to make its plunger pressure responsive at any desired point, instantly adjustable, and applicable to the widest range of ware." [41] After 1900 one kind of power press or another invaded shops making jar lids,[42] pressed tumblers, nappies, sherbets, and similar items.

Nearly all the machines, equipped with several molds, involved a circular table which rotated intermittently beneath a plunger. According to the model built by M. J. Owens and his engineers in 1912–1915,

[39] By contrast, it should be noted, a worker could blow and tool glass into almost any shape.

[40] Automatic feeders displaced hand gatherers, and the remaining workers (the pressers) became simply semiskilled machine operators.

[41] *National Glass Budget*, XIV, September 24, 1898.

[42] Frank O'Neill, who later started a glass-machinery factory at Toledo, Ohio, and built some of the best semiautomatic bottle machines, had constructed his jar-lid power press in Ontario around 1901. The machine pressed and automatically ejected about 27 lids per minute. — *Ibid.*, XVI, April 20, 1901.

each mold received a gob of glass and then passed beneath a power-driven plunger. After the glass was pressed into shape, it hardened in the mold until the table almost completed one revolution. A worker then lifted the pressed object from the mold and carried it to an annealing oven. Another gob of glass dropped into the same mold as it came once again before the gatherer, and the process continued. The machine's operation may be described in another way. At any instant of time, one mold on the table was receiving its gob of metal, another was in contact with the plunger, several others were holding pressed pieces of glass while they cooled, and one was being emptied so that it would be ready to receive another charge.[43]

Depending upon the nature of their output, the semiautomatic presses made anywhere from 50 to 110 per cent more ware per man-hour of work than did the side-lever presses and lowered the direct labor costs of production anywhere from 39 to 46 per cent.[44] Compared with those equipped with side levers, the fully automatic presses increased worker productivity from three fold to twelve fold and reduced direct labor costs almost proportionately.

Pressed glassware nearly always bore rather sharp and undesirable mold seams when it left the machine; hence the pieces were usually put through a fire-finishing process which melted and smoothed the edges and gave the surface a brilliant finish. One "boy" carried the glass from the press to the reheating hearth, another handled the preliminary heating, and a third took charge during the final polishing stages. Beginning in 1886 various men built machines for finishing tableware which employed the same basic principles. The glass was subjected to a series of gas flames and then delivered to a worker who stacked the pieces in an annealing lehr.[45] Eventually automatic take-out devices and conveyor belts made it possible for the glass to pass from the presses to the finishing machines and finally through the lehr without human handling.

5. Technological Developments in Plate-Glass Production

The production of plate glass was the only major branch of the American glass industry which did not experience truly revolutionary changes between 1880 and 1920. Output expanded rapidly, but manufacturing techniques changed little. Manufacturers still cast plates on tables and then annealed, ground, polished, and cut them into appropriate sizes. It was only after 1920 that they adopted the continuous process whereby glass was rolled or formed as it flowed from the melting furnace and then

[43] See *Productivity of Labor in the Glass Industry*, p. 93.

[44] *Ibid.*, pp. 95–99.

[45] See Davis, "American Glass Industry," p. 340; *Pottery and Glassware Reporter*, December 3, 1891, in Gaffield Collection, Scraps, III, 147; *National Glass Budget*, XIV, November 19, 1898.

was carried slowly through the lehr and underneath the grinding wheels and polishing pads.[46] American manufacturers, nevertheless, increased the efficiency of their equipment and gradually became world leaders in large-scale production and cost reduction.[47]

In 1891 the *Pottery and Glassware Reporter* stated that "the American mechanic has just begun to pay a little attention to the grinding and polishing machinery used in plate glass manufacture, and already several sets of polishers have been relegated to the scrap pile, though they are still used in European factories. Grinders of larger capacity, and polishers of improved construction are being introduced each year, so that already American machinery is being sought abroad, and has been introduced by the plate glass manufacturers of Europe."[48] The factory at Crystal City, Missouri, around 1890 replaced its square and octagonal grinding tables with round ones and about ten years later added removable decks. Plates of glass were plastered to the deck and ground on one side; then the deck was removed and a new one, covered with other plates, was screwed to the grinding table. In this way it was possible to use the tables and grinding machinery practically all the time and to avoid leaving them idle while the plates were being pried loose, turned over, and reset directly on the table's top. Once round instead of square tables were used in polishing as well as in grinding, workers could remove the deck from the grinder to the polisher and thereby completely finish one side of the glass before they turned the plates over.[49]

The Marsh Plate Glass Company at Floreffe, Pennsylvania, installed the first continuous lehr for annealing plate glass around 1897. To anneal plates by means of the old-type oven required about three days; the continuous lehr reduced this time to three hours.[50] Furthermore, factories began to use electricity to turn the grinding and polishing wheels and also to move the overhead crane which carried the pots (each holding, approximately, one ton of molten glass) from the furnace to the casting table. Although they were improved and enlarged, pot furnaces were employed instead of melting tanks as long as discontinuous casting remained the mode. Pearce Davis has concluded that as a result of these improvements the time taken to produce a sheet of polished plate

[46] Henry Ford, of automobile fame, in 1922 introduced continuous rolling in the manufacture of plates suitable for windshields.

[47] Cf. Davis, "American Glass Industry," pp. 377 n. 1, 390.

[48] In Gaffield Collection, Scraps, III, 136.

[49] The number of times the glass had to be handled during the finishing process accordingly was reduced from 9 to 3. — Davis, "American Glass Industry," pp. 379–380.

[50] Tariff Commission, *Flat Glass*, p. 82. It also encouraged the use of stationary casting tables. Each lehr was able to handle as many plates as could be cast on any one table. Formerly, the casting table had had to be wheeled from one oven to another.

declined between 1889 and 1923 from about ten days to thirty-six hours.[51]

6. Other Developments

Container, window-glass, and tableware factories, like those making plate glass, benefited from various innovations which lowered costs and enhanced the quality of their product, although they were less spectacular technological advances than the contemporary development of glassmaking machinery. The lehr, for example, underwent improvement, and largely displaced the wasteful and inefficient annealing oven. Over a hundred patents for lehrs were issued to individuals between 1880 and 1900. Infringements, of course, were rife, and the resulting litigation extended over a decade.[52] The muffle lehr, of American origin, appeared around 1916. It reheated and gradually cooled glass without directly exposing it to fire. Hence smoke and oil did not smudge the glass nor did fire mark it with "blooms"[53] before it emerged. Manufacturers developed various mechanical ways of conveying glassware from one part of the workroom to another — for example, from the flattening oven or forming machine to the lehr.[54] Some establishments had trolleys for transferring raw materials from their bins to a weighing machine, and power-driven batch mixers performed the work formerly done by hand. The Chambers Glass Company at New Kensington, Pennsylvania, in 1897 was reputedly the most modern window-glass factory in this respect. It even employed mechanical means for conveying batch from the mixer to the furnace and then for dumping it into the continuous melting tank.[55]

An experienced manufacturer in 1893 ranked the introduction of the continuous tank along with burning petroleum and the first bottlemaking machine as "the three great and only improvements that have been made in glassmaking for a long period of time."[56] "The tank furnace," he said,

[51] "American Glass Industry," p. 381.

[52] *Ibid.*, p. 166.

[53] Blooms were highly glazed patches on the surface of glass.

[54] "In 1904 it was reported that bottle plants had adopted mechanical conveyance of formed glassware; from the press and blow machines the bottles moved, on an automatic carrier, to and through annealing lehrs pyrometrically regulated and, after inspecting and packing, to warerooms or railway cars." — *Ibid.*, p. 312.

[55] *National Glass Budget*, XII, May 8, 1897 (as cited in Davis, "American Glass Industry," pp. 269–270). It is difficult to appraise the extent and efficiency of the early attempts to mechanize the pre-melting operations. Throughout the first two decades of the twentieth century, improvements were being made in the layout of plants and in the construction of mixers, weighers, and conveyors. What may have been described as the most modern plant in 1897, therefore, would possibly have been considered as outmoded ten and fifteen years later.

[56] Quoted by Henderson, "The Glass Industry," *Popular Science Monthly*, XLII (1893), 592–593.

"is rapidly superseding the old-fashioned pot furnaces, and in a very few years I do not think a pot furnace will be in operation [in bottle and window-glass shops] in the entire country." The first one for melting window glass was built at Jeannette, Pennsylvania, in 1888.[57] A manufacturer told the Tariff Commission in 1918 that he had seen only one small pot furnace making window glass during his entire sixteen years in business.[58] All bottle plants not using the Owens bottle machine in 1916 had in operation 201 continuous tanks, 9 day tanks, and only 24 pot furnaces; [59] and the Owens machine always worked from tanks. Ninety-two factories making blown and pressed tableware, lighting goods, and lamp chimneys out of lime glass in 1916 employed 48 continuous tanks, 47 day tanks, and 159 pot furnaces.[60]

In addition to being more economical than pot furnaces,[61] the continuous tank had certain other advantages which rendered it especially serviceable wherever glass was worked mechanically. All completely automatic machines could handle large quantities of molten glass each day; and because of the heavy capital investment in the machines it was highly desirable to keep each one in constant operation. It would have been a serious handicap, therefore, if the machines had had to gather from pots. The largest pot furnace in the country around 1877 held only twenty tons of glass.[62] After each pot was emptied, furthermore, workers had to refill it with batch and leave it in the furnace for the glass to be melted and refined. Many tanks in 1910–1920, on the other hand, had a capacity of over 1,000 tons [63] and continuously melted and refined glass at the same time that men or machines were working it. The level of glass in the furnace, furthermore, could be kept more or less constant. This last advantage was most essential whenever machines were used to dip into the furnace and gather sufficient glass for blowing tumblers, light bulbs, or bottles. Manufacturers could use pots with semiautomatic pressing and blowing machines without any great inconvenience, and they sometimes did. Workers who fed the first Owens paste-mold ma-

[57] Tariff Commission, *Flat Glass*, p. 30. By 1895 tank furnaces accounted for well over half the total melting capacity of all window-glass factories. — J. Gillinder in *One Hundred Years of American Commerce*, p. 282.

[58] Tariff Commission, *Glass Industry as Affected by War*, p. 89.

[59] *Glass Industry*, p. 212. The day tank was much smaller than the continuous tank; and, as its name suggests, glass was melted as night and worked during the following day. See above, p. 28.

[60] *Glass Industry*, p. 217.

[61] See above, p. 28. The U. S. Department of Commerce has estimated that the average daily output of bottle blowers in hand factories increased by as much as 25 per cent after the tank was introduced. — *Glass Industry*, p. 299.

[62] See above, p. 28, n. 90.

[63] The first tanks built in 1888 at Jeannette held 600 or 800 tons and could melt 30 tons per day. — J. Gillinder, in *One Hundred Years of American Commerce*, p. 281; Atwater, "Glass-Making in America," *Engineering Mag.*, IV (1893), 887.

chines, for example, collected glass on their blowpipes by dipping them into pots. Until the decade following 1910 all incandescent lamp bulbs and tubing were made of lead glass because its coefficient of expansion was similar to that of the metal used for the lamp's base. Since open tanks could not melt lead glass satisfactorily, pot furnaces were indispensable. Eventually, however, technicians improved the metal lamp base and also the expansion coefficient of lime glass.[64] As a result, it was no longer necessary to use lead glass in the manufacture of bulbs. It was not entirely fortuitous, therefore, that the automatic bulb machine developed by the Libbey Glass Company and Westlake Machine Company was not a commercial success until lime glass, melted in continous tanks, could be substituted for the other kind.

On the whole, American entrepreneurs before 1920 paid slight attention to the chemistry and physics of glass. Germany, the outstanding leader in this field, made the best optical, photographic, telescopic, and chemical glass in the world. When the first World War interrupted the import of such goods into the United States, domestic producers tried valiantly to fill the gap.[65] But their ignorance of scientific matters proved a serious handicap. The U. S. Bureau of Foreign and Domestic Commerce in 1916 stated that "the average manufacturer's chemical knowledge of the materials that enter into the manufacture of glass is very vague and indefinite." [66] Many still thought that glass was pure silica and that all other batch ingredients went up the chimney in smoke after helping the sand to fuse.[67] Some manufacturers did not even weigh the materials they mixed in the batch but only guessed at the proper proportions.[68] Not more than one chemist was discovered for about every twenty plants visited.[69] The Corning Glass Works, however, had begun its experiments with low-expansion glasses and had introduced "Pyrex" to consumers in 1915.[70] Little more than this was accomplished by any firm until the 1920's and 1930's.

[64] Lime glass had long been melted successfully in continuous tanks.

[65] See, for example, the testimony compiled in 1918 by the Tariff Commission in its *Glass Industry as Affected by War.* "In the production of glass for ordinary purposes American invention, equipment, and methods surpass those of all other countries, but . . . in the higher branches of glassmaking the United States is the most backward of countries." — *Ibid.*, pp. 11–12.

[66] *Glass Industry*, p. 316.

[67] *Ibid.*, p. 316, citing R. L. Frink in the November 1, 1915, issue of the *Metallurgical and Chemical Engineering Journal.*

[68] *Ibid.*, p. 68.

[69] The "dean" of American glass chemists was George Estes Barton, who was first employed at the Whitall-Tatum plant at Millville, New Jersey, in 1896. By 1902 there were only seven chemists in the entire industry. — Davis, "American Glass Industry," p. 418.

[70] *National Glass Budget*, XXXI, April 22, 1916, p. 1. About 15 years earlier it had experimented with melting and casting processes in order to provide better railway signal colors and transmission lenses.

APPENDIX C

SOME STATISTICS ON THE FINANCIAL SUCCESS OF THE TOLEDO FIRMS

TABLE 26

SELECTED ASSETS OF THE OWENS BOTTLE COMPANY, 1904–1920

Date	Patents and Invention's Account	Plant and Equipment Account	Investments in Controlled and Other Companies	Total Assets
Sept. 1, 1904	$2,747,977.52	$ 93,028.28	—	$3,107,117.75
Aug. 1, 1905	2,747,977.52	91,101.34	—	2,980,311.40
Aug. 1, 1906	2,747,977.52	105,923.63	$ 5,000.00	3,036,808.50
Sept. 30, 1907	2,747,977.52	116,519.96	7,427.04	3,196,032.29
Sept. 30, 1908	2,600,000.00	159,967.75	116,500.00	3,310,121.64
Sept. 30, 1909	1,888,500.00	164,756.89	616,500.00	3,232,718.93
Sept. 30, 1910	1,613,500.00	164,187.96	791,500.00	3,347,942.62
Oct. 1, 1911	1,516,667.49	162,175.16	866,500.00	3,643,470.56
Oct. 1, 1912	1,517,577.84	1,549,941.82	893,965.00	4,900,163.85
Sept. 30, 1913	1,519,022.59	1,728,698.32	1,493,850.00	6,154,665.13
Sept. 30, 1914	1,524,172.08	1,770,438.01	2,419,315.82	7,712,506.98
Sept. 30, 1915	1,526,187.09	2,525,366.52	2,588,328.33	8,709,858.15
Sept. 30, 1916	1,593,739.09	2,384,108.46	11,501,905.00	19,647,802.26
Sept. 30, 1917	1,699,792.37	4,281,849.12	12,031,836.83	24,994,244.15
Dec. 31, 1918	1,692,257.23	7,027,271.72	11,749,477.91	27,121,361.67
Dec. 31, 1919	2,608,230.20	7,593,985.63	12,864,215.99	31,062,956.03
Dec. 31, 1920*	3,143,212.54	10,753,522.97	4,558,499.97 **	35,879,476.34

* Consolidated statement for Owens Bottle, American Bottle, Graham Glass, and Peytona Gas Company.

** The decrease in this asset over the previous year was a result of merging the assets of the American Bottle with those of Owens Bottle.

Table 27

NET WORTH OF THE OWENS BOTTLE COMPANY, 1904–1920

Date	Preferred Stock Outstanding	Common Stock Outstanding	Surplus and Undivided Profits	Total Net Worth
Sept. 1, 1904	$ 400,000.00	$2,500,000.00	$ 188,797.53	$3,088,797.53
Aug. 1, 1905	450,000.00	2,500,000.00	*20,856.63* *	2,929,143.37
Aug. 1, 1906	464,000.00	2,500,000.00	*36,530.56* *	2,927,469.44
Sept. 30, 1907	464,000.00	2,500,000.00	99,296.59	3,063,296.59
Sept. 30, 1908	464,000.00	2,500,000.00	186,109.67	3,150,109.67
Sept. 30, 1909	464,000.00	2,500,000.00	235,135.83	3,199,135.83
Sept. 30, 1910	464,000.00	2,500,000.00	220,739.31	3,184,739.31
Oct. 1, 1911	464,000.00	2,500,000.00	624,852.30	3,588,852.30
Oct. 1, 1912	464,000.00	2,500,000.00	1,699,716.94	4,663,716.94
Sept. 30, 1913	500,000.00	3,750,000.00	1,692,924.88	5,942,924.88
Sept. 30, 1914	500,000.00	5,000,000.00	1,915,370.83	7,415,370.83
Sept. 30, 1915	500,000.00	6,250,000.00	1,367,447.30	8,117,447.30
Sept. 30, 1916	7,257,400.00	9,050,000.00	2,685,817.90	18,993,217.90
Sept. 30, 1917	7,757,400.00	9,584,875.00	5,829,908.12	23,172,183.12
Dec. 31, 1918	7,336,700.00	9,546,525.00	6,291,720.96	23,174,945.96
Dec. 31, 1919	9,587,500.00	10,391,900.00	8,990,607.47	28,970,007.47
Dec. 31, 1920	9,450,200.00	11,037,575.00	9,720,564.02	30,208,339.02

* *Deficits.*

Table 28

CASH DIVIDENDS PAID BY THE OWENS BOTTLE COMPANY, 1906–1920

Period	Cumulative Preferred Stock		Common Stock		Total Cash Dividends
	Amount	Rate	Amount	Rate	
Year ending Sept. 30, 1907	$115,567.81	7%	$ 25,000.00	1%	$ 140,567.81
" " Sept. 30, 1908	32,480.00	7	100,000.00	4	132,480.00
" " Sept. 30, 1909	32,480.00	7	150,000.00	6	182,480.00
" " Sept. 30, 1910	32,480.00	7	200,000.00	8	232,480.00
" " Sept. 30, 1911	32,480.00	7	250,000.00	10	282,480.00
" " Sept. 30, 1912	32,480.00	7	300,000.00	12	332,480.00
" " Sept. 30, 1913	34,988.00	7	450,000.00	12	484,988.00
" " Sept. 30, 1914	35,012.00	7	600,000.00	12	635,012.00
" " Sept. 30, 1915	35,000.00	7	750,000.00	12	785,000.00
" " Sept. 30, 1916	282,425.50	7	1,715,500.00	20	1,997,925.50
" " Sept. 30, 1917	510,343.36	7	1,814,920.85	23	2,325,264.21
Oct. 1, 1917– Dec. 31, 1918	650,756.75	7	2,199,673.30 *	27 *	2,850,430.05 *
Year ending Dec. 31, 1919	614,149.32	7	1,178,254.25	12	1,792,403.57
" " Dec. 31, 1920	668,069.50	7	1,289,206.50	12	1,957,276.00

* The major portion of this was paid in war bonds.

Table 29

EARNINGS OF THE OWENS BOTTLE COMPANY, 1906–1920

Period	Royalty and License Income	Dividends, Profits from Investments, Capital Gains	Profit from Bottle Sales	Gross Income *	Total Net Earnings *
July 31, 1906–Sept. 30, 1907	$ 317,972.96	$ 2,520.68	$ 12,687.35	$ 333,180.99	$ 270,008.34
Year ending Sept. 30, 1908 **	?	?	?	?	308,447.40
" " Sept. 30, 1909	497,717.05	2,634.47	16,701.66	500,315.12	442,295.26
" " Sept. 30, 1910	520,290.19	48,121.47	25,138.27	582,007.68	493,083.48
" " Sept. 30, 1911	789,855.24	153,606.60	65,474.83	1,496,394.62	919,500.03
" " Sept. 30, 1912	998,397.54	130,763.68	307,632.81	2,734,384.31	1,308,650.64
" " Sept. 30, 1913	978,251.31	279,783.14	713,651.39	3,256,494.80	1,756,162.72
" " Sept. 30, 1914	1,191,980.26	301,520.54	856,369.43	2,349,870.23	2,107,457.95
" " Sept. 30, 1915	1,054,382.62	436,791.55	481,685.61	1,972,859.78	1,722,950.69
" " Sept. 30, 1916	1,361,790.80	1,080,618.45	982,945.55	3,436,198.76	3,104,081.48
" " Sept. 30, 1917	1,647,514.52	2,121,158.59	968,254.52	4,936,004.81	3,833,068.22
Oct. 1, 1917– Dec. 31, 1918	1,802,312.28	1,823,133.46	1,837,011.40	5,531,690.19	3,795,571.64
Year ending Dec. 31, 1919	1,293,979.49	1,199,506.93	1,450,200.80	4,039,795.99	2,966,388.97
" " Dec. 31, 1920	2,125,614.11	936,689.22	4,343,490.08	7,788,802.91	4,222,205.41

* These figures include earnings in addition to those listed separately in this table.

** I found no profit-and-loss statement for the year ending on September 30, 1908. The total net earnings for that year were interpolated.

Table 30

TOTAL ASSETS AND NET WORTH OF THE OWENS EUROPEAN BOTTLE MACHINE COMPANY, 1906–1920

Date *	Total Assets	Net Worth
Nov. 1, 1906	$2,722,011.58	$2,712,734.70
Oct. 1, 1907	2,761,225.08	2,759,116.58
Sept. 30, 1908	2,481,415.38	2,381,855.99
Sept. 1, 1909	3,113,912.44	2,074,701.33
Sept. 1, 1911	2,690,356.43	1,405,044.99
Sept. 1, 1912	1,566,133.36	1,171,809.53
Aug. 31, 1913	1,124,933.50	886,851.02
Aug. 31, 1914	834,978.82	584,380.70
Aug. 31, 1915	703,095.08	304,840.19
Aug. 31, 1917	129,778.48	40,270.35
Aug. 31, 1918	120,729.67	98,432.10
Aug. 31, 1920	114,710.45	73,885.60

* Balance sheets for 1910, 1916, and 1919 were unavailable.

Table 31

SELECTED ASSETS AND NET WORTH OF THE LIBBEY-OWENS SHEET GLASS COMPANY, 1916–1920

Date	Value of Patents	Plant, Land, and Equipment	Investments in Other Companies	Total Assets
December 31, 1916	$4,289,544.82	$ 653,480.02	—	$6,018,991.88
December 31, 1917	4,292,564.82	1,391,434.06	—	6,264,516.72
September 30, 1918	4,292,641.82	1,459,795.32	—	6,169,316.65
September 30, 1919	3,475,641.82	1,321,216.37	$507,237.50	6,773,425.98
September 30, 1920	3,258,641.82	4,216,590.43	677,980.00	10,936,454.51
	Preferred Stock Outstanding	**Common Stock Outstanding**	**Surplus**	**Net Worth**
December 31, 1916	$1,926,500.00	$4,000,000.00	$ *22,549.92* *	$5,903,950.08
December 31, 1917	2,000,000.00	4,000,000.00	*70,286.87* *	5,929,713.13
September 30, 1918	2,000,000.00	4,000,000.00	13,280.10	6,013,280.10
September 30, 1919	2,000,000.00	4,000,000.00	418,805.35	6,418,805.35
September 30, 1920	2,000,000.00	4,000,000.00	2,090,526.20	8,090,526.20

* *Deficits.*

TABLE 32

ASSETS AND NET WORTH OF THE LIBBEY GLASS COMPANY, 1889–1920

Date *	Manufacturing Plant, Land, and Equipment	Total Assets	Preferred Stock Outstanding	Common Stock Outstanding	Surplus	Net Worth
January 1, 1889	$ 72,481.30	$ 128,652.61	—	$ 100,000.00	$ *5,000.00* **	$ 95,000.00
January 1, 1890	78,409.81	151,154.75	—	100,000.00	*3,000.00* **	97,000.00
August 11, 1891	74,464.88	202,942.04	—	100,000.00	50,000.00	150,000.00
January 1, 1892	70,322.77	216,517.27	—	100,000.00	75,000.00	175,000.00
January 1, 1893	107,864.98	283,000.00	—	125,000.00	109,500.00	234,500.00
January 1, 1894	82,309.51	279,026.79	—	125,000.00	147,026.79	272,026.79
January 1, 1895	78,220.24	281,944.58	—	125,000.00	149,858.66	274,858.66
January 1, 1896	89,638.74	262,091.50	—	125,000.00	129,977.82	254,977.82
January 1, 1897	81,879.74	275,335.87	—	125,000.00	139,870.67	264,870.67
January 1, 1898	87,137.78	255,898.72	—	125,000.00	123,819.57	248,819.57
January 1, 1899	91,465.55	251,737.23	—	125,000.00	119,699.21	244,699.21
January 1, 1900	102,819.11	279,750.00	—	125,000.00	147,750.00	272,750.00
January 1, 1901	130,783.08	336,529.17	—	125,000.00	204,529.17	329,529.17
January 1, 1902	140,668.93	372,805.33	—	125,000.00	240,805.33	365,805.33
January 1, 1903	184,140.78	429,754.65	—	125,000.00	297,754.65	422,754.65
January 1, 1904	250,713.56	585,147.53	—	125,000.00	432,321.71	557,321.71
January 1, 1905	287,385.91	596,650.28	—	125,000.00	447,025.03	572,025.03
January 1, 1906	279,230.98	663,411.30	—	500,000.00	140,034.65	640,034.65
January 1, 1907	266,123.78	786,133.79	—	500,000.00	252,753.76	752,753.76
January 1, 1908	294,828.44	949,654.73	—	500,000.00	412,326.31	912,326.31
January 1, 1909	287,965.38	940,629.33	—	500,000.00	416,052.84	916,052.84
January 1, 1910	277,313.50	926,125.41	—	500,000.00	402,325.45	902,325.45
January 1, 1911	296,149.80	957,394.67	—	500,000.00	427,101.31	927,101.31
January 1, 1912	291,662.79	938,894.35	—	500,000.00	410,807.74	910,807.74
January 1, 1913	506,560.28	1,089,033.35	—	500,000.00	540,507.74	1,040,507.74
January 1, 1914	556,873.25	1,330,346.60	—	1,000,000.00	297,107.74	1,297,107.74
January 1, 1915	577,099.62	1,271,951.74	—	1,000,000.00	243,357.74	1,243,357.74
January 1, 1916	678,826.45	1,454,451.38	—	1,000,000.00	396,907.74	1,396,907.74
January 1, 1918	1,142,000.00	2,831,746.54	$1,000,000.00	1,000,000.00	689,889.13	2,689,889.13
January 1, 1919	1,249,500.00	3,327,507.91	1,000,000.00	1,000,000.00	850,889.13	2,850,889.13
January 1, 1920	1,180,300.00	3,387,346.47	880,000.00	1,000,000.00	1,358,889.13	3,238,889.13

* I was unable to find a balance sheet for 1917.

** *Deficits*

TABLE 33

ANNUAL SALES OF THE LIBBEY GLASS COMPANY, 1894–1919

Year *	Bulbs and Tubes	Other Goods	Total Sales
1894	?	?	$ 300,113.04
1895	?	?	375,436.99
1896	?	?	354,267.02
1897	?	?	381,660.83
1898	?	?	467,768.52
1899	?	?	583,148.69
1900	?	?	685,144.43
1901	?	?	728,809.85
1902	?	?	726,775.28
1903	?	?	1,151,568.91
1904	$ 506,985.19	$ 454,304.19	961,289.38
1905	500,083.02	565,940.91	1,066,023.93
1906	634,197.53	711,378.05	1,345,575.58
1907	936,026.54	835,353.37	1,771,379.91
1908	549,392.25	643,433.04	1,192,825.29
1909	763,423.70	816,187.65	1,579,611.35
1910	916,728.58	919,671.80	1,836,400.38
1911	1,108,026.67	900,450.09	2,008,476.76
1912	1,082,394.12	1,069,168.33	2,151,562.45
1913	1,086,414.97	903,264.96	1,989,679.93
1914	971,773.11	611,586.29	1,583,359.40
1915	1,169,001.68	499,925.44	1,668,927.12
1916	1,671,339.89	728,607.23	2,399,947.12
1917	2,511,189.96	653,500.78	3,164,690.74
1919	?	?	3,388,634.47

* I was unable to find data on the sales during 1918.

NET EARNINGS AND CASH DIVIDENDS OF THE LIBBEY GLASS COMPANY, 1889–1919

Year	Net Earnings	Cash Dividends				Total
		On Common		On Preferred		
		Amount	Rate	Amount	Rate	
1889	?					
1890	?					
1891	?					
1892	?	$10,000.00	10%			$10,000.00
1893	?	12,500.00	10			12,500.00
1894	$51,500.00	50,000.00	40			50,000.00
1895	90,000.00	93,750.00	75			93,750.00
1896	70,000.00	62,500.00	50			62,500.00
1897	79,800.00	93,750.00	75			93,750.00
1898	121,500.00	125,000.00	100			125,000.00
1899	175,000.00	143,750.00	115			143,750.00
1900	180,000.00	125,000.00	100			125,000.00
1901	155,200.00	125,000.00	100			125,000.00
1902	146,000.00	87,500.00	70			87,500.00
1903	180,000.00	43,750.00	35			43,750.00
1904	65,000.00	50,000.00	40			50,000.00
1905	140,140.00	72,500.00	14.5 *			72,500.00
1906	210,109.55	100,000.00	20			100,000.00
1907	334,750.00	175,000.00	35			175,000.00
1908	167,365.00	165,000.00	33			165,000.00
1909	370,190.00	300,000.00	60			300,000.00
1910	444,350.00	400,000.00	80			400,000.00
1911	454,200.00	470,000.00	94			470,000.00
1912	544,700.00	415,000.00	83			415,000.00
1913	362,780.72	110,000.00	22			110,000.00
1914	197,900.00	250,000.00	25 **			250,000.00
1915	302,630.00	150,000.00	15			150,000.00
1916	405,704.73	310,000.00	31			310,000.00
1917	269,060.70	67,500.00	6.75	$51,668.61	7%	119,168.61
1918	?	120,000.00 ***	12 ***	70,000.00	7	190.000.00
1919	524,686.37	180,000.00	18	70,000.00	7	250,000.00

* As of $500,000. ** As of $1,000,000. *** Partly estimated; a portion was paid in war bonds.

Table 35

TOTAL ASSETS AND NET WORTH OF THE TOLEDO GLASS COMPANY FOR SELECTED YEARS, 1898–1920

Date *	Patent Account	Total Assets	Capital Stock Out-standing	Surplus	Net Worth
Jan. 1, 1898	$60,000.00	$ 333,902.21	$150,000	$ 85,313.97	$ 235,313.97
Feb. 1, 1900	60,000.00	151,795.53	150,000	*1,413.16* **	148,586.84
Feb. 2, 1904	590,323.24	768,605.00	150,000	600,784.90	750,784.90
Feb. 1, 1905	29,504.69	196,527.29	150,000	44,527.29	194,527.29
Feb. 1, 1906	13,721.01	2,406,451.45	150,000	1,581,451.45	1,731,451.45
Feb. 1, 1907	36,963.11	2,175,077.24	150,000	1,082,649.89	1,232,649.89
Feb. 1, 1908	2,300.00	2,026,156.66	150,000	2,026,156.66	2,176,156.66
Jan. 30, 1909	2,300.00	949,788.45	150,000	795,297.10	945,297.10
Jan. 31, 1910	4.00	831,572.17	150,000	675,309.60	825,309.60
Oct. 31, 1910	4.00	718,801.57	150,000	564,081.64	714,081.64
Nov. 1, 1911	4.00	625,539.49	150,000	469,982.46	619,982.46
Nov. 1, 1912	22,949.20	584,940.27	150,000	397,026.04	547,026.04
Oct. 31, 1913	34,501.14	636,371.94	150,000	404,951.78	554,951.78
Oct. 31, 1914	176,302.76	678,533.77	150,000	395,454.88	545,454.88
Oct. 31, 1915	174,674.20	635,058.97	150,000	357,661.04	507,661.04
Oct. 31, 1919	?	2,053,118.45	150,000	1,903,091.19	2,053,091.19
Oct. 31, 1920	?	2,044,133.96	150,000	1,893,927.16	2,043,927.16

* I was unable to find financial statements for the missing years.

** *Deficit*

TABLE 36

DIVIDENDS PAID BY THE TOLEDO GLASS COMPANY, 1895–1920 *

Date	Amount	Rate on Stock Outstanding	How Paid
1895–1898	None		
January 1, 1899	$ 150,000	100 %	American Lamp Chimney common
October 1, 1899	75,000	50	Macbeth-Evans preferred
February 6, 1900	11,500	7.66	Macbeth-Evans common
March 21, 1900	3,450	2.33	Macbeth-Evans common
November 7, 1903	50,000	33.33	Owens Bottle preferred
November 7, 1903	1,550,000	1033.33	Owens Bottle common
May 15, 1907	75,000	50	Owens Bottle common
January 7, 1908	50,000	33.33	Owens Bottle common
March 25, 1908	150,000	100	Cash
March 31, 1908	15,000	10	Owens Bottle common
March 31, 1908	37,500	25	Cash
December 30, 1908	15,000	10	Cash
1909	120,000	80	Cash
1910	120,000	80	Cash
1911	90,000	60	Cash
1912	75,000	50	Cash
1917	82,500	55	Cash
1919	27,000	18	Cash
1920	157,500	105	Cash
TOTAL	$2,854,450	1903 %	

* So far as I know, all dividends are included in this table.

BIBLIOGRAPHY

Manuscript Materials
(*including typewritten*)

Bacon, George S. "Whitall-Tatum Company, 1836–1938." Millville, N. J.

Bining, William. "The Glass Industry of Western Pennsylvania, 1797–1860." Unpublished master's thesis, University of Pittsburgh, 1936. Carnegie Library, Pittsburgh, Pa.

Corning Glass Works (Corning, N. Y.)
Minutes of the Meetings Held by the Board of Directors. Vols. I (1863–1868), II (1868–1911).
George B. Hollister. "Historical Notes: Corning Glass Works."

Davis, Pearce. "The American Glass Industry and the Tariff." Unpublished doctoral dissertation; Harvard University, 1938. Cambridge, Mass.

Gaffield, Thomas, Collection. (Massachusetts Institute of Technology Library, Cambridge, Mass.) In manuscript unless otherwise stated.
Copies of Letters Written and Received. 1 vol. (1868–1892).
Correspondence on Glass and Glass Making. 1 vol.
Correspondence on Sunlight Experiments. 1 vol.
Experiments in Photography and the Action of Sunlight on Glass. 1 vol.
Glass and Glass-Making, etc. 3 vols. Contains extracts and clippings from newspapers, magazines, and trade journals.
Glass Journal. Vols. I (1858–1864), II (1864–1874), III (1874–1881), and IV (1881–1894). Observations and notes in manuscript.
Glass-Making, etc. 8 vols. Contains printed articles, extracts, and clippings pertaining to the industry and art.
Letters of E. S. Drone and Joseph D. Weeks.
Notes on Glass. 5 vols.
Notes on Glass-Making. 1 vol.
Prices Current of Window Glass. Printed.
Scraps. 3 vols. Contains articles and newspaper clippings on glass and glasshouses.

Historical Society of Western Pennsylvania (Pittsburgh, Pa.).
Letters written by James O'Hara regarding his glass industry interests. 1 vol. (covering March 2, 1805, to December 1, 1819).
Daybook, (covering 1814–1820).
Small folder containing material on Bakewell and Page, glass manufacturers in Pittsburgh (covering 1826–1837).

Hollister, George B. "Historical Notes: Corning Glass Works." See Corning Glass Works (Corning, N. Y.).

Libbey Estate (Justin Wilcox, executor; Toledo). Minutes of the Meetings Held by the Board of Directors and Stockholders of the Westlake European Machine Company.

Libbey Glass Division of the Owens-Illinois Glass Company (Toledo, Ohio). Printed material, typescripts, and manuscripts.
Minutes of the Meetings Held by the Libbey Glass Board of Directors and Stockholders. Vols. I (1888–1909), II (1909–1913), III (1913–1917), and IV (1917–1922). Vols. I and II are in the possession of Joseph W. Robinson at Toledo.
Minutes of the Meetings Held by the Board of Directors and Stockholders of the Westlake Machine Company.
Court hearings and proceedings involving the Libbey Glass Company and the Owens Bottle Company.
Three old safes filled with documents relating to licenses, patents, lawsuits, auditor's reports, balance sheets, patent interferences, and blueprints.

Libbey-Owens-Ford Glass Company (Toledo, Ohio). Printed material, typescripts, and manuscripts.
Minutes of the Meetings Held by the Board of Directors and Stockholders of the Owens European Bottle Machine Company. Vols. I–III (1905–1920).
Minutes of the Meetings Held by the Board of Directors and Stockholders of the Toledo Glass Company. Vols. I (1895–1904), II (1905–1908), III (1908–1912), IV (1913–1926), and V (1927–1932).
Minutes of the Meetings Held by the Board of Directors and Stockholders of the Libbey-Owens Sheet Glass Company. Vols. I (1916–1919) and II (1919–1924).
Toledo Glass Company. Licenses and Correspondence Thereto. 2 vols.
Owens European Bottle Machine Company. Licenses and Correspondence Thereto. 1 vol.
Colburn Records, including several albums of photographs of the

Colburn Machine, seven volumes of newspaper and trade-journal clippings, and a report on the validity of the Colburn patents.

Owens-Illinois Glass Company (Toledo, Ohio). Printed material, typescripts, and manuscripts.

Annual Reports of the President of the Owens Bottle Company to the Stockholders. 1908–1935.

Clarence Brown's Private Files.

Europäischer Verband Lawsuit Files.

"History of the Owens-Illinois Glass Company."

"The Intimate Lives of Edward Drummond Libbey and Michael Joseph Owens as We Knew Them." December 1, 1933.

John H. McNerney's Private Files.

Letters of the American Lamp Chimney Company. 1 vol.

Letters of E. D. Libbey. 3 vols.

Letters of the Kent Machine Company. 2 vols.

Letters of William S. Walbridge. 1 vol.

Memorandum prepared by Leonard D. Soubier on April 17, 1943.

Minutes of the Meetings Held by the Board of Directors and Stockholders of the Owens Bottle Machine Company of New Jersey. I (1903–1908).

Minutes of the Meetings Held by the Board of Directors and Stockholders of the Owens Bottle Machine Company of Ohio. 7 vols. (1908–on).

Minutes of the Meetings Held by the Executive Committee of the Owens Bottle Company. 2 vols.

M. J. Owens' Private Files.

Opinion of James Whittemore, October 30, 1906. Patents Department.

Soubier, Leonard D. "The Development of the Owens Suction Machine." December 15, 1935.

"Suggested Revision of the Biographical Sketch of M. J. Owens, on Proofs Submitted by the National Cyclopedia of American Biography, under Date of June 3, 1936."

Walbridge, William S. "Revision for American Bottles, Old and New."

——. "The Romance of an Invention: The Owens Bottle Machine."

Pennsylvania Historical Society (Philadelphia).

John Dickinson Collection. Two ledgers and one journal of Stiegel's Manheim glasshouse. Account book of Caspar Wistar and several letters of the Wistar family.

Item 379. The Logan Papers, containing some Henry W. Stiegel documents (1766–1775) and papers relating to the Elizabeth furnace.

Robinson, Ralph J. "Glass." Industrial Bureau, Baltimore Association of Commerce.

Rodefer Glass Company (Bellaire, Ohio). Miscellaneous Records: Hundreds of letter-file boxes, minute books, and journals dating back to about 1877.

Swindell Brothers, Inc. (Baltimore, Md.).
Cash books (1882–on).
Monthly comparative stock of window glass (1880–1920).
Monthly total sales records (1873–1937).

United States Glass Company (Tiffin, Ohio). Miscellaneous Records: Journals, ledgers, balance sheets, profit-and-loss statements, annual reports, auditor's reports, stock and bond records, some correspondence, inventories, sales, by-laws, and rules (dating from 1891).

Printed Materials

Acton, R. M. "Short History of Glass Manufacture in Salem County, N. J." *Pennsylvania Magazine of History and Biography*, IX (1885).

Aldrich Report. See United States. *Wholesale Prices . . .*

American Glass Review. Pittsburgh, 1882 on. In earlier years called successively *American Glass Worker*, *Commoner and Labor Herald*, *Commoner*, *Commoner and American Glass Worker*, *Commoner and Glassworker*, and *Glassworker.* Contains innumerable articles on problems and labor conditions relating to the trade. Especially useful was the *Directory of the Glass Trade* (1905) published by the Commoner Publishing Company of Pittsburgh for the *Commoner and Glassworker.*

Antiques. New York, 1922 on. Contains many articles written by antique dealers and those interested in regional histories on glassmakers and local glasshouses.

Atwater, R. M. "The Glass-Making Industry in America." *Engineering Magazine*, IV (1893).

Austin, Mrs. Jane G. "Cullet." *Atlantic Monthly*, XIV (1864).

Barber, Edwin Atlee. *American Glassware Old and New.* Philadelphia: Patterson and White Co., 1900.

Barnett, George E. *Chapters on Machinery and Labor.* Cambridge, Mass.: Harvard University Press, 1926.

Bezanson, Anne, Gray, Robert D. and Hussey, Miriam. *Wholesale Prices in Philadelphia, 1784–1861.* Philadelphia: University of Pennsylvania Press, 1937.

Bidwell, W. H. *The American Plate Glass Manufactory, at Platin Creek, near St. Louis, Missouri.* New York, 1870.

Bining, William. "The Glass Industry of Western Pennsylvania, 1797–1857." *Western Pennsylvania Historical Magazine*, XIX (1936).

Bishop, J. Leander. *A History of American Manufactures from 1608 to 1860.* 3 vols. Philadelphia, 1866.

Bontemps, Georges. *Essays on Glass and Glass Making.* Paris, 1845–1872. Available in the Gaffield Collection in the Massachusetts Institute of Technology Library.

——. *Guide du Verrier.* Paris, 1868.

Brothers, J. Stanley, Jr. *Thumbnail Sketches.* Kalamazoo, Mich: J. S. Brothers, Jr., Publications, 1940.

Burbank, Leonard H. "The Keene Glass Works." *Antiques*, IX (1930).

Cady, Edward H. "Builders of Business." *System*, June, 1919.

Chipman, Frank W. *The Romance of Old Sandwich Glass.* Sandwich, Mass.: Sandwich Publishing Co., 1932.

Clark, J. B. *The Distribution of Wealth.* New York: The Macmillan Co., 1931.

Clark, Victor S. *History of Manufactures in the United States, 1607–1928.* 3 vols. New York: McGraw-Hill Book Co., 1929.

Coggeshall, William T., ed. *Five Black Arts: A Popular Account of the History, Processes of Manufacture, and Uses of Printing, Gas-Light, Pottery, Glass, Iron.* Columbus, Ohio, 1861. Condensed from the *Encyclopaedia Britannica.*

Cole, Arthur H. "An Approach to the Study of Entrepreneurship: A Tribute to Edwin F. Gay." *The Tasks of Economic History* (supplemental issue of *The Journal of Economic History*), VI (1946).

——. "Business History and Economic History." *The Tasks of Economic History* (supplemental issue of *The Journal of Economic History*), V (1945).

——. *Wholesale Commodity Prices in the United States, 1700–1861. Statistical Supplement.* Cambridge, Mass.: Harvard University Press, 1938.

Colné, Charles. *Paris Exposition, 1878. Glass and Glassware.* Washington (?), 1881 (?).

Cooper, William. *The Crown Glass Cutter & Glazier's Manual.* Edinburgh, 1835.

Coxe, Tench. *A Statement of the Arts and Manufactures of the United States for the Year 1810.* Philadelphia, 1814.

Curtiss, W. H. "Pyrex: A Triumph for Chemical Research in Industry." *Journal of Industrial and Engineering Chemistry*, XIV (1922).

Daglish, George H. "On Machinery for the Manufacture of Plate Glass." *Journal of the Franklin Institute of the State of Pennsylvania.* LXXVIII (1864).

Dictionary of Arts and Sciences. Gregory, G., ed. 2d American edition. 3 vols. New York, 1822. Article on "Glass."

Dimock, Marshall E. *The Executive in Action.* New York: Harper and Brothers, 1945.

Directory of the Glass Trade. See *American Glass Review.*

Dodd, George. *The Curiosities of Industry: Glass and Its Manufacture.* London, 1853.

Douglas, Paul H. *Real Wages in the United States, 1890–1926.* New York: Houghton Mifflin Co., 1930.

——. *The Theory of Wages.* New York: Houghton Mifflin Co., 1934.

Drugs, Oils and Paints. Philadelphia, VIII–XXXVI (1893–1921).

Dubbs, Joseph H. "Baron Stiegel." *Pennsylvania Magazine of History and Biography*, I (1877).

Durfee, W. F., and Gobert, August, fils. *Plate-Glass: Its Manufacture in the United States.* Philadelphia, 1876.

Dyott, T. W. *An Exposition of the System of Moral and Mental Labor, Established at the Glass Factory of Dyottville, in the County of Philadelphia.* Philadelphia, 1833.

Edward D. Libbey v. *Mount Washington Glass Company et al.* U. S. Circuit Court, District of Massachusetts, in Equity, No. 2189.

Ellsworth, P. T. *International Economics.* New York: The Macmillan Co., 1938.

Evans, Thomas. "A Brief History of the Chimney Business." See Gillinder, James and Evans, Thomas, *Papers.*

Fettke, Charles Reinhard. *Glass Manufacture and the Glass Sand Industry of Pennsylvania.* Harrisburg, Pa.: J. L. L. Kuhn, printer, 1919. (Report No. XII of the Topographic and Geologic Survey of Pennsylvania.)

Finn, Dr. Horatio N. "On the Manufacture of Glass." *American Journal of Science and Arts.* XVI (1829).

Flamm, P. *Un Chapitre sur la Verrerie ou Transformation Complète de la Fabrication Actuelle du Verre.* Paris, 1866.

Fourth Census. See United States, *American State Papers.*

Fowle, Arthur E. *Flat Glass.* Toledo, Ohio: The Libbey-Owens Sheet Glass Company, 1924.

Friends of Domestic Industry. *General Convention of the Friends of Domestic Industry, Assembled at New York, October 26, 1831.* Baltimore, 1832.

Gillinder, James. "American Glass Interests." *One Hundred Years of American Commerce.* Chauncey M. Depew, ed., New York, 1895.

——. and Evans, Thomas. *Papers Read before the Members of the American Association of Flint and Lime Glass Manufacturers at the Annual Meetings at Atlantic City, July, 1900 and 1901.*

Gillinder, William. *Treatise on the Art of Glass Making.* 2d edition. Birmingham, England, 1854.

Gillingham, Harrold E. "Pottery, China, and Glass Making in Philadelphia." *Pennsylvania Magazine of History and Biography,* LIV (1930).

Gilpin, Joshua. "Journal of a Tour from Philadelphia Thro the Western Counties of Pennsylvania in the Months of September and October, 1809." *Pennsylvania Magazine of History and Biography,* LII (1928).

Glass Industry. Report on the Cost of Production of Glass in the United States. (Bureau of Foreign and Domestic Commerce, Misc. Series No. 60.) Washington, 1917.

"Great American Industries: VIII, A Piece of Glass." *Harper's New Monthly Magazine,* LXXIX (1889).

Harding, William G. "Glass Manufacture in the Berkshires." *Collections of the Berkshire Historical and Scientific Society.* II, Part I (1894).

Hartford–Empire Company et al., v. *United States of America.* Supreme Court; October term, 1943.

Heiges, George L. *Henry William Stiegel: The Life Story of a Famous American Glass-Maker.* Manheim, Pa.: The Author, 1937.

Henderson, C. Hanford. "Glass-Making." *Popular Science Monthly,* XXXIV–XXXIX (1888–1891).

——. "The Glass Industry." *Popular Science Monthly,* XLII (1893).

History of the Processes of Manufacture and Uses of Printing, Glass and Iron. J. Bradburn, publisher. New York, 1864. From the *Encyclopaedia Britannica.*

Hollister, George B. "Manufacture of Early Incandescent Lamp Bulbs at Corning." *Bulletin of the American Ceramic Society*, XV (1936).

Hower, Harry S. "Some Scientific and Technological Contributions to the Glass Industry in the Pittsburgh District." *Western Pennsylvania Historical Magazine*, XVIII (1935).

Hull, M. P. *Early Glass-Making in Virginia.* Richmond, Va.: Jones Printing Co., 1933.

Hunter, Frederick William. *Stiegel Glass.* Boston: Houghton Mifflin Company, 1914.

Irwin, Frederick T. *The Story of Sandwich Glass and Glass Workers.* Manchester, N. H.: Granite State Press, 1926.

Jarves, Deming. *Reminiscences of Glass-Making.* 2d edition. New York, 1865.

——. "The Manufacture of Glass." *Hunt's Merchants' Magazine and Commercial Review*, XXVII–XXVIII (1852–1853).

Jerome, Harry. *Mechanization in Industry.* New York: National Bureau of Economic Research, 1934.

Journal of the Franklin Institute of the State of Pennsylvania, XXIV (1833).

Juergens, Carl H. "Movement of Wholesale Prices in New York City, 1825–1863." *American Statistical Association Quarterly Publications*, XII (1910–1911).

Knight, Frank H. "Profits and Entrepreneurial Functions." *The Tasks of Economic History* (supplemental issue of *The Journal of Economic History*), II (1942).

Knittle, Rhea Mansfield. *Early American Glass.* New York: The Century Company, 1927.

Lanier, Sterling. "A Cloth of Glass." *Technology Review*, XLIV (1942).

Lee, Mrs. Ruth Webb. *Sandwich Glass: The History of the Boston and Sandwich Glass Company.* 2d edition. Framingham Centre, Mass.: The Author, 1939.

Long, Clarence D., Jr. *Building Cycles and the Theory of Investment.* Princeton: Princeton University Press, 1940.

Macbeth-Evans Glass Company. *Fifty Years of Glass Making, 1869–1919.* Pittsburgh: Macbeth-Evans Glass Company, 1920.

McKearin, George S. and Helen. *American Glass.* New York: Crown Publishers, 1941.

McLane, Louis. *Report on Manufactures* (House Document No. 308, 22 Congress, 1 session). 2 vols., Washington, 1833.

MacManus, Theodore F. *A Century of Glass Manufacture, 1818–1918.* Toledo, Ohio: Libbey Glass Company, 1918.

Marshall, T. H. *James Watt.* London: Leonard Parsons, Ltd., 1925.

Marson, Percival. *Glass and Glass Manufacture.* London: Sir I. Pitman and Sons, 1918.

Martin, Robert F. *National Income in the United States, 1799–1938.* New York: National Industrial Conference Board, 1939.

Mechanics Magazine, The. November 18, 1871.

Miller, Bruce. "Glass . . . and Edward D. Libbey." *Batch.* A Libbey-Owens-Ford Glass Company publication. Carl Matson, ed. No. 22 (November, 1940).

——. "'It Can Be Done.'" *Batch,* No. 24 (January, 1941).

——. "The Great Pioneer of Our Business — Capt. John B. Ford." *Batch,* No. 19 (August, 1940).

Mitchell, W. C. *A History of Greenbacks.* Chicago: University of Chicago Press, 1903.

Monro, William L. *Window Glass in the Making: An Art, A Craft, A Business.* Pittsburgh: American Window Glass Company, 1926.

Moore, N. Hudson. *Old Glass, European and American.* New edition. New York: Tudor Publishing Company, 1935.

Morrison, James. "A 60-Year History of Glass Blowing." *Ceramic Industry,* XXXIX (1942).

——. *The Romance of Glass.* (Published by the Owens Bottle Co.)

National Glass Budget. Especially vols. XIV–XXXV (Pittsburgh, 1898–1920).

National Industrial Conference Board. Division of Industrial Economics. *Studies in Enterprise and Social Progress.* New York: National Industrial Conference Board, 1938.

New American Cyclopaedia. George Ripley and Charles A. Dana, eds. "Glass." VIII. New York: D. Appleton and Co., 1859.

Newman, William H. "The Building Industry and Business Cycles." *Studies in Business Administration.* Vol. V, No. 4 (1935). Chicago: *Journal of Business of the University of Chicago.*

Nichols, I. T. *The City of Bridgeton, New Jersey.* Philadelphia: Burk & McFetridge, printers, 1889.

Northend, Mary Harrod. *American Glass.* New edition. New York: Tudor Publishing Company, 1936.

Nourse, Edwin G. and Drury, Horace B. *Industrial Price Policies and Economic Progress.* Washington: The Brookings Institution, 1938.

Ohio Bureau of Labor Statistics, *Annual Reports,* XXVII–XXXVI (1904–1913).

Ormsbee, Thomas H., and Allen, Florence C. "Glassmaking at Lake Dunmore, Vermont." *American Collector,* VII (1937).

Owens Bottle Machine Company. *The Owens-Automatic Bottle Machine* (a prospectus). Toledo, 1903.

Peligot, M. E. *Douze Leçons sur l'Art de la Verrerie.* Paris, 1862.

Pellatt, Apsley. *Curiosities of Glass Making.* London, 1849.

Persons, Warren M. *Forecasting Business Cycles.* New York: J. Wiley and Sons, 1931.

Phillips, C. J. *Glass: The Miracle Maker. Its History, Technology and Applications.* New York: Pitman Publishing Company, 1941.

Pilkington, Richard Jr. "On the Processes and Mechanical Appliances in the Manufacture of Polished Sheet Glass." *Journal of the Franklin Institute of the State of Pennsylvania.* LXXX (1865).

Pittsburgh Plate Glass Company. *Glass: Paints, Varnishes and Brushes: Their History, Manufacture and Use.* Pittsburgh: Pittsburgh Plate Glass Company, 1923.

Porter, George R. *A Treatise on the Origin, Progressive Improvement, and Present State of the Manufacture of Porcelain and Glass.* Philadelphia, 1834.

Pound, Arthur. *Salt of the Earth. The Story of Captain J. B. Ford and Michigan Alkali Company.* Boston: Atlantic Monthly Company, 1940.

Powell, Harry J. and others. *The Principles of Glass-Making.* London, 1883.

Productivity of Labor in the Glass Industry. U. S. B. L. S. *Bulletin* 441. Washington, 1927.

Réaumur, René-Antoine de. "Expériences et Réflexions sur la Prodigieuse Ductilité de Diverses Matières." *Histoire de l'Académie Royale des Sciences, avec les Mémoires de Mathématique et de Physique, année 1713.* Paris, 1739.

Robinson, James Harvey. *The Mind in the Making.* New York: Harper and Brothers, 1921.

Rosanoff, M. K. "Edison in His Laboratory." *Harper's Magazine,* Vol. 165 (1932).

Rosenhain, Walter. *Glass Manufacture.* London: A. Constable & Co., Ltd., 1908.

Saunders, Allen. "The Genius of the Bottle." *Toledo News-Bee,* October 31, 1934.

Schumpeter, Joseph A. *The Theory of Economic Development.* Translated by Redvers Opie. Cambridge, Mass.: Harvard University Press, 1936.

Scoville, Warren C. "Labor and Labor Conditions in the French Glass Industry, 1643–1789." *Journal of Modern History,* XV (1943).

——. "Large-Scale Production in the French Plate-Glass Industry, 1665–1789." *Journal of Political Economy,* L (1942).

——. "Technology and the French Glass Industry, 1640–1740." *Journal of Economic History,* I (1941).

Siemens, C. W., F. R. S., D. C. L., and F. *Siemens' Patent Continuous Glass Melting Furnace.* London, 1871. Available in the Gaffield Collection in the Massachusetts Institute of Technology Library.

Snow, Julia D. Sophronia. "The Franklin Glass Factory – Warwick's Venture." *Antiques,* II (1933).

Snyder, Carl. *Business Cycles and Business Measurements.* New York: The Macmillan Company, 1927.

Soubier, L. D. See Owens-Illinois Glass Company.

Stainsby, William. *Diseases and Disease Tendencies of Occupations. The Glass Industry and the Hatting Industry.* In "Monographs on Economic Subjects." The New Jersey Bureau of Statistics. Trenton, N. J., 1901.

Thurston, George H. *Pittsburgh as It Is; Or, Facts and Figures Exhibiting the Past and Present of Pittsburgh, Its Advantages, Resources, Manufactures, and Commerce.* Pittsburgh, 1857.

Tock, Charles. "Note sur les Différents Systèmes de Fours Employés dans la Fabrication des Verres à Vitres." *Revue Universelle des Mines, de la Métallurgie, des Travaux Publics, des Science et des Arts Appliqués à l'Industrie.* XXXIX (1876).

Toledo News-Bee, October 31, 1934. See Saunders, Allen.

Trowbridge, F. T. "Among the Glass-Makers." *Our Young Folks,* V (1869).

Turner, W. E. S. "Twenty-one Years: A Professor Looks Out on the Glass Industry." *Journal of the Society of Glass Technology,* XXII (1938).

United States.

American State Papers. Finance, Vol. IV, Document No. 29. (Commonly called the *Fourth Census*, 1820.)

Bureau of Labor Statistics, *Course of Wholesale Prices.*

1890–1901	*Bulletin*	39 (1902).
1890–March 1910	*Bulletin*	87 (1910).
1890–1910	*Bulletin*	93 (1911).
1890–1921	*Bulletin*	320 (1922).
1890–1925	*Bulletin*	415 (1926).

——. *History of Wages in the United States from Colonial Times to 1928. Bulletin* 604 (1934).

Bureau of the Census, *Census Reports on Manufactures*, 1860, 1870, 1880, 1890, 1900, 1905, 1910, 1914, 1919, 1921.

Commissioner of Labor, *Regulation and Restriction of Output.* 1904.

Department of Commerce, *Statistical Abstracts of the United States.*

Tariff Commission, *Cylinder, Crown, and Sheet Glass.* Second Series No. 33, 1932.

——. *The Glass Industry as Affected by the War.* Tariff Information Series, No. 5, 1918.

——. *Flat Glass and Related Glass Products.* Second Series No. 123, 1937.

Tariff Commission of 1882, *Report.* (47 Cong., 2 sess., House Misc. Doc. No. 6, Part I.

Temporary National Economic Committee, *Investigation of Concentration of Economic Power. Hearings.* 75 Cong., 3 sess.

Wholesale Prices, Wages and Transportation, 1893. Aldrich Report: 52 Cong., 2 sess., Senate Report 1394.

See also

Coxe, Tench, *A Statement . . . of Manufactures . . .*

Glass Industry.

McLane, Louis, *Report on Manufactures.*

Productivity of Labor in the Glass Industry.

Wages and Hours of Labor.

Wages and Hours of Labor in Manufacturing Industries, 1890–1907.

Weeks, Joseph D., *Report . . . on Glass . . .*

Valerio, M. "Industrie des Glaces." *Revue Universelle des Mines, de la Métallurgie, des Travaux Publics, des Science et des Arts Appliqués à l'Industrie.* I–II (1857–1859).

Van Rensselaer, Stephen. *Early American Bottles and Flasks.* Revised edition. Peterborough, N. H.: Transcript Printing Co., 1926.

Veblen, Thorstein. *Absentee Ownership and Business Enterprise in Recent Times.* New York: Viking Press, 1938.

Wages and Hours of Labor. 19th Annual Report of the U. S. Commissioner of Labor. Washington, 1905.

Wages and Hours of Labor in Manufacturing Industries, 1890–1907. U. S. B. L. S. Bulletin 77. Washington, 1908.

Walbridge, William S. *American Bottles Old and New.* Toledo, Ohio: Caslon Press, 1920.

——. See Owens-Illinois Glass Company.

Warren, George F. and Pearson, Frank A. *Physical Volume of Production in the United States, The.* Memoir 144 of the Cornell University Agricultural Experiment Station. Ithaca, N. Y., 1932.

——. *Prices.* New York: J. Wiley and Sons, 1933.

——. *World Prices and the Building Industry.* New York: J. Wiley and Sons, 1937.

Watkins, Lura Woodside. *Cambridge Glass, 1818 to 1888.* Boston: Marshall Jones Company, 1930.

Weeks, Joseph D. *Directory of the Glass Works of the United States.* Washington, 1880.

——. *Report on the Manufacture of Glass in the United States.* Washington, 1883. Also available in Vol. II of the *Tenth Census.*

White, Harry Hall. "Early Pittsburgh Glass Houses." *Antiques,* X (1931).

——. "The Lancaster Glass Works." *Antiques,* XII (1933).

Wright, Chester W. *Economic History of the United States.* New York: McGraw-Hill Book Company, 1941.

INDEX

INDEX

THE EVOLUTION OF CAPITALISM

Allen, Zachariah. **The Practical Tourist,** Or Sketches of the State of the Useful Arts, and of Society, Scenery, &c. &c. in Great-Britain, France and Holland. Providence, R.I., 1832. Two volumes in one.

Bridge, James Howard. **The Inside History of the Carnegie Steel Company:** A Romance of Millions. New York, 1903.

Brodrick, J[ames]. **The Economic Morals of the Jesuits:** An Answer to Dr. H. M. Robertson. London, 1934.

Burlamaqui, J[ean-] J[acques]. **The Principles of Natural and Politic Law.** Cambridge, Mass., 1807. Two volumes in one.

Capitalism and Fascism: Three Right-Wing Tracts, 1937-1941. New York, 1972.

Corey, Lewis. **The Decline of American Capitalism.** New York, 1934.

[Court, Pieter de la]. **The True Interest and Political Maxims, of the Republic of Holland.** Written by that Great Statesman and Patriot, John de Witt. To which is prefixed, (never before printed) Historical Memoirs of the Illustrious Brothers Cornelius and John de Witt, by John Campbell. London, 1746.

Dos Passos, John R. **Commercial Trusts:** The Growth and Rights of Aggregated Capital. An Argument Delivered Before the Industrial Commission at Washington, D.C., December 12, 1899. New York, 1901.

Fanfani, Amintore. **Catholicism, Protestantism and Capitalism.** London, 1935.

Gaskell, P[eter]. **The Manufacturing Population of England:** Its Moral, Social, and Physical Conditions, and the Changes Which Have Arisen From the Use of Steam Machinery; With an Examination of Infant Labour. London, 1833.

Göhre, Paul. **Three Months in a Workshop:** A Practical Study. London, 1895.

Greeley, Horace. **Essays Designed to Elucidate the Science of Political Economy,** While Serving to Explain and Defend the Policy of Protection to Home Industry, As a System of National Cooperation for the Elevation of Labor. Boston, 1870.

Grotius, Hugo. **The Freedom of the Seas,** Or, The Right Which Belongs to the Dutch to Take Part in the East Indian Trade. Translated with Revision of the Latin Text of 1633 by Ralph Van Deman Magoffin. New York, 1916.

Hadley, Arthur Twining. **Economics:** An Account of the Relations Between Private Property and Public Welfare. New York, 1896.

Knight, Charles. **Capital and Labour;** Including *The Results of Machinery*. London, 1845.

de Malynes, Gerrard. **Englands View, in the Unmasking of Two Paradoxes:** With a Replication unto the Answer of Maister John Bodine. London, 1603. New Introduction by Mark Silk.

Marquand, H. A. **The Dynamics of Industrial Combination.** London, 1931.

Mercantilist Views of Trade and Monopoly: Four Essays, 1645-1720. New York, 1972.

Morrison, C[harles]. **An Essay on the Relations Between Labour and Capital.** London, 1854.

Nicholson, J. Shield. **The Effects of Machinery on Wages.** London, 1892.

One Hundred Years' Progress of the United States: With an Appendix Entitled Marvels That Our Grandchildren Will See; or, One Hundred Years' Progress in the Future. By Eminent Literary Men, Who Have Made the Subjects on Which They Have Written Their Special Study. Hartford, Conn., 1870.

The Poetry of Industry: Two Literary Reactions to the Industrial Revolution, 1755/1757. New York, 1972.

Pre-Capitalist Economic Thought: Three Modern Interpretations. New York, 1972.

Promoting Prosperity: Two Eighteenth Century Tracts. New York, 1972.

Proudhon, P[ierre-] J[oseph]. **System of Economical Contradictions:** Or, The Philosophy of Misery. (Reprinted from *The Works of P. J. Proudhon,* Vol. IV, Part I.) Translated by Benj. R. Tucker. Boston, 1888.

Religious Attitudes Toward Usury: Two Early Polemics. New York, 1972.

Roscher, William. **Principles of Political Economy.** New York, 1878. Two volumes in one.

Scoville, Warren C. **Revolution in Glassmaking:** Entrepreneurship and Technological Change in the American Industry, 1880-1920. Cambridge, Mass., 1948.

Selden, John. **Of the Dominion, Or, Ownership of the Sea.** Written at First in Latin, and Entituled *Mare Clausum.* Translated by Marchamont Nedham. London, 1652.

Senior, Nassau W. **Industrial Efficiency and Social Economy.** Original Manuscript Arranged and Edited by S. Leon Levy. New York, 1928. Revised Preface by S. Leon Levy. Two volumes in one.

Spann, Othmar. **The History of Economics.** Translated from the 19th German Edition by Eden and Cedar Paul. New York, 1930.

The Usury Debate After Adam Smith: Two Nineteenth Century Essays. New York, 1972. New Introduction by Mark Silk.

The Usury Debate in the Seventeenth Century: Three Arguments. New York, 1972.

Varga, E[ugen]. **Twentieth Century Capitalism.** Translated from the Russian by George H. Hanna. Moscow, [1964].

Young, Arthur. **Arthur Young on Industry and Economics:** Being Excerpts from Arthur Young's Observations on the State of Manufactures and His Economic Opinions on Problems Related to Contemporary Industry in England. Arranged by Elizabeth Pinney Hunt. Bryn Mawr, Pa., 1926.